D1260930

Reinventing Public Service Communication

Also by Petros Iosifidis

EUROPEAN TELEVISION INDUSTRIES (with Jeanette Steemers and Mark Wheeler)

PUBLIC TELEVISION IN THE DIGITAL ERA: TECHNOLOGICAL CHALLENGES AND NEW STRATEGIES FOR EUROPE

Reinventing Public Service Communication

European Broadcasters and Beyond

Edited by

Petros Iosifidis
Reader in Media and Communications, City University London, UK

palgrave
macmillan

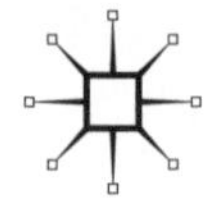

First published 2010 by
PALGRAVE MACMILLAN

Palgrave Macmillan in the UK is an imprint of Macmillan Publishers Limited, registered in England, company number 785998, of Houndmills, Basingstoke, Hampshire RG21 6XS.

Palgrave Macmillan in the US is a division of St Martin's Press LLC, 175 Fifth Avenue, New York, NY 10010.

Palgrave Macmillan is the global academic imprint of the above companies and has companies and representatives throughout the world.

Palgrave® and Macmillan® are registered trademarks in the United States, the United Kingdom, Europe and other countries.

ISBN 978–0–230–22967–9 hardback

This book is printed on paper suitable for recycling and made from fully managed and sustained forest sources. Logging, pulping and manufacturing processes are expected to conform to the environmental regulations of the country of origin.

A catalogue record for this book is available from the British Library.

A catalog record for this book is available from the Library of Congress.

10 9 8 7 6 5 4 3 2 1
19 18 17 16 15 14 13 12 11 10

Printed and bound in Great Britain by
CPI Antony Rowe, Chippenham and Eastbourne

To the twins Yiannis and Haris

Contents

List of Figures

List of Tables

Foreword

Carole Tongue

In this volume a distinguished and rare combination of international audiovisual expert commentators discuss uniquely and presciently the future of Public Service Media (PSM) within a European prism on the one hand and national public service broadcasting experiences on the other.

A range of audiences, particularly in academia and policy-making, can benefit from this theoretical critical analysis and empirical national case studies at a time when PSM faces challenges to its very existence, definition and funding, and where debate rages as to what role it should play in protecting and enhancing democracy, supporting national culture, identity and creativity in an era of fast technological change and globalisation.

While PSM remains largely imprisoned within the nation-state in a global economy, it is invaluable that the first group of essays take a critical look at Public Service Media in a European context. They throw a welcome spotlight into rarely illuminated corners of the European public service debate.

The EU Amsterdam Protocol, as well as Treaty provisions to uphold cultural and linguistic diversity and the 2005 UNESCO Convention on cultural diversity of expression sit alongside, and in unresolved tension with, free market principles where the European Commission's task is to uphold competition and trade laws. Quite rightly a range of authors here illuminate and discuss the implications of these often paradoxical and contradictory obligations, commitments and loyalties.

In the wake of the mixed public reception to the Lisbon Treaty, European political elites anxiously debate how best to build citizen support for the EU in the absence of transnational public service broadcasting. Best practice and innovation from abroad are barely considered by national governments when framing new laws.

Most welcome therefore are chapters on how PSM could play a part in creating a European demos. In transcending national boundaries and disciplines and in describing the actual PSM environment in a range of countries, including Central and Eastern Europe as well as non-EU territories, these essays could be an important contribution to sound policy-making that draws on multiple diverse experiences.

With few exceptions, media debates are dominated by a technological determinism, claiming multiple digital channels will automatically deliver a cornucopia of diverse programming. Actual principles and regulation necessary to support local culture and the creation of high quality indigenous drama, documentary and film have received less attention in many countries. National examples here, particularly from France and New Zealand, fill this gap.

Our times demand an unfettered imagination of different audio-visual futures where the fragmentation and often lack of diversity produced by market forces are balanced with a diverse PSM that thrives within nation-states and also extends across frontiers to enhance democracy that can match European decision-making, multinational markets and new media.

In this respect, this book helpfully explores the many different public service possibilities in audio-visual media across the globe. It should thus be part of the lexicon and essential toolkit of everyone, particularly policy-makers, to encourage greater understanding of PSM's role in fostering values, meaning, identity and cultural diversity both within nation-states and across frontiers where the screen is so important for mutual understanding.

Acknowledgements

This volume could not have been written without the generous sabbatical scheme of the Department of Sociology, City University London, which allowed me research time. I would like to express my gratitude to the reviewer of the book proposal for welcome comments and suggestions at the beginning of this effort. Most of all, I would like to thank the authors of the essays included in this volume for responding to my call and submitting pieces which provoke fresh thinking, understanding and inquiry about Public Service Media in Europe and beyond. Special thanks are due to my family for their patience and for supporting me throughout this project, and especially to my daughter and twin boys who have been an endless source of inspiration. Thanks also to Christabel Scaife and Renée Takken at Palgrave Macmillan for believing in this project and providing helpful advice. Every effort has been made to trace all copyright holders, but if any have been inadvertently overlooked, the publisher will be pleased to make the necessary arrangements at the first opportunity.

Notes on Contributors

Walter S. Baer is Visiting Fellow in International Communication at the Annenberg School for Communication at the University of Southern California. He has previously held positions as deputy vice president of the RAND Corporation and professor of policy analysis at the RAND Graduate School; chief technology officer for the Times Mirror Company; and assistant to the director of the Office of Science and Technology in the Executive Office of the President. He has served on the Governor's Council on Information Technology for the state of California, and as chair of the Telecommunications Policy Research Conference.

Jean K. Chalaby is currently Reader at the Department of Sociology, City University London. He has published *The Invention of Journalism* (Macmillan, 1998) and *The de Gaulle Presidency and the Media: Statism and Public Communications* (Palgrave Macmillan, 2002). He has edited *Transnational Television Worldwide* (2005) and is the author of *Transnational Television in Europe* (2009). Chalaby has been a visiting lecturer at the University of Geneva for the past five years.

Farrel Corcoran is Professor of Communication at Dublin City University in Ireland, where he previously served as Head of the School of Communication and Dean of the Faculty of Humanities and Social Sciences. In the 1990s he was Chairman of the Irish PSB, RTE. He is a founding member of the editorial board of the *International Journal of Media and Foreign Affairs*. Recent publications include *RTE and the Globalisation of Irish Television* (2004) and various book chapters on topics such as censorship and cultural nationalism in Ireland and digital television.

Alessandro D'Arma is a Lecturer in Media Industry and Policy at the Department of Journalism and Mass Communications, University of Westminster. His research has focused on Italian and British broadcasting policy. In 2007–8 Alessandro worked as postdoctoral researcher on an AHRC-funded research project on pre-school television in the UK.

Trisha Dunleavy is a Senior Lecturer in Media Studies at Victoria University of Wellington, New Zealand. She has published extensively on television systems and dominant TV forms, concentrating on three national paradigms and industries – New Zealand, the United States and the United Kingdom. Her single-author books include *Ourselves in Primetime: a History of New Zealand Television Drama* (2005), and *Television Drama: Form, Agency, Innovation* (Palgrave Macmillan, 2009).

Gay Hawkins is a Professor of Media and Social Theory in the School of English, Media and Performing Arts at the University of New South Wales, Sydney, Australia. Key books she has published are *From Nimbin to Mardi Gras: Constructing Community Arts* (1993) and, with Ien Ang and Lamia Dabboussy, *The SBS Story: the Challenge of Cultural Diversity* (2008). She has also published numerous articles and book chapters on Australia's public service media, the ABC and SBS, and on diasporic and transnational media flows into Australia.

Petros Iosifidis is Reader in Media and Communications in the Department of Sociology, City University London. He is the author of *Public Television in the Digital Era* (Palgrave Macmillan, 2007) and co-author of *European Television Industries* (2005). He has published articles in peer-reviewed journals, book chapters, and has guest-edited a special issue of *Javnost/The Public* on *Digital Television and Switchover* (April, 2007). He has acted as a national expert at the European Commission. In 2008–9 he was Co-Principal Investigator in an EPSRC funded project on the theme of the Digital Economy.

Karol Jakubowicz is a member of the Council of the Independent Media Commission in Kosovo, as well as of the Intergovernmental Council of the UNESCO Information for All Programme. He heads Working Group 2 of the COST A30 ACTION 'East of West: Setting a New Central and Eastern European Media Research Agenda' of the European Science Foundation. He was Director of the Strategy and Analysis Department, the National Broadcasting Council of Poland, the broadcasting regulatory authority (2004–6). He has been active in the Council of Europe and as a member of the Digital Strategy Group of the European Broadcasting Union. As a Council of Europe, UNESCO, European Union and OSCE expert, he has taken part in many missions to advise on the development of broadcasting legislation in a number of countries and has published widely.

Raymond Kuhn is Professor of Politics at Queen Mary, University of London. His publications include *The Politics of Broadcasting* (1985, editor), *Broadcasting and Politics in Western Europe* (1985, editor), *The Media in France* (1995), *Political Journalism* (2002, co-editor) and *Politics and the Media in Britain* (Palgrave Macmillan, 2007). He has published widely on French media policy and political communication in peer-reviewed journals.

Márk Lengyel joined the Office of the National Radio and Television Commission in Hungary in 1998 and in 2002 he established Körmendy-Ékes & Lengyel Consulting while in 2005 he set up his law firm. He has taken part in a series of projects including the Digital Switchover. He is also a lecturer of law at the Moholy-Nagy University of Art and Design Budapest. Márk Lengyel has been an expert on various bodies of the Council of Europe. Currently he is an expert of the CDMC's Group of Specialists on PSM in the information society. In 2006 he was elected vice chairman of the group.

Bienvenido León is Associate Professor at the School of Public Communication of the University of Navarra, Spain. He also teaches regularly at other academic institutions of Spain, Portugal, Argentina, Costa Rica and Ecuador. He has published several books, including *Science on Television: the Narrative of Scientific Documentary* (2007), and various articles for journals and edited books.

Gregory Ferrell Lowe is Professor of Media Management at the University of Tampere in Finland. He has held academic appointments at the University of Kentucky and at the George Washington University in the US. He has worked with executive managers for the Finnish Broadcasting Company, YLE, as Senior Adviser for Corporate Strategy and Development (1997–2007). Lowe is the Continuity Director of the RIPE initiative, an acronym for Re-visionary Interpretations of the Public Enterprise. This initiative has produced four conferences and four books to date.

Maria Michalis is Principal Lecturer and course leader for the Masters programmes in Communication, Communication Policy and Global Media at the University of Westminster, London. She is the author of *Governing European Communications* (2007) and has contributed several book chapters and articles in peer-reviewed journals. She is deputy head of the Communication Policy and Technology Section of the International Association for Media and Communication Research (IAMCR) and a founding member of the Media, Communications and Cultural Studies Association (MeCCSA) policy group.

Cinzia Padovani is Assistant Professor in the School of Journalism, Southern Illinois University at Carbondale. Her publications include *A Fatal Attraction: Public Television and Politics in Italy* (2005), also published as *Attrazione Fatale: Televisione pubblica e politica in Italia* (2007). She has also published various articles in peer-reviewed journals.

Teemu Palokangas is a doctoral candidate in the Department of Journalism and Mass Communication at the University of Tampere, Finland. He has previously worked for YLE, the Finnish public service broadcasting company, as a current affairs journalist and as a market analyst in programme development (YLEdge).

Stylianos Papathanassopoulos is Professor in Media Organisation and Policy and Head of the Faculty of Communication and Media Studies at the National and Kapodistrian University of Athens. His recent books include *Television in the 21st Century* (2005, in Greek), *Politics and the Media* (2004, in Greek) and *Television in the Digital Age: Issues, Dynamics and Realities* (2002). He is also co-editor of *The Professionalization of Political Communication* (2007) and editor of *Communication and Society from the 20th to 21st Century* (2000).

Katharine Sarikakis is Senior Lecturer in Communications Policy and Director of the Centre for International Communications Research at the Institute of Communications Studies, University of Leeds. She has co-edited *Feminist Interventions in International Communication* (2008), co-authored *Media Policy and Globalization* (2006), authored *Powers in Media Policy* (2004) and guest edited *Media and Cultural Policy in Europe* for the European Studies issue (2007).

Philip Savage is Assistant Professor in the Department of Communication Studies and Multimedia, McMaster University. His recent contribution 'The Audience Massage: Audience Research and Canadian Public Service Broadcasting' was published in *From Public Service Broadcasting to Public Service Media* (2008). He has acted as Head of Research for CBC Radio (through the 1990s) and then Senior Manager of Planning and Regulatory Affairs for CBC Radio, CBC Television and CBC.ca (prior to his return full-time to the university in 2005). He is a board member and active participant of RIPE.

Jeanette Steemers is Professor of Media and Communications at the Communications and Media Research Institute (CAMRI), University of Westminster in London. Her book publications include *Changing Channels: the Prospects for Television in a Digital World* (1998), *Selling Television: British Television in the Global Marketplace* (2004) and *European Television Industries* (2005, together with P. Iosifidis and M. Wheeler). She is European Editor of *Convergence*. Her work on television trade and children's television has been funded by the Leverhulme Trust, the British Academy and the Arts and Humanities Research Board.

Paweł Stępka is a media expert and works in the Department of European Policy and International Relations of the National Broadcasting Council of Poland, the Polish regulatory authority in the broadcasting sector. In March 2005 he became a member of the Group of Specialists on Media Diversity (MM-S-MD) established by the Council of Europe. He is a member of the Steering Committee on the Media and New Communication Services (CDMC) in the Council of Europe. He is author of several articles and many analyses prepared for the National Broadcasting Council of Poland.

Barbara Thomass is Professor for International Comparison of Media Systems at the Institute for Media Studies, Ruhr-University in Bochum, Germany. She also held posts at the universities of Hamburg, Göttingen, Lüneburg and Bremen and at the universities in Vienna and Paris. She has been working with several international organisations in courses on journalism standards and ethics in different parts of Eastern and South Eastern Europe. In her former professional life she was a journalist.

Carole Tongue is a former Member of the European Parliament, spokesperson on Public Service Broadcasting, and Chair of the UK Coalition for Cultural Diversity and Visiting Professor, University of the Arts.

Josef Trappel is the head of IPMZ transfer, at the University of Zurich in Switzerland. Since 2008, he has been Visiting Professor for Media Policy and Media Economy at the University of Vienna, Austria. He is member of the Board of Directors of SwissGIS, the Swiss Centre for Studies on the Global Information Society and co-chair of the Euromedia Research Group. He has also acted as media policy expert in the Austrian Federal Chancellery, as national expert at the European Commission, as well as a consultant.

Jeremy Tunstall is Professor Emeritus of Sociology at City University London. He is the author, co-author and editor of twenty books. His sole authored books include *Journalists at Work: Special Correspondents, Their News Organizations, News-Sources and Competitor Colleagues* (1971), *The Media Are American: Anglo-American Media in the World* (1977), *The Media in Britain* (1983), *Communications Deregulation: Unleashing of America's Communications Industry* (1986), *Television Producers* (1993), *Newspaper Power: the New National Press in Britain* (1996) and *The Media Were American: US Mass Media in Decline* (2007). Co-authored books include *Media Made in California* (with David Walker, 1981), *Liberating Communications: Policy Making in France and Britain* (with Michael Palmer, 1990), *Media Moguls* (with Michael Palmer, 1991) and *The Anglo-American Media Connection* (with David Machin, 1999). His edited books include *Media Occupations and Professions* (2001). He is currently writing a book about the revolution in British television, 1990–2010.

Mark Wheeler is a Reader in Politics in the Department of Law, Governance and International Relations, London Metropolitan University. He has published *Politics and the Mass Media* (1997), *Hollywood: Politics and Society* (2006) and he is the co-author of *European Television Industries* (2005). He has contributed several articles for peer-reviewed research journals.

Runar Woldt is a journalist writing for *Media Perspektiven*, a monthly journal published by ARD in Frankfurt, Germany. Between 1991 and 2001 he was Head of Research at the European Institute for the Media, Manchester/Dusseldorf and was involved in numerous research projects. He has published books and articles, including *Perspectives of Public Service Television in Europe* (1998).

Introduction

Petros Iosifidis

This edited collection addresses one of the most challenging debates in contemporary media studies: the transition of the traditional Public Service Broadcasters (PSBs) into Public Service Media (PSM) – that is, widening their remit to be available in more delivery platforms for producing and distributing public service content. Cross-platform strategies help PSM retain audience share, reach new audiences and develop on-demand services, while enabling them to create a stronger partnership with civil society and serve an extended form of citizenship.

The principal challenges facing PSBs include the pressures generated by rapid technological change; the dilemma between the obligation to safeguard citizenship ambitions and support market principles; the legitimacy and performance of PSBs in a multimedia ecology characterised by convergence and fragmentation; and their transition to PSM. Technological change and growing competition in broadcasting has opened up media markets and enhanced media choice, thus challenging the performance of publicly funded PSBs. Furthermore, the shape of society and the mass media have become more internationally oriented and this shift has brought into question the very legitimacy of national communities, identities and ideologies. This has impacted greatly on Europe, which is evolving dynamically and where historically language, ideology, politics, economics and religions complicate the crafting of a unified sense of Europe. The continent's emerging cultural mix, facilitated by emigration and the accession of twelve new nations since 2004 with nearly 120 million people, has complicated the formation of a common cultural conversation and identity and exacerbated the difficulty of framing a constitution.

In the light of the above changes PSBs are struggling to come to terms with Europeanisation and globalisation of media ownership, production, programming and distribution, the 'marketisation' of media output, technological convergence and audience fragmentation, as well as the shift from analogue to digital transmission. While the prevailing nation-state frameworks for cultural and political identity are gradually fading, some PSBs are finding it hard to serve and promote national culture and identity, and meet the challenges of the growing uncertainties in light of a cosmopolitan Europe. But these are considered among the central institutions that can help European citizens make sense of such developments by bearing traces of collective identities and therefore creating an expanded, pan-European cultural space. Can PSBs be 'multicultural', mobilise a new sense of Europeanness, while at the same time transform into PSM and deliver public service content that would meet audience needs in a digital age?

The scholars in this volume discuss the contemporary relevance of PSM as a cultural and political enterprise and as a forum in which a variety of cultural demands are best met. The idea of an edited collection on PSM was born when I was finishing my monograph, *Public Television in the Digital Era* (2007) which discussed PSB in six EU countries. Already then I had a strong sense that further research on public service institutions must be carried out more systematically and on a broader scale. Meanwhile, other titles like Lowe and Bardoel's 2007 edited collection *From Public Service Broadcasting to Public Service Media* dealt with this issue, but this was mostly Europe-centred with particular emphasis on the North. A Council of Europe report prepared by Lowe (2007) and a special issue of the journal *Convergence* edited by d'Haenens et al. (2008) also contributed to the debate on the role of PSM, the former tackling issues of citizen participation and the latter media policy matters. Other edited collections such as Sarikakis' *Media and Cultural Policy in the European Union* (2007), de Bens' *Media Between Culture and Commerce* (2007), Bondebjerg and Madsen's *Media, Democracy and European Culture* (2008) and Uricchio's *We Europeans? Media, Representations, Identities* (2008) address some of the issues explored here (public sphere, public policy, culture, identity and democracy, the European integration project), though none of them focus specifically on how PSBs evolve in this ecology.

The title *Reinventing Public Service Communication: European Broadcasters and Beyond* may seem rather vague and ambitious, but it reflects how social change and new technologies require these public institutions to evolve from basic broadcasting services into an engine that provides information and useful content to all citizens using various platforms. What distinguishes this edited collection is that it blends theoretical critical analysis with empirical national case studies. A diverse group of scholars has been brought together to provide a range of well-argued, independent yet critical perspectives on the issue of PSM. The main questions posed in this book are:

- What strategies would public service enterprise need to renew and reshape while maintaining public service principles?
- How can PSBs take account of the different media platforms for PSM (online, on-demand, mobile) and the changing relationship with the audience (as content generators and a community of users)?

- Are European media and cultural policies developing satisfactorily within the context of enlargement and alleged European integration?
- To what extent and why is there a European public sphere and identity and how is this documented by public media?
- How do non-EU PSM systems perform and how do these broadcast economies link with the EU area?

The authors address these questions not merely through the lens of European integration, cultural policy and the public sphere (or the absence of it), but also through wider concerns relating to the PSM position within highly competitive national marketplaces, in particular by touching on issues of PSM strategy, branding, content and, crucially, PSM's relationship with its audiences, who may expect more interaction and also generate content themselves. These questions are approached more specifically in the two parts of the book.

Book structure

Part I gives the context for the theory of PSM in political and regulatory thinking that aims to inform and promote debate around public service communication. Jakubowicz (Chapter 1) discusses broadly the key issues affecting PSM and opens up interesting questions about content, PSM's relationship with the audience (as content generators and a community of users), funding and PSM 'renewal'. The author suggests that there is a need for a structural reorganisation of the public institutions to promote fully integrated, digital multi-platform production processes. Iosifidis (Chapter 2) outlines the changing PSB environment in the UK and beyond and considers questions of funding, content and regulatory arrangements. Having discussed in detail the PSB systems in the UK and across Europe, Iosifidis argues that *plurality of institutions (providers)* and *plurality of funding* cannot necessarily ensure *plurality and diversity of content* and instead favours an adequately funded, consolidated PSB system. Michalis' Chapter 3 on European broadcasting governance and the Audiovisual Media Services Directive and Wheeler's Chapter 4 on competition law and state aid provide good overviews on key areas of EU involvement. Michalis stresses that although the EU has explicitly recognised the significance of socio-cultural aims, public services and citizens' rights, its substantive output remains centred on economic and competition considerations. In the same vein, Wheeler implies that the EU's rigid employment of state aids fails to take into account the social, cultural and democratic functions of PSBs.

Thomass (Chapter 5) challenges the idea of a single public sphere and opts for a multiplicity of European public spheres in which PSM can contribute politically and culturally. She suggests the extension of the national mandates to a European scale and the use of online media by PSBs to address audiences beyond national borders as a possible PSM contribution to Europeanising the public sphere(s). Corcoran's Chapter 6 is useful in underlining the problems of looking at PSM through an EU lens – namely that there is widespread indifference to EU affairs; that there is no European public sphere; and that there is little interest in Brussels' regulatory powers by the national media. Sarikakis (Chapter 7) explores the setting of international standards for PSM and the relationship of PSM to culture and democracy. She argues that in line with technological, socio-cultural and political developments, a European redefined PSM system would protect and promote an extended form of citizenship that encourages the continuous critical approach to established practices and dominant values. Chalaby's Chapter 8 on transnational broadcasting underlines how difficult it is to maintain effective international channels by pinpointing that it is only the BBC that understands today's market conditions (but of course has the benefits of broadcasting in the English language which opens up the US market).

By focusing on children's media, D'Arma and Steemers (Chapter 9) offer an interesting entry point to debates about PSBs as they reconfigure as PSM organisations. Having examined the remit and obligations of the BBC, Italian RAI and US PBS in relation to children's provision and funding, the authors are convinced that *funding,* and not *regulation,* is the critical issue for the continuing provision of children's programming. The tenth and final chapter in Part I by Lowe and Palokangas focuses on the public service brand and explores the strategic reasons for brand management as PSB transitions into PSM. The authors advocate the option of developing the strategic brand management capability, with its emphasis on understanding and nurturing PSB as a heritage brand from a customer-centric perspective. Chapters 9 and 10 include specific national case studies and therefore are deliberately positioned last to provide a link between Parts I and II. Chapter 10 stands out in presenting the business perspective, in contrast to the rest of Part I, which is highly normative in tone and focuses on the political, policy and citizenship framework.

Part II is the empirical section and covers five large (UK, France, Germany, Italy, Spain) and two small EU countries (Austria, Greece), a non-EU territory (Switzerland), a large and a small Eastern/Central EU country (Poland, Hungary), and four countries beyond Europe (US, Australia, Canada, New Zealand). Academic localisation enables the authors to highlight nationally distinct patterns of PSM focusing on particular ideas, trends, values and systems existing in different territories. By taking a comparative approach to the topic, this volume is organised around a set of common questions, themes and methodological reflections. The contributors reflect on key issues in the countries concerned and adapt a multi-vocal and multi-disciplinary perspective, by covering the following basic areas: organisation, funding, political independence, editorial autonomy and regulation of the national PSB; the distinctiveness of PSB services and content in the digital era and its transformation to PSM; and reflection on the national PSBs' contribution to the creation of a genuine multicultural society.

Tunstall (Chapter 11) discusses extensively UK broadcasting policy, how UK PSB is gradually and continuously redefined, and why UK broadcasting policy and its regulation has in recent years evolved from an amateur to a more professional approach, while PSB has been quantified and is therefore more receptive to quantitative measurements. Kuhn (Chapter 12) discusses the distinctive contribution of the French public service provider in a system characterised by market competition and audience fragmentation. What makes the situation particularly difficult for France Télévisions is the combination of the hostility of an interventionist president and the lobbying influence of commercial broadcasters on government policy. In Chapter 13, Woldt discusses the problems facing PSBs in Germany, paying particular attention to digital and online developments and the relating political tensions between German public broadcasters and the EC. Padovani (Chapter 14) explores RAI's plans towards becoming a multimedia public service corporation. She is sceptical of its ability to contribute to a new multicultural society and to the formation of a European public sphere in a media ecology influenced by the so-called 'Berlusconi factor'. León's Chapter 15 concludes the overview of large European countries by focusing on the largely ineffective transformation process of the Spanish public broadcaster RTVE in both organisational and programming terms.

Trappel delves into the public debate on PSM in Austria and Switzerland (Chapter 16), where uniquely television offerings are strongly influenced by non-national players. In both these small countries expectations of ownership diversity and programming autonomy are unrealistic so media governance requirements focus on the conduct of PSM organisations. Chapter 17 by Papathanassopoulos presents the PSB system in Greece which, as in many other Southern European states, has been used as a vehicle for negotiating with and pressuring the government of the day, rather than facilitating a public conversation and discussion. The next two chapters focus on Eastern/Central EU countries. Stępka (Chapter 18) argues that while market size has not 'protected' Polish PSBs from the difficulties typical for this region (that is, pressure exerted by the main political forces and ineffectiveness of the funding mechanism), public institutions enjoy a fairly strong position, which makes this system unique among other Eastern/Central European countries. Chapter 19 by Lengyel stresses that the term 'PSB' in Hungary started to gain real meaning only after the country's transition from the former communist regime to democracy, but there are still a number of shortcomings, including an insufficient definition of the public service remit and a weak PSB connection with the rest of society.

The last four chapters consider other national situations beyond the EU for comparative purposes and to inform the debate internationally. Chapter 20 by Baer argues that the US system of PSB (which differs markedly from those in Europe in that the US chose to grant local broadcast licences to non-government entities) needs to undergo substantial structural and cultural changes if it wants to stay relevant and thrive in today's media ecology. Savage (Chapter 21) pinpoints that while the CBC was assumed to help form and protect the Canadian identity by broadcasting Canadian Content ('CanCon') the digital revolution has made such a cultural project much tougher in the context of the world's highest levels of integration with foreign – read American – markets. Hawkins (Chapter 22) shows how Australian ABC and SBS ignore the impact of massive changes in technology, audience practices and social composition, while the last contribution by Dunleavy (Chapter 23) interrogates the New Zealand political experience in changing the broadcasting funding model over the last fifteen years and provides useful lessons for the current UK debate on a top slicing of the licence fee revenue for use by private companies in making public service programmes.

Final word

I am a firm supporter of a PSM system that would provide a wide range of high quality, universally accessible content, free at the point of consumption. In the midst of a global economic crisis it is becoming increasingly apparent that the *public sector*, rather than the *free market*, is the answer to the continuing supply of high quality public service output. Policy-makers, politicians, academics and the media industry have much to learn from the practical experience of studying diverse national public service media landscapes.

References

de Bens, E. (2007) *Media Between Culture and Commerce* (Bristol, UK: Intellect).

Bondebjerg, I. and P. Madsen (2008) *Media, Democracy and European Culture* (Bristol, UK: Intellect).

d'Haenens, L., H. Sousa, W. A. Meier and J. Trappel (2008) 'Editorial' in *Convergence: The International Journal of Research into New Media Technologies*, 14(3): 243–7.

Iosifidis, P. (2007) *Public Television in the Digital Era: Technological Challenges and New Strategies* (Basingstoke: Palgrave Macmillan).

Lowe, G.F. (2007) *The Role of Public Service Media for Widening Individual Participation in European Democracy* (Strasburg: Council of Europe), November.

Lowe, G. F. and J. Bardoel (2007) *From Public Service Broadcasting to Public Service Media*, RIPE@2007 (Göteborg, Sweden: NORDICOM).

Sarikakis, K. (2007) *Media and Cultural Policy in the European Union* (Amsterdam and New York: Rodopi).

Uricchio, W. (2008) *We Europeans? Media, Representations, Identities* (Bristol, UK: Intellect).

Part I

1
PSB 3.0: Reinventing European PSB

Karol Jakubowicz

Introduction

The history of European Public Service Broadcasting (PSB) has so far encompassed two main periods: the time up until the 1980s, before it faced commercial competition (PSB 1.0) and the period of great upheavals and change since then (PSB 2.0). PSB also needed to find its bearings in a multi-channel broadcasting landscape, leading to 'a significant level of commercialisation, where differences with commercial television are, in general, relatively small' (León, 2007: 98). Now is the time for PSB 3.0 – the twenty-first-century version that we would probably invent if we were to create PSB today, necessarily very different from the one we have inherited.

PSB is challenged by neoliberal and postmodern sentiments, convergence, internationalisation and globalisation, privatisation and commercialisation (Syvertsen, 2003). The key challenge is general social, cultural and ideological change (see Ofcom, 2004), above all the ascendancy of neoliberalism, opposed to the existence of public sector institutions where market forces should, in this view, operate without hindrance. The main rationale for PSB has been that it serves the public interest, defined as 'informational, cultural and social benefits to the wider society which go beyond the immediate, particular and individual interests' (McQuail, 1992: 3), but it has been dethroned by commercial and individual interests.

General democratisation and rising affluence in European societies have levelled social divisions and stratification. The old paternalism of PSB as the voice of authority (or, even worse, of the authorities), or of the social elite, is thus no longer acceptable. Cultural change has undercut the view that some cultural products are more valuable than others. A more postmodern attitude rejects traditional taste and cultural hierarchies. The boundary line between art and commerce has become blurred. Consumers are seen as the only relevant arbiters of taste. The postmodern aesthetics cherishes playfulness, irony and intertextuality, best expressed in advertising and commercial forms, such as music videos (Syvertsen, 2003).

Failure to follow these cultural trends is why PSB is losing the young audience, in any case being weaned by new technologies away from traditional radio and television altogether: 'Consumers (especially young people) are actively looking for ... specific content and use multiple platforms, such as TV, games console, radio, PC, mobile phone and MP3 players' (European Commission, no date: 4). Another factor is a sense of entitlement shared by many individuals. They claim opportunities of access to creative and cultural content (Ewing and Thomas, 2008), and they are no longer willing to be passive. For many, the traditional mode of PSB communication – one-way and top-down – is no longer tenable.

Because of individualisation and fragmentation of society, PSB must redefine its service to social integration and cohesion and go beyond collective experience (that is, generalist channels) and cater to group and individual interests, for example by providing thematic services and online services. The 'crisis of democracy' (Political Affairs Committee, 2007) and rising political disengagement among the public require a serious rethinking of the traditional role of PSB in democracy and its dedication to serving citizenship.

Globalisation and international integration disorient PSB, typically closely bound with the nation-state. Societies are increasingly multi-ethnic and multi-lingual. Respect for cultural diversity is recognised as a key policy objective, again forcing a revision of the traditional PSB task to preserve national and cultural identity.

Change in communication technology and consequently in patterns of social communication (Cardoso, 2006) calls for an upheaval in the way PSB has produced and disseminated content. The traditional PSB role as a communicator engaged in linear 'push' communication is out of touch with the interactive, multimedia technological reality of today.

In short, the entire context within which PSB operates has changed fundamentally. So must PSB: its remit should be extended and revised to fit the new societal and global circumstances and its organisation, production methods and relations with the audience, society and with the outside world in general must also be completely overhauled (Bardoel and d'Haenens, 2008; Jakubowicz, 2007, 2008). The same goes for policy and regulatory frameworks relating to PSB (Group of Specialists, 2008a). Nothing less than a Copernican revolution is likely to achieve the desired effect.

This chapter reviews how European PSB should evolve in response to these challenges.

Do we still need PSB?

Defining PSB is notoriously difficult. For technological reasons, new names for public service *broadcasting* are being proposed: 'public service *media*' (Committee of Ministers, 2007), or 'full-service public *communicator*' (Raboy, 2008: 3). PSB is a broad concept, seen as part of 'the European model of society' (European Commission, 1998; Parliamentary Assembly, 2004), but here we can use Syvertsen's (2003: 156) technology- and institution-neutral definition of PSB as a particular model of communication governance, a set of political interventions into the media market to ensure that content 'valuable to society' is available to all. It is a public institution set up for that specific purpose and offered an (often unsatisfactory and insufficient) institutional and financial framework needed to meet that purpose.

PSB would no longer be needed if without it the public could have equally easy access to a comparable volume and quality of such content, offered by the market. In fact, paradoxically, online media fail to deliver much new quality content (*The Economist*, 2006), and commercial broadcast media may in future, due to rising competition and movement of advertising spending to the internet, be even less able to do that (Ward, 2006), especially via free-to-air generalist radio or television channels. The UK government has recognised that 'in television and radio, as with news, we may no longer be able to rely on the provision in the future of the wider range of public service programming from varied sources to which we have been used' (DCMS/DBERR, 2009: 45).

PSB will be more important than ever in the digital age in order to counterbalance the declining scope for broadcasting law and regulation so as to ensure that all society's communication needs are met; to maintain quality programme standards in the context of widespread commercialisation; to act as a counterweight to powerful private media companies; and to promote national and societal cultural identities in the face of globalisation (Humphreys, 2008). PSB relevance for the democratic process and the policy discourse grows as big corporate media lose – due to commercialisation, deregulation, concentration and internationalisation – their identification with, and commitment to, any specific political system (Trappel, 2008a). Enli (2008a) adds that serving as an alternative to a global hyper-commercial market for children's programming is probably among the most important public service remits of the next decades.

However, PSB can no longer rely on good will and the axiomatic recognition of the need for its existence. Governments, parliaments and other policy-makers seem to be engaged in a waiting game, unsure where to go next. A crucial factor here is EU media policy, where PSB is like 'a square peg in a round hole' (Jakubowicz, 2004), an object of concern primarily in terms of competition and state aid policy that are extraneous to what PSB is about.

Do we still need PSB institutions?

It is often argued that in the digital era, public intervention to guarantee a supply of 'socially valuable' content could take the form of direct funding for content producers (Foster, 2007; Ofcom, 2008) rather than PSB institutions. This is known as 'distributed public service', or 'deinstitutionalisation of PSB'. However, there is no evidence of a general policy move to this as a replacement for PSB institutions. The British government is 'committed to a strong, fully funded BBC at the core of delivering public purposes in Britain's media' (DCMS/DBERR, 2009: 45). At a time when the media are entering a 'post-objectivity' period and, given that the internet is a source of highly partisan content, the importance of PSB as a provider of impartial, high-quality news is seen to grow (Humphreys, 2008).

The shape and number of PSB institutions is another matter. The British government believes that 'we need at least one other provider of scale as well as the BBC' (DCMS/DBERR, 2009: 45; see also Foster, 2008). In this view, plurality of public service provision could also mean a decentralised production system, producing regional programming in the regions it is meant for, as well as reliance on independent producers.

Another tack is taken by those who believe that PSB institutions could (perhaps should) be replaced by community media, social networks or alternative, internet-based media (Harrison and Wessels, 2005; Kearns, 2003; Rozanova, 2007), or, as in Coleman's proposal, by a 'publicly-funded, independently-managed online civic commons', with a 'key role' for PSB (cited in Moe, 2008). It should encompass diverse forms of communication – from dialogic to disseminating modes.

Another view is that 'commercial convergence' should be countered by 'public sector convergence': PSB should become 'the central node in a new network of public and civil institutions that together make up the digital commons, a linked space defined by its shared refusal of commercial enclosure and its commitment to free and universal access, reciprocity, and collaborative activity' (Murdock, 2004). This could encompass various educational, cultural and other public institutions, libraries, universities, museums, community and alternative media, user-generated content and other elements of the non-commercial public forum and public-spirited digital commons.

This could also be the structural and organisational answer to the issue of 'plurality' and to the obvious mismatch between, on the one hand, the network society, characterised by 'creative chaos' and 'relentlessly variable geometry' on the national and global scale (see Castells, 1996), and, on the other, a 'Fordist', centralised PSB organisation.

Policy intervention to support this form of 'PSB plurality' could ensure the availability of socially valuable content from a diversity of sources (including commercial entities, but without weakening PSB institutions, or their funding). To be able to play this role, PSB needs to have considerable institutional, organisational and financial critical mass.

It should, therefore, be well-funded (Papathanassopoulos, 2007; Picard, 2005). To ensure this for the future, licence fees are charged for possession of any TV signal-receiving equipment, or replaced by budgetary appropriations. Licence fees may not survive for ever, and probably funding will eventually come mostly from budgetary appropriations, but a commitment to maintenance of PSB must also mean provision of adequate funding.

The 'commons' concept would have to be mandated by law, as has happened in Portugal, where the second channel of RTP has been 'opened up to institutions of "civil society" which participate through production and exhibition of programmes' (Correia and Martins, 2007: 268). Plans announced in September 2008 to create a new Dutch public broadcaster for Muslims show that an open PSB system, capable of admitting new voices, is possible.

The public service remit

The PSB remit is usually defined in law in terms of qualitative requirements (PSB must broadcast certain programme types: news, education, culture, and so on); or quantitative requirements: an obligation to broadcast a minimum percentage or amount of certain programme types or genres. Requirements may focus on the audience: for instance, PSB must reach certain (shares of) audiences, serve specific target groups, and broadcast certain programmes at peak time.

The remit is usually defined in terms of an 'all-embracing' approach and PSB has the legal obligation to serve a large general audience as well as specific target groups. PSB is normally entitled to broadcast programmes that are also offered by commercial broadcasting (Betzel, 2007).

The debate concerning the 'distinctiveness' of PSB tends to be very hypocritical. PSB is expected 'to differ from the private channels in its programming and . . . be similar at the same time' (Atkinson, 1997: 25). The European Union and some national policy-makers, often acting under pressure from the commercial sector and motivated by competition policy, are pushing PSB to concentrate on redressing market failure and thus on content that is not attractive to commercial content providers (Jakubowicz, 2004). Meanwhile, 'the "Reithian trinity" ... has been in a process of redefinition ... public service broadcasting in the digital era may best be described as "entertainment, education and participation"'. Also, 'in tandem with the increased focus on participation, there has been a general popularisation of PSB entertainment, with less focus on academic knowledge and more focus on general knowledge, and even *play*' (Enli, 2008b: *passim*; emphasis added).

It is time for a reality check, an approach going beyond the nostalgic vision of the 'enlightenment' role of PSB and 'high culture distinctiveness', and acknowledging cultural change, postmodern tastes and standards and new audience/user expectations. PSB should be distinct by virtue of the *functions* it performs and the *value* it brings to society – regardless of the genres and types of content used to perform those functions. If, in the words of the Amsterdam Protocol, it continues to be 'directly related to the democratic, social and cultural needs of each society and to the need to preserve media pluralism' (see also Committee of Ministers, 2007, for a more developed definition of the PSB remit), then it will be distinctive enough, even if entertainment, enjoyment and play are involved.

Whatever else happens, however, it is clear, as we show below, that tasks subsumed under the general rubric of 'socially valuable content' need to be modernised and adjusted to new circumstances. Thus, obligations relating to support for political citizenship and democracy require a new approach, especially as regards promoting the democratic participation of individuals (Group of Specialists, 2008b) and encouraging participatory forms of democratic involvement (Lowe, 2007). The role of PSB in democracy should be extended to the international arena, to serve as a watchdog also of the EU and other European organisations and to help create a 'European public sphere'. To do this and reduce the EU democratic deficit, PSB should disseminate a European news agenda, and help develop European publics with background knowledge, interest and interpretive skills to make sense of the EU and its policy options and debates (see Schlesinger, cited in Golding, 2006).

In terms of its cultural tasks, PSB must respond to globalisation, migration, the increasingly multicultural nature of many societies and the need to maintain or promote social cohesion and facilitate intercultural and interreligious dialogue and understanding among peoples (Committee of Ministers, 2008). As regards education, new tasks for PSB result from educational challenges and skills needed in the new societal, technological and cultural circumstances, such as the 'new literacies' of the twenty-first century: technology literacy; information literacy; media creativity; and global literacy: literacy with responsibility (Varis, no date).

Given the ongoing fragmentation and individualisation trends, PSB's role in maintaining social cohesion must reconcile adequate service to various groups and individuals with maintenance of a necessary level of social cohesion, and reduce the digital divide and exclusion (Bennett, 2008), among other things by playing a leading role in the digital switchover; being available on all digital platforms; supporting traditional broadcasting content with internet and interactive resources; providing multimedia interactive services, independent and complementary web services; actively promoting digital media literacy, and so on.

The need to regain the youth audience requires recourse to its tastes and aesthetics, as shown by public service radio channels for young audiences dedicated, for example, to CHR (Contemporary Hit Radio), Pop/Rock, Alternative and Urban music formats (Strategic Information Service, 2008). It is not the kind of programming fare to set them apart from commercial channels, but here the broadcaster has to adjust to the audience's tastes, and not vice versa.

PSB and the new technologies

The concept of Public Service Media (PSM) can be briefly summed up as 'PSB + all relevant platforms + Web 2.0', representing a technology-neutral definition of the remit. The goal is multimediality due to technological, media and organisational convergence, changing the way in which programming/content is made (Erdal, 2007), and requiring market, product and process innovation (Bechmann, 2008). This makes possible cross-media strategies, such as newsroom convergence (Dailey et al., 2003).

This requires a structural reorganisation to promote fully integrated, digital multi-platform production processes, making possible the application of the COPE strategy, 'create once, play everywhere' (Huntsberger, 2008), and retraining of staff, enabling it to engage in multi-skilling. Cross-platform strategies pursued by PSB institutions help retain audience share in multichannel households; reach new audiences, including young people, with branded content; provide an additional offer for niche audiences and for typical PSB programming (children, culture and documentary); and develop on-demand services (Leurdijk et al., 2008). Special interest channels and on-demand services become part of the cross-media strategy linked to websites, communities, user-generated content, civil journalism, archive material, mobile services and so on, including cross-referencing, branding and marketing. PSB offers podcasting and/or catch-up streaming radio services. Mobile services, online forums and other value-added services are also widely implemented. They take advantage of new media trends such as social networking, audio and video sharing, blogging and so on.

In all this, 'the future of Public Service hangs in the balance . . . The greatest single threat may be the risk of inactivity by being too slow or too late in engaging online audiences' (Strategic Information Service, 2007: *passim*).

Not all aspects of PSB evolution in the twenty-first century are the subject of equally heated policy controversy (Trappel, 2008b). Again, the EU, by attempting to set limits on what PSB can do with the use of the new technologies, is having a strong impact on national policy in member states; hence a wide variety of policy and regulatory approaches are adopted in different countries. Moe (2008) identifies three main ones:

- 'Extending broadcasting' – fitting new services under the umbrella of 'broadcasting'.
- 'Adding to broadcasting' – new activities are appended to the traditional ones as complementary and secondary, offering only programming-related content. This hinders the use of new platforms for PSB purposes (see Dörr, 2008, on German policies of this nature, adopted under pressure from the European Commission).
- 'Demoting broadcasting' – this leaves 'broadcasting' behind as the principal defining term, with broadcasting no longer seen as the key component of public service provision.

The third approach, which is the case in the UK, is the most future-oriented one, though PSB has a duty not to abandon those who prefer traditional radio and television. PSB should be legally supported and required to offer mass, specialised and internet content as the default and core content offer, as 'legal and financial empowerment is the prerequisite for making the innovation of online distribution a sustaining one' (Strategic Information Service, 2007: 132).

Policy and regulation should also redefine some of the basic features of PSB by changing the 'unit of account'. Universality of content – yes, but across the full range of services, including thematic channels or online services. Universal access – yes, but not universal use, as that cannot be expected in the case of thematic or online services. Internal diversity – yes, but in many cases PSB can also provide external diversity via different services, or indeed different institutions.

Re-embedding PSB in society

PSB has so far largely failed to respond, in its organisation, management structures and relations with civil society, to the rise of networked, non-hierarchical forms of multi-stakeholder governance and social relations. No wonder: it even has a limited view of itself as a *public* institution: 'many PSBs have kept the people and civil society at a distance, while politics and the government proved to be the preferred partner' (Bardoel and d'Haenens, 2008: 340).

Of course, many PSB organisations have outreach programmes and public accountability systems. Referring to the BBC, Born (2002) listed, among others: the 'BBC Programme Complaints Unit'; the 'Board of Governors'; 'The BBC Listens', four-yearly review of services; BBC Online inviting feedback on the Annual Report; Governors' seminars; and Annual Statement of Promises. Many of the above have strong links with civil society. German PSB institutions have large Administrative Councils with extensive representation of 'socially relevant groups' (though these Councils often turn out to be highly politicised (Hoffmann-Riem, no date). Simple forms of feedback and participation (phone-in programmes, Short Message Service and internet voting) are common. In their policy documents some PSBs 'emphasise [such] audience participation as a strategic response to the challenges in the digital age' (Enli, 2008b: 109; Jackson, 2008). This is meant to provide a new source of legitimacy by replacing 'passive viewing' with active participation, indeed 'public service participation'. 'User-generated content' and collaboration with 'citizen journalists' are experimented with, though in limited ways (Heinrich, 2008).

PSB must, however, redefine its place in society, seeking and enabling genuine participation by (and partnership with) civil society (Lowe, 2008). Access to air time and participatory programming would respond to a growing desire of many to be 'producers of cultural content' (Ewing and Thomas, 2008) and to use the internet and other ICTs (information and communication technologies) 'as a tool for the multimodal, individualised and specialised contribution to technology-mediated communication processes . . . a window for the world to self-broadcasting' (Tabernero et al., 2008: 288).

Participatory programming can also mean public scrutiny of editorial policy and a broadcaster–audience dialogue about it. Public scrutiny is made possible, for example, by 'Open Newsroom', a project of Aktuellt, a daily news programme of the Swedish public television SVT which publishes on its website film recordings of staff meetings and other discussions among journalists, only minutes after they occur (Elia, 2008). Dialogue is made possible by the 'The Editors', the blog of BBC News editors, used to explain editorial policy and engage in debate with viewers. Many public broadcasters also maintain online communities connected to particular programmes.

As for social re-embedding of management, and policy-making, Kleinsteuber (2008) calls for meetings of the Administrative Councils of German PSB institutions to be held in public (and broadcast online), with all the documentation to be uploaded on the internet as a matter of routine.

In the UK, research conducted before the creation of the BBC Trust showed that respondents wanted the then BBC Governors to be far more representative of (and more directly accountable to) the licence fee-paying public. They argued that the Governors should be elected by the general public and called for greater transparency: online webcasts of meetings, with 'cyber-seats' available for licence fee payers online (Ubiqus Reporting, 2004).

The Facebook social network has been forced to introduce a new charter, giving members voting rights over the firm's future policies regarding how the site is governed (Waters, 2009). Radio and television stations will sooner or later have to follow suit.

Conclusion

Let us sum up, then. While the fundamental rationale of PSB to deliver socially valuable content and protect and promote the public interest remains the same, almost everything in the way it performs its mission should change. What is lacking, though, is properly informed and motivated policy to realise this change.

PSB functions and the value it provides are not (and may not any time soon be) offered in equal volume and quality by commercial electronic media, or by online content providers. Nevertheless, a new definition of PSB distinctiveness is needed: one that goes beyond the 'enlightenment' role of PSB and takes into consideration cultural change, postmodern tastes and standards, and new audience/user expectations. To regain the young audience, PSB should adjust its content to the younger population's needs, aesthetic tastes, forms of expression and favourite platforms.

There is a continued need for strong, well-funded PSB institutions, capable of delivering socially valuable content (also as commissioners of content produced externally). PSB could perform as the central node (and co-organiser) of a broad network of public and civic institutions and groups, as well as individuals, capable of contributing to the provision of such content and keeping public debate alive. Policy can provide for a plurality of PSB institutions and for provision of socially valuable content by commercial entities, but this should not weaken or undercut PSB institutions, or their funding.

There is a need for a technology-neutral definition of the remit, with broadcasting and the new platforms treated equally, each in terms of how they can best be used to deliver a public service. The new technologies offer PSB a chance to perform its role better and to serve the audience in more varied ways than before. This is why PSB should be transformed into PSM – multimedia institutions restructured to produce and distribute content digitally and to take full advantage of opportunities offered by the new platforms.

It is imperative to re-embed PSB institutions in society, by means of 'participatory programming', open and accountable management, opportunities for the public to participate in editorial decision-making, and finally systems of governance in line with the way the network society operates.

Policy today lags far behind the needs of society as concerns PSB because it is made under growing pressure from the commercial sector and from the European Union, which perceives PSB increasingly as a problem, and not, as it should be, part of the solution to societal problems. If that attitude does not change, the future of PSB is not assured.

The rising tendency to treat PSB as an 'anomaly' and a threat to the interests of the commercial sector may perhaps be reversed as, in the wake of the financial and economic crisis of 2008–9, nation-states and the international community re-evaluate the neoliberal model of society and the role of the state and the public sector in social life in protecting the public interest.

EU member states should insist on full respect for the principle, enshrined in the Amsterdam Protocol, that it is their competence to confer, define, fund and organise the fulfilment of the public service remit.

References

Atkinson, D. (1997) 'Public Service Television in the Age of Competition', in D. Atkinson and M. Raboy (eds), *Public Service Broadcasting: the Challenges of the Twenty-First Century. Reports and Papers on Mass Communication*, No. 111 (Paris: UNESCO), pp. 17–74.

Bardoel J. and d'Haenens, L. (2008) 'Reinventing Public Service Broadcasting in Europe: Prospects, Promises and Problems', *Media, Culture & Society*, 30(3): 337–55.

Bechmann P. (2008) 'Cross Media as Innovation Strategy: Digital Media Challenges in the Danish Broadcasting Corporation', paper for the RIPE@2008 Conference, Mainz.

Bennett, J. (2008) 'Interfacing the Nation: Remediating Public Service Broadcasting in the Digital Television Age', *Convergence*, 14(3): 277–94.

Betzel, M. (2007) 'Public Service Broadcasting in Europe: Distinctiveness, Remit and Programme Content Obligations', in E. De Bens et al. (eds), *Media between Culture and Commerce* (Bristol, UK: Intellect Books), pp. 142–50.

Born, G. (2002) 'Accountability and Audit at the BBC', paper prepared for an expert meeting 'Auditing Public Broadcasting Performance' organised by the Amsterdam School of Communications.

Cardoso, G. (2006) *The Media in the Network Society: Browsing, News, Filters and Citizenship* (Lisbon: Centre for Research and Studies in Sociology).

Castells, M. (1996) *The Rise of the Network Society* (Cambridge: Blackwell Publishers).

Committee of Ministers (2007) *Recommendation Rec(2007)3 on the Remit of Public Service Media in the Information Society* (Strasbourg: Council of Europe).

Committee of Ministers (2008) *White Paper on Intercultural Dialogue: 'Living Together as Equals in Dignity'*, CM(2008)30 final 2 (Strasbourg: Council of Europe).

Correia, F. and C. Martins (2007) 'The Portuguese Media Landscape', in G. Terzis (ed.), *European Media Governance: National and Regional Dimension* (Bristol, UK: Intellect), pp. 263–77.

Dailey, L., L. Demo and M. Spillman (2003) 'The Convergence Continuum: a Model for Studying Collaboration Between Media Newsrooms', paper for the Newspaper Division of the Association for Education in Journalism and Mass Communication, Kansas City, Missouri, July–August, http://web.bsu.edu/ldailey/converge.pdf, accessed 25 February 2009.

DCMS/DBERR (2009) *Digital Britain: the Interim Report* (London: Department for Culture, Media and Sport; Department for Business, Enterprise and Regulatory Reform).

Dörr, D. (2008) 'The Inner German Debate on Online Actions of Public Service Broadcasting and the Realization of the "Public Value Test" in Germany', paper for the RIPE@2008 Conference, Mainz.

Elia, C. (2008) 'Open Newsroom', *Neue Zürcher Zeitung*, 22 August, http://www.ejo.ch/index.php?option=com_content&task=view&id=1572&Itemid=138, accessed 15 February 2009.

Enli, G. (2008a) 'Serving the Children in Public Service Broadcasting: Exploring the TV-channel NRK SUPER', paper for the RIPE@2008 Conference, Mainz.

Enli, G. (2008b) 'Redefining Public Service Broadcasting: Multi-Platform Participation', *Convergence*, 14(1): 105–20.

Erdal, I. (2007) 'Researching Media Convergence and Crossmedia News Production: Mapping the Field', *Nordicom Review*, 28(2): 51–61.

European Commission (1998) *The Digital Age: European Audiovisual Policy. Report from the High-Level Group on Audiovisual Policy* (Brussels: European Union).

European Commission (no date) *Review of the Broadcasting Communication: Summary of the Replies to the Public Consultation*, Commission Staff Working Paper, Brussels, European Union, http://ec.europa.eu/competition/state_aid/reform/comments_broadcasting/summary.pdf, accessed 10 February 2009.

Ewing, S. and J. Thomas (2008) 'Broadband and the "Creative Internet": Australians as Consumers and Producers of Cultural Content Online', *Observatorio (OBS*) Journal*, 6: 187–208.

Foster, R. (2007) *Future Broadcasting Regulation* (London: Department for Culture, Media and Sport), http://www.culture.gov.uk/images/publications/FutureBroadcasting-Regulation.pdf, accessed 8 February 2009.

Foster, R. (2008) 'Plurality and the Broadcasting Value Chain – Relevance and Risks?' in T. Gardam and D. A. L. Levy (eds), *The Price of Plurality: Choice, Diversity and Broadcasting Institutions in the Digital Age* (Oxford: Reuters Institute for the Study of Journalism, University of Oxford), pp. 25–35.

Golding, P. (2006) 'European Journalism and the European Public Sphere: Some Thoughts on Practice and Prospects', paper presented at the conference on European media, democracy and Europe, Copenhagen.

Group of Specialists (2008a) *Report on How Member States Ensure the Legal, Financial, Technical and other Appropriate Conditions Required to Enable Public Service Media to Discharge their Remit: Compilation of Good Practices*, prepared by the Group of Specialists on Public Service Media, MC-S-PSM(2008)004 rev. (Strasbourg: Council of Europe).

Group of Specialists (2008b) *Report on Good Practices of Public Service Media as Regards Promoting a Wider Democratic Participation of Individuals*, prepared by the Group of Specialists on Public Service Media, MC-S-PSM(2008)005 rev. (Strasbourg: Council of Europe).

Harrison, J. and B. Wessels (2005) 'A New Public Service Communication Environment? Public Service Broadcasting Values in the Reconfiguring Media', *New Media & Society*, 7(6): 834–53.

Heinrich, A. (2008) 'Network Journalism: Moving towards a Global Journalism Culture', paper for the RIPE@2008 Conference, Mainz.

Hoffmann-Riem, W. (1996) 'Germany: the Regulation of Broadcasting', in M. Raboy (ed.), *Public Broadcasting for the 21st Century* (Luton: John Libbey Media), pp. 64–86.

Humphreys, P. (2008) 'Redefining Public Service Media: a Comparative Study of France, Germany and the UK', paper for the RIPE@2008 Conference, Mainz.

Huntsberger, M. (2008) 'Create Once, Play Everywhere: Convergence Strategies for Public Radio in the US', paper for the RIPE@2008 Conference, Mainz.

Jackson, L. (2008) 'Facilitating Participatory Media at the BBC', paper for the RIPE@2008 Conference, Mainz.

Jakubowicz, K. (2004) 'A Square Peg in a Round Hole: the EU's Policy on Public Service Broadcasting', in I. Bondebjerg and P. Golding (eds), *European Culture and the Media* (Bristol, UK: Intellect Books), pp. 277–302.

Jakubowicz, K. (2007) 'Public Service Broadcasting in the 21st Century: What Chance for a New Beginning?' in G. F. Lowe and J. Bardoel (eds), *From Public Service Broadcasting to Public Service Media* (Göteborg: NORDICOM), pp. 29–49.

Jakubowicz, K. (2008) 'Participation and Partnership: a Copernican Revolution to Re-engineer Public Service Media for the 21st Century', paper for the RIPE@2008 Conference, Mainz.

Kearns, I. (2003) 'A Mission to Empower: PSC. From Public Service Broadcasting to Public Service Communications', speech presented on behalf of the Institute for Public Policy Research, Westminster e-Forum.

Kleinsteuber, H. J. (2008) 'Participation in the Management of Public Service Media. Broadcasting Councils in Germany: Making them Fit for the Future', paper for the RIPE@2008 Conference, Mainz.

León, B. (2007) 'Commercialisation and Programming Strategies of European Public Television: a Comparative Study of Purpose, Genres and Diversity', *Observatorio (OBS*) Journal*, 2: 81–102.

Leurdijk, A., G. Bodea and J. Esmeijer (2008) 'Following the Audience: a Comparative Analysis between PSM's Cross-Platform Strategies in Four European Countries', paper for the RIPE@2008 Conference, Mainz.

Lowe, G. F. (2007) *The Role of Public Service Media for Widening Individual Participation in European Democracy* (Strasbourg: Council of Europe), http://www.coe.int/t/dghl/standardsetting/media/Doc/H-Inf(2008)012_en.pdf, accessed 8 February 2009.

Lowe, G. F. (2008) 'Public Service Broadcasting in "Partnership" with the Public: Meanings and Implications', paper for the RIPE@2008 Conference, Mainz.

McQuail, D. (1992) *Media Performance: Mass Communication and the Public Interest* (London: Sage).

Moe, H. (2008) 'Defining Public Service beyond Broadcasting: the Legitimacy of Different Approaches', paper for the RIPE@2008 Conference, Mainz.

Murdock, G. (2004) *Building the Digital Commons: Public Broadcasting in the Age of the Internet*, The 2004 Spry Memorial Lecture, http://www.com.umontreal.ca/spry/spry-gm-lec.htm, accessed 6 February 2009.

Ofcom (2004) 'Looking to the Future of Public Service Television Broadcasting', *Ofcom Review of Public Service Television Broadcasting: Phase 2* (London: Office of Communications).

Ofcom (2008) *Second Public Service Broadcasting Review. Phase 2: Preparing for the Digital Future*, Consultation (London: Office of Communications).

Papathanassopoulos, S. (2007) 'Financing Public Service Broadcasters in the New Era', in E. De Bens et al. (eds), *Media between Culture and Commerce* (Bristol, UK: Intellect Books), pp. 151–66.

Parliamentary Assembly (2004) *Public Service Broadcasting*, Doc. 10029 (Strasbourg: Council of Europe), http://assembly.coe.int/Documents/WorkingDocs/doc04/EDOC10029.htm, accessed 6 February 2009.

Picard, R. G. (2005) 'Audience Relations in the Changing Culture of Media Use: Why Should I Pay the License Fee?' in G. F. Lowe and P. Jauert (eds), *Cultural Dilemmas in Public Service Broadcasting* (Göteborg: NORDICOM), pp. 277–92.

Political Affairs Committee (2007) *State of Human Rights and Democracy in Europe: State of Democracy in Europe*, Doc. 11203 (Strasbourg: Council of Europe), http://assembly.coe.int/Main.asp?link=/Documents/WorkingDocs/Doc07/EDOC11203.htm, accessed 6 January 2009.

Raboy, M. (2008) 'Dreaming in Technicolor: the Future of PSB in a World Beyond Broadcasting', *Convergence*, 14(3): 361–5.

Rozanova, J. (2007) 'Public Television in the Context of Established and Emerging Democracies: Quo Vadis?' *The International Communication Gazette*, 69(2): 129–47.

Strategic Information Service (2007) *Broadcasters and the Internet* (Geneva: European Broadcasting Union), www.ebu.ch/en/sis, accessed 16 February 2009.

Strategic Information Service (2008) *Public Youth Radio in Europe: Executive Summary* (Geneva: European Broadcasting Union), www.ebu.ch/en/sis, accessed 6 February 2009.

Syvertsen, T. (2003) 'Challenges to Public Television in the Era of Convergence and Commercialization', *Television & New Media*, 4(2): 155–75.

Tabernero, C., J. Sánchez-Navarroand and I. Tubella (2008) 'The Young and the Internet: Revolution at Home: When the Household becomes the Foundation of Socio-Cultural Change', *Observatorio (OBS*) Journal*, 6: 273–91.

The Economist (2006) 'Don't Write off Hollywood and the Big Media Groups Just Yet', 19 January, http://www.economist.com/opinion/displaystory.cfm?story_id=5411930, accessed 6 February 2006.

Trappel, J. (2008a) 'Dividends of Change: Can Deregulation, Commercialisation and Media Concentration Strengthen Public Service Media?', paper for the RIPE@2008 Conference, Mainz.

Trappel, J. (2008b) 'Online Media Within the Public Service Realm? Reasons to Include Online into the Public Service Mission', *Convergence*, 14(3): 313–22.

Ubiqus Reporting (2004) *BBC Royal Charter Review: an Analysis of Responses to the DCMS Consultation*, http://www.bbccharterreview.org.uk/pdf_documents/ubiques_analysis_bbccr_responses.pdf, accessed 3 February 2009.
Varis, T. (no date) *New Literacies and e-Learning Competences*, http://www.elearningeuropa.info/index.php?page=doc&doc_id=595&doclng=6, accessed 2 February 2009.
Ward, D. (2006) 'Can the Market Provide? Public Service Media Market Failure and Public Goods', in C. Nissen (ed.), *Making a Difference: Public Service Broadcasting in the European Media Landscape* (Eastleigh: John Libbey Publishing), pp. 51–64.
Waters, D. (2009) *Facebook Offers Control to Users*, http://news.bbc.co.uk/2/hi/technology/7913289.stm, accessed 27 February 2009.

2
Pluralism and Funding of Public Service Broadcasting across Europe

Petros Iosifidis

Introduction

On 21 January 2009 the Office of Communications (Ofcom), the UK regulator for broadcasting and telecommunications, published its final statement of a long-running Public Service Broadcasting (PSB) review, titled 'Putting Viewers First', setting out recommendations for the future of PSB. Ofcom is required to undertake such a review at least every five years under the UK's Communications Act 2003, and the 2009 document follows several phases and reviews of PSB, all starting with the 2004 review (Is Television Special?) which had opened up the debate on whether the existing UK PSBs (the BBC channels, Channel 3/ITV, Channel 4 and S4C in Wales and Five) are delivering the range and breadth of programming and audience needs that constitute PSB, and how it is to be delivered in the future.

The Ofcom review identified a number of challenges and opportunities concerning the current PSB system, including: the transition from analogue to digital; that audiences value public service (PS) content and they want it sustained; and they want choice beyond BBC. Having considered that the public continues to value the benefits of PSB and that plurality (defined as competition in the provision of public service content) is critically important, Ofcom's main recommendations to government were to: maintain the BBC's role and funding at the heart of the system; free-up ITV and Five as commercial networks with a limited PS commitment; and create a strong, alternative PS voice to BBC with Channel 4 at its heart. Ofcom acknowledged that new models of replacement funding will be needed and brought forward a wide range of possible funding sources, including Existing Funding (the licence fee, direct government funding and regulatory assets), Extra Value (BBC Partnerships, other partnerships, digital switchover surplus), and New Funding (direct government funding, industry levies – levies charged on revenue from organisations such as Internet Service Providers (ISPs), mobile phone operators, broadcasters, video labels, and so on).

Some of the suggestions include 'institutional' competition for PS provision to end the BBC's near monopoly in the area (the Ofcom's analysis is that commercial pressures will make it harder for commercially funded broadcasters to sustain their public service obligations), competition in the provision of PS programming, and 'contestable' funding (that is, income top-sliced from the licence fee). The suggestion for the BBC to lose a portion of its licence fee funding to help subsidise local and regional news and possibly older children's programming was echoed in the June 2009 *Digital Britain* report (DCMS/DBERR, 2009), which like Ofcom added to the growing pressures on PSB by arguing for a more metrically defined PSB, that is, measuring the service delivered to the public, and implied that 3.5 per cent of licence fee revenue that is ring-fenced for digital TV switchover be top-sliced.

This chapter is divided into two parts. Part one critically assesses Ofcom's recommendations with regard to provision of PSB in the digital age. It argues that *plurality of content* is more important than *plurality of providers* (or *institutional plurality* in Ofcom's analysis) for the provision of PS content. To demonstrate this, the chapter discusses the state of plurality of PS providers and PS programming in other European PSB systems. The European dimension to the current PSB debate in the UK, presented in the second part, is significant as the debates on PSB in European countries take place within a shared regulatory framework which is subject to the same constraints as every other EU member state.

Institutional competition and the (abandoned) idea of a Public Service Publisher

Ofcom first expressed a concern in the 2004 review 'Is Television Special?' that there should be more than one PSB – the 'Public Service Publisher' or the 'Arts Council of the airwaves' as outlined initially by the Peacock Report (1986). In view of an emerging deficit in the provision of PSB as television moves to digital, Ofcom proposed the establishment of a new entity, provisionally called a Public Service Publisher (PSP), to ensure that the necessary level of competition for quality in PSB continues through the transition to digital. Ofcom's analysis of the responses of the relevant consultation process on assessing the establishment of a PSP, an entity rooted in the ideas, creativity and ethos of the new media, was that there is 'significant interest in the idea of a PSP' (Ofcom, 2007), although in 2008 the regulator abandoned plans for a PSP.

Meanwhile the idea of an Arts Council of the airwaves (that would commission individual PS programmes to compete with the BBC) has been mooted with some regularity down the years as a way to fund PS programming outside the BBC, but in 2008 it was ruled out by then Culture Secretary James Purnell who argued that 'it would itself become the commissioner of programmes but without the necessary relationship with the audience' (see Tryhorn, 2008).

Ofcom's analysis considers whether the digital switchover (that is, the progressive migration of television households, from analogue to digital-only reception) and the intensified competition that will follow will force commercial PSBs to water down or give up their PS remit. There are at least three sorts of pressures that make it difficult for commercial channels to deliver PS content in the digital world: pressure of audience fragmentation; pressure of alternative media; and pressure of advertising revenues. These pressures will only intensify as today's cartography is seen by Ofcom as transitional to the fully digital era envisaged for the next decade. Alongside Ofcom's review of PSB, the 2006 BBC's Charter review process and a discussion over ITV's PS obligations after the 2003 Carlton–Granada merger[1] have all fuelled the debate over the role of publicly funded programming in the digital, multi-channel future. While UK viewers have so far benefited from provision by five PS television broadcasters, changes in the market may mean it is no longer realistic to expect commercial broadcasters to deliver significant PS obligations around regional, religious, children and arts content. This is now government policy. The 2009 *Digital Britain* report explicitly made a case for progressive liberalisation for ITV and Five, so that they can move towards becoming fully commercial networks, while for Channel 4 it envisaged a new, modernised and more online focused remit.

However, relieving commercial broadcasters from some of their PSB obligations has major implications for the BBC which may emerge as a PSB quasi-monopoly. The basic idea of establishing a Public Service Publisher was to provide competition to the BBC and to avoid the country being left with just one PSB. Although the PSP proposal has faded out, one of the models suggested by Ofcom – the new or competitive funding model, which involves funding to provide PS content beyond the BBC – sounds like a relaunch of the PSP proposal. Further, the idea behind the *Digital Britain* report's suggestion that Channel 4 could be encouraged to take more steps into multi-media content would be to make it a strong competitor to the BBC. Therefore contestability both in PS provision and PS funding is still on the agenda.

One could ask what is wrong with having just one PSB catering for internal pluralism. Schlesinger (2004) identifies a number of far-reaching undesirable consequences of having PSB production largely or exclusively limited to one institution such as the BBC. First, the analytical separation between PSB and its particular institutional incarnations would be largely undermined. As the quasi-monopolist of PSB, the BBC would be overwhelmingly identified with it. Second, this would make the future of PSB more vulnerable by largely equating it with one institution's output and profile. Third, it would impair the capacity of British television to develop alternative ideas about public service outside the BBC. Lastly, given that institutions are far from perfect, a BBC near-monopoly would mean that the corporation largely became its own measure in matters of performance, thereby reinforcing an inward-looking culture. In sum the future of PSB would be less sustainable and more vulnerable because everything would hang on the fate of the BBC. There is therefore a need to ensure that more than one institution is centrally tasked with providing PSB. In Schlesinger's words (2004: 4), 'competition between organisations whose purposes are focused on public service broadcasting, within a market dominated by a commercial imperative, is a desirable counterweight to the unmediated impact of commercial imperatives to a quasi-monopoly'.

However, the creation of institutional competition to deliver public service is being met with scepticism by various scholars. For a start, one should look for the main motivation behind this proposal. According to Karol Jakubowicz (personal communication), who has been active in the Council of Europe and whose chapter is being used as a framework in this volume, Ofcom's motivation is ideological and has to do with a desire to promote competition, also in the public sector, and to prevent PSB from dominating the market, rather than with ensuring a plurality of PSBs. In their response to the first phase of the Ofcom review of PSB former member of the European Parliament Carol Tongue (who provides the Foreword to this volume) and professor of broadcasting policy Sylvia Harvey argued that a new body would only be required if the BBC were to be abolished and this, they think, would be an unacceptable waste of resources, reputation, brand name and accumulated cultural capital (Tongue and Harvey, 2004). Jean Seaton, Professor of Media History at the University of Westminster and the Official Historian of the BBC, responding to the 2009 Digital Britain report, said that 'top-slicing a bit of our international worth [that is, the BBC], seems unwise.'[2]

Along these lines the Broadcasting Entertainment Cinematograph and Theatre Union (BECTU, 2008) and the Voice of the Listener and Viewer (VLV, 2008) representing the citizen and consumer interest in broadcasting, in their responses to Ofcom's second consultation review – phase one – strongly opposed any such development, for it would weaken the BBC. In its own response, the National Union of Journalists (NUJ, 2008) believes the impact of institutional competition would be the creation of a new layer of administration, less money may be directed to programme making, and there would be more fragmentation. On the issue of possible models for PSB in the future, BECTU, the VLV, the NUJ and the Campaign for Press and Broadcasting Freedom (CPBF, 2008) all argue that the BBC should be kept at the heart of PSB and commercial PSBs should retain a designated PS role.

It can be seen that some scholars and think-tanks favour competition in PS provision, while others are sceptical about the establishment of bodies other than the BBC to provide PS content, especially to the idea of a PSP, pointing to the fact that the creation of such a body might introduce a superfluous layer of chaotic competition, weaken a valuable public service and therefore lead to an inferior service for viewers. Most of the commentators, though, share the view that commercial broadcasters should retain their PS obligations, thereby challenging Ofcom's recommendation to free up ITV and Five from their PS commitments.

But the debate about institutional contestability has far-reaching consequences for the ecology of the broadcasting system in the UK in general and for the existing public broadcaster more particularly since it calls for a system of contested funds. The existence and track record of the BBC gives confidence that standards can be maintained and that the BBC should continue to act as a benchmark of quality across the entire broadcasting system. This can be achieved as long as the BBC's current method and level of funding is maintained. The idea about 'contestability of institutions' implies 'contestability of funding'; therefore it proposes an end to the integrity of the licence fee as an exclusive resource for the BBC. Diverting or 'top-slicing' BBC licence fee income to other PS broadcasters has initiated a hot debate. This is discussed in the next section.

Top-slicing

Top-slicing is the suggestion that a part of each licence fee should go to a body that would use the money to subsidise PS content from broadcasters other than the BBC. Under this proposition a portion of the licence fee should be given to broadcasters other than the BBC (including, presumably, the main terrestrial TV channels) in return for PS content. The idea of top-slicing the BBC's licence fee and creating a separate fund has a resonance as a means of preserving programme plurality on ITV, Channel 4 and Five as the digital world erodes the traditional incentives for making them. Splitting licence fee money with other broadcasters also allows independent producers, undoubtedly an important part of the mix to deliver effective PSB, to bid directly for finance.

Indeed there are many predators who eye the licence fee and would be happy with a system of contested funds. ITV, ailing and hard hit by advertising decline,[3] would certainly bid for a slice if it was given the chance. Channel 4 also wants public subsidy and this is perhaps a more compelling case for allocating a portion of the licence fee. As said, the government encourages Channel 4's ambition to extend its PSB role across new channels and media platforms. Given that viewers care most about news and children's TV (Ofcom, 2004), *Channel 4 News* could be publicly funded to ensure the broadcaster's funding difficulties do not adversely affect the quality of the programme. However, to justify this Channel 4 should aim at airing fewer reality shows, such as *Big Brother*, and a return to its original remit: diverse, innovative and experimental programming, in conjunction with a more online focused remit, as envisaged by the *Digital Britain* report.

But is top-slicing a good idea? It certainly presents a very fundamental change in the ecology of PSB. In his speech to the IPPR Oxford Media Convention in January 2008 Sir Michael Lyons, Chairman of the BBC Trust, stressed that the commercial PSBs and the BBC 'compete for audiences, not for revenue and this has resulted in incentives for all players to invest in high quality content'. But as the system is coming under strain given the downturn in TV advertising and the tight licence fee settlement[4] there is good reason to question possible changes regarding the fundamental nature of the licence fee. The BBC Trust is open to an energetic debate on the future funding of PSB provision and its chairman sees the following as emerging questions: Should the clear relationship between the BBC and the licence fee be diluted? Could the BBC deliver public purposes with less money? Would it be a good idea to weaken the BBC's ability to deliver PSB mission in order to enable other broadcasters to deliver theirs?

These are some of the fundamental issues that need addressing in the current debate, but the reader should be reminded that the debate over the role of publicly funded programming has been open for years and top-slicing is in fact not a new idea as it has been on the agenda for at least a decade, promoted by a group of academic economists and pro-market regulators. Stephen Carter, the former Ofcom chief executive, originally conceived the idea of top-slicing and the very same idea is now put forward by the current chief executive officer of Ofcom Ed Richards. Evidence of current thinking is Lord Carter's *Digital Britain* report proposal to top-slice 3.5 per cent of the television licence fee after 2013.

On the other hand, there is a camp that defends the licence fee and opposes breaking the link between the licence fee and the BBC, as this would be, to borrow Sir Michael's words, 'an act of bad faith', or like 'cutting an umbilical cord with viewers and listeners' (Jones, 2008). Maggie Brown (2008), a regular columnist in the *Guardian*, says that top-slicing is not the answer to TV's problems and takes a critical stance of then Culture Secretary James Purnell's speech at the IPPR Oxford Media Convention which was associated with resurrecting of the policy of sharing the BBC licence fee income around other worthy users (see Gibson, 2008).

Polly Toynbee (2008) is categorically against giving a portion of the licence fee to broadcasters other than the BBC in return for PS content and here is why: the BBC reaches well over 90 per cent of the population with its many services, and independent studies show that the licence fee is acceptable. Toynbee agrees with Jones that once the link between the BBC and the licence fee is breached then the way is opened to go much further and reduce the organisation to a US-style niche subscription service offering only education and information. The BBC is Britain's most powerful global brand, capable of providing quality and diversity of content. Toynbee argues that this will be jeopardised if the BBC channels were drained of funds.

Top-slicing will not be the end of the BBC, but it may be *the beginning of the end*. But is the case of contestability for public funding a uniquely British consideration or is it an issue debated beyond Britain's borders? The rest of this chapter presents the European dimension to the current PSB debate in the UK.

The EU experience of PS institutional competition for plurality provision

Ofcom's concern that there should be more than one PSB (assuming that the new body can be considered a broadcaster, given that Ofcom initially envisaged the PSP as a content provider and commissioning house rather than a conventional TV channel) seems to be unique in Europe – providing of course that one refers to broadcasting serving and competing in the same market. For example, in countries such as Belgium, Switzerland and Spain there are several types of PSB which exist because of historical, cultural or linguistic reasons to serve different communities/regions. France Télévisions, the holding group for the national PSBs and Arte, the Franco–German cultural channel, do not testify to a plurality of PSBs, as Arte was always meant to be a niche broadcaster. France 2 and France 3 fit the bill better, but have of course been folded into the France Télévisions holding.

Germany has two PSBs (ARD and ZDF) serving the same national market, but as Jakubowicz (personal communication) put it, 'that is an accident of history, given that the federal structure of ARD was imposed by the occupying countries in Western Germany, and ZDF was then created separately'. Today there are nine broadcasting corporations in the Länder (states) that cooperate under the ARD, which is the first channel, and each of them broadcasts a third channel in their own Länder. This complicated system ensuring plurality of PS institutions is attributed to the fact that Germany is a federal state and broadcasting issues are by definition cultural issues which are by constitution the responsibility of the Länder. While in Germany the regional output is safeguarded by the constitution, in the UK some fear that regional programming is in peril as terrestrial commercial broadcasters have been released from some of their obligations around regional (as well as religious and arts) content.

In France, despite the existence of a national provider and regionally focused channels, it has always been hard to secure provision of regional news and political coverage. Plurality will be even harder to keep going, particularly as provision increases generally across television and audiences fragment. The PS television sector is in very poor shape and is reeling from Sarkozy's recent announcement to take advertising away from PS channels (see Kuhn's contribution to this volume). This move is likely to increase advertising revenues for the commercial channels, but at the same time it looks as if it might create a BBC-type funding situation, but with significantly less resources!

In Italy, there are no provisions for PS obligations on the part of non-PSBs. The Bill proposed by the Centre Left government of Romano Prodi in 2006 for the reformation of PSB (which never made it through the phase of parliamentary discussions as Berlusconi returned to power in May 2008) in fact aimed at diminishing RAI's PS responsibilities with its provision that RAI's two main channels be privatised and the third remain as the only publicly funded channel, without, however, expanding its PS obligations to other players (see Padovani's chapter in this volume). Likewise in Germany the commercial broadcasters are not subject to PS obligations. The German constitutional court even ruled on several occasions that private channels are allowed to be truly market-oriented as long as PSB exists. It seems as though commercial broadcasting is only allowed to exist in Germany as long as the existence of PSB is guaranteed!

In the Mediterranean country of Spain PS programming is hard to secure too. For a start, commercial broadcasters are not subject to PS obligations. However, the main concern is to ensure that public broadcaster RTVE truly offers a PS output since for the last decade the public institution has been very commercialised (at least the first channel TVE-1) (Iosifidis, 2007: 126–9; see also León's chapter in this volume).

In the smaller European territory of Denmark, in spring 2002 the government opted to privatise the nationwide television channel TV2, which until its foundation in 1988 has been partially funded by the licence fee. Despite its privatisation the channel must still abide by certain PS obligations with respect to news and current affairs and a continued financial commitment to Danish film. Advertising-funded TV2 currently sees itself as a hybrid PSB-commercial broadcaster competing mainly with licence fee-funded DR (Danmarks Radio) which has clear PS obligations imposed on its two TV channels and four radio stations.

The Dutch PSB NPO (Nederlandse Publieke Omroep), which is part of the Broadcasting Corporation NOS, the umbrella organisation for public broadcasters, is a member-based broadcasting association whose members share common facilities. This arrangement has its origins in the polarisation of the previous century when various religious and political streams in Dutch society (Catholics, Protestants, Socialists) all had their own separate associations, newspapers, sports clubs, educational institutions and broadcasting organisations. The aim of the NPO is to provide a voice for each social group in a multicultural society.[5] This new model of organising PSB in the Netherlands demonstrates that public broadcasting pluralism in the media should no longer be the responsibility of a single medium or sector, but should include the full supply of content and its use via other media. According to Bardoel and d'Haenens (2008) this resembles new concepts such as 'distributed public service' and 'deinstitutionalising' PSB, though for the BBC Trust (2008) the award of a single licence to NPO shows evidence of consolidation (see below).

Denmark, the Netherlands and Germany aside there is little evidence that European countries aim for competition between broadcasters for the production and distribution of programmes in key PS genres.[6] In Denmark and the Netherlands 'institutional competition' and 'convergent media policies' appear to have gained some ground, but the debates on alternative PSB policies go on and have by no means been finalised. Even in Germany fierce competitors ARD and ZDF sometimes coordinate their scheduling in order to avoid programming duplication. In fact, the kind of formalised cooperation which used to exist in the old days (for example, regular meetings to discuss programming issues or conflicts of interest) does not exist any more. However, since ARD and ZDF are partners in several channels (Arte, 3sat, Kinderkanal, Phoenix) and still jointly acquire rights in premium content such as sports (Olympics, Football World Cup), there is a need to coordinate in these areas.

Furthermore and as Runar Woldt, who contributes to this volume, put it (personal communication) there are occasionally 'gentlemen's agreements' concerning for instance very expensive own productions. If, for example, ARD schedules a high-budget TV programme which has the potential of a large appeal (typically TV movies or documentaries), ZDF will not schedule another blockbuster against it, but maybe a repeat of an older film. To sum up, even though ARD and ZDF compete with each other as they do with the private channels, there is still evidence of cooperation.

Ironically, if predictions about the demise or decline of free-to-air broadcasting prove correct, PSB may regain monopoly on both free-to-view programmes and on PS content, at least on terrestrial mass audience channels, as commercial broadcasters are forced to compete for dwindling advertising revenues. The Ofcom answer to this, that another body beyond the established PSB could be created, is a peculiarly British, if not Ofcom idea, as the UK government has not so far subscribed to this proposal. There is no similar policy development elsewhere in Europe, though evidence of such policies can be found in non-European territories such as New Zealand (see Dunleavy's chapter in this volume).

Internal versus external pluralism and independent producers

Plurality in European PSB systems is seldom conceptualised in the same terms as is the case in the current PSB debate in the UK. European nations have mainly focused on pluralism within the PSB, rather than between different providers. PSB is still primarily defined in terms of internal pluralism and where a plurality of PSB institutions does exist, it tends to be considered a 'legacy' feature of the PSB system, and the focus of the debate is on consolidation rather than on the preservation of this type of plurality for its own sake.

For instance, the Netherlands has witnessed substantial consolidation of its PSBs over the last decade, during which twenty-four (mostly regional) PSBs, each of which held their own broadcasting licence within a loose group structure, have been replaced by a single licence awarded to NPO for ten years and centralised scheduling of the three national TV channels and radio stations (BBC Trust, 2008: 6). Sweden provides another example that European countries aim for a unified public service system, for up to the end of 1995 the two national public channels were competing openly with each other, with SVT1 (then named Kanal 1) showing Stockholm-based programmes and

SVT2 (named TV2 at the time) broadcasting programmes from other parts of Sweden. In January 1997 the two channels were reorganised under a common administration and have since cooperated closely in the areas of production and broadcasting (Iosifidis, 2007: 157).

The Italian way to secure internal pluralism within the PSB has been reflected in the practice called *lottizzazione*, according to which each RAI channel, each news bulletin and public affairs programme, had its layers of political affiliation. Although this practice still continues today its legitimisation and effectiveness have shifted. Today's internal pluralism is mainly related to the various scheduling and programming strategies for the different audience targets of the various channels. Only RAI3 continues to provide an element of internal pluralism. As far as the debate on internal versus external pluralism is concerned in Spain, there are no clear rules for the participation of independent content providers in public TV. In fact, a few production companies take most of the cake and in many cases content providers for public broadcaster RTVE are 'friends' of the political party in power. Regarding the commercial channels, in many cases the production companies are wholly owned by broadcasters.

In the federal state of Germany PS plurality is considered as a cultural issue and by constitution is the responsibility of the Länder. When post-war German PSB was modelled after the BBC the prime aim was to prevent political influence on programming. However, the issue of access for independent producers has never acquired the same degree of political salience that it has in Britain, where independent producers are operating on a raised production quota and there appears to be a widely shared set of values relevant to PSB in the independent sector. Whereas internal pluralism prevails in Germany, the situation is rather different in France. France Télévisions, like Channel 4, commissions most of its production from external producers and pluralism is provided via an increase in diversity of supply.

But again that is an 'accident of history', as the old ORTF production resources, hived off into a separate company upon the break up of ORTF in the mid-1970s, were finally privatised, leaving France Télévisions with almost no production capacity, except in the regions. Production obligations, which find concrete expression in a complex set of quotas, broadcasting time limits, and multiple contributions to audiovisual and cinematographic production, aim at preserving national culture through the programming of French and European works. But internal pluralism has always been hard to secure in the French case, especially in news and political coverage. While the BBC is taken for granted in the UK, for it is perceived as a cornerstone of PSB, the public service ethos is less well implemented and more susceptible to political attack in France (as in Italy and Spain), where little national discussion has taken place on PS purposes, especially citizenship and content (Iosifidis, 2008: 188).

In Italy, the notion of pluralism has been associated with quantitative issues. For example, the introduction of digital terrestrial television is being seen by many as the solution to the lack of external pluralism, as if more channels would automatically result in a more plural TV market! The 1996 Bill of the Prodi government supported a more independent and diversified pool of independent producers, but never became law. In the Netherlands there was a plan to leave PSB organisations with the task of producing only the news and current affairs, with all the rest commissioned from external producers, but it was never implemented. Otherwise the independent production sector exists primarily because of the ruling of the TWF (now Audio-Visual Media Services) Directive that obliges European broadcasters to commission at least 10 per cent of their programming from this sector. It is a shame really, given that the expansion of independently produced supply has the potential to shift the internal culture of PSBs and, provided the workforce is trained, reconfigure the wider culture of TV production.

Conclusion

The first part of this chapter showed that traditionally the UK has had a multiplicity of PS providers/programming, but as Ofcom's analysis demonstrates it will be hard for commercially funded broadcasters to sustain their PS obligations. This has initiated a debate on whether something should be done about it. This debate has gathered pace in the UK without anyone asking whether the historic situation was unique to the UK, whether it is coming under pressure everywhere, and whether the current UK preoccupation is generally held or peculiar to the UK context. Therefore the second part looked for a European perspective on the current UK debate about the prospects for a plurality of PS provision.

Several conclusions can be drawn from the above analysis. First, despite the development of different models of intervention and funding, existing broadcasting institutions will certainly matter in delivering PS programming for the transition to digital. Second, institutional competition for PS provision and top-slicing to subsidise other broadcasters in the UK at least risks becoming unacceptable if this implies a weaker BBC, which commands worldwide respect and remains Britain's most powerful global brand. Politicians should start thinking seriously of other forms of funding such as industry levies, which may not appear politically popular, but as Patrick Barwise, Emeritus Professor of Management and Marketing at the London Business School said, they could help bridge the funding gap which PSBs will face after digital switchover in 2012.[7] Third, the vigorous UK debate on PS plurality of institutions and plurality of funding has not so far featured in many other European discussions. Plurality in European PSB systems other than the UK is seldom conceptualised in the same terms, for what prevails is *plurality of content* rather than *plurality of providers*. Where PSB plurality is observable in the European countries beyond the UK (Germany), it is more likely to be a function of underlying political and constitutional structures rather than an actively managed regulatory outcome. Otherwise PSB ecologies in European countries are tending towards greater consolidation, rather than increased plurality of providers. Finally, there is no precedent in Europe for licence fee revenue being used to fund the creation of content or channels beyond the existing PSB.

Notes

1. In 2003 the two big London-based independent television companies Granada and Carlton agreed a £2.6 billion merger.
2. See http://www.guardian.co.uk/commentisfree/2009/jun/16/digital-britain-bbc-licence-fee1, accessed 17 June 2009.
3. In March 2008 British broadcaster ITV reported a 37 per cent drop in profit, with profits reaching £137 million in 2007, down from £219 million in 2006 (see http://www.cnbc.com/id/23478160/for/cnbc, accessed 17 March 2008).
4. January 2007 witnessed the BBC's funding settlement for the next six years, which will see the licence fee rise from £131.50 from 1 April 2007 and increase steadily to £151.50 by 2012. The money provided the BBC with £1.2 billion of investment in new services, but left the corporation £2.4 billion short of what it had asked the government for.
5. See http://www.culturalpolicies.net/web/netherlands.php?aid=425, accessed 11 June 2009.
6. There is uncertainty as to which programmes should be considered as offering a public service and when a TV channel is fulfilling any kind of public service obligation. A traditional line of thought is that genres are relevant for a definition of PSB and therefore it is largely news, documentaries, current affairs, children, religious and arts outputs that satisfy multiple PS commitments. In a significant break with this doctrine an emerging view is that 'creativity', 'innovation' and 'risk' are important notions to encapsulate PS purposes. Born's (2004) ethnographic study of the BBC confirms the validity of this emerging view. The book's principal argument is that there has been a declining creative culture at the BBC which in recent years tends to create more standardised or imitative programmes and commissions significantly less risky and innovative pieces. So long as PSB is not a well-defined output, commercial broadcasters can mislead the regulator by claiming they deliver PS content, questioning publicly funded channels and demanding a slice of the public money.
7. See http://www.bectu.org.uk/news/315, accessed 25 June 2009.

References

Bardoel J. and d'Haenens, L. (2008) 'Reinventing Public Service Broadcasting in Europe: Prospects, Promises and Problems', *Media, Culture & Society*, 30(3): 337–55.

BBC Trust (2008) 'BBC Response to Ofcom's Second PSB Review – Phase One, An International Perspective', http://www.bbc.co.uk/thefuture/pdf/international_perspective.pdf, accessed 27 September 2008.

BECTU (Broadcasting Entertainment Cinematograph and Theatre Union) (2008) 'Ofcom: Second PSB Review, Phase One – Consultation Response', 13 June, http://www.bectu.org.uk/policy/pol123.html, accessed 23 September 2008.

Born, G. (2004) *Uncertain Vision: Birt, Dyke and the Reinvention of the BBC* (London: Secker & Warburg).

Brown, M. (2008) 'Top-slicing Comeback is Deeply Depressing', *Media Guardian*, 18 January, http://blogs.guardian.co.uk/organgrinder/2008/01/please_spare_us_that_hoary_old.html, accessed 18 February 2008.

CPBF (Campaign for Press and Broadcasting Freedom) (2008) 'Response to Ofcom's Second PSB Review, Phase One', June, http://www.cpbf.org.uk/files/PSB_review_June_2008_Final.doc, accessed 14 September 2008.

DCMS/DBERR (Department of Culture, Media and Society and Department for Business Enterprise and Regulatory Reform) (2009) *The Digital Britain White Paper*, 16 June (London: Crown Copyright).

Gibson, O. (2008) 'BBC Licence Fee may be Shared, says Purnell', *The Guardian*, 18 January, http://www.guardian.co.uk/media/2008/jan/18/television.media?gusrc=rss&feed=media, accessed 22 February 2008.

Iosifidis, P. (2007) *Public Television in the Digital Era: Technological Challenges and New Strategies* (Basingstoke: Palgrave Macmillan).

Iosifidis, P. (2008) 'Plurality in Public Service Provision: a European Dimension', in D. Levy and T. Gardam (eds), *The Price of Plurality: Choice, Diversity and Broadcasting Institutions in the Digital Age* (London: Reuters Institute/Ofcom), pp. 185–9.

Jones, N. (2008) 'Begin the Fight Back: How Corporate Strategies Neutered the BBC', online article, 24 January, http://www.cpbf.org.uk/body.phtml?category=&id=1980, accessed 3 March 2008.

Lyons, M. (2008) Chairman of the BBC Trust, 'Putting Audiences at the Heart of the PSB Review', Speech to the IPPR Oxford Media Convention, 17 January, http://www.bbc.co.uk/bbctrust/news/speeches/ml_ippr.html (accessed 19 January 2008).

NUJ (National Union of Journalists) (2008) 'Ofcom: Second PSB Review, Phase One – Consultation Response', June, http://www.ofcom.org.uk/tv/psb_review/psb_2review/responses/nuj.pdf, accessed 10 June 2009.

Ofcom (2004) 'Is Television Special?', First Statutory Review of Public Service Television Broadcasting (London: Office of Communications), http://www.ofcom.org.uk/consult/condocs/psb/psb/psb.pdf, accessed 8 June 2009.

Ofcom (2007) 'A New Approach to Public Service Content in the Digital Media Age: the Potential Role of Public Service Publisher' (London: Office of Communications), 24 January, http://www.ofcom.org.uk/consult/condocs/pspnewapproach/newapproach.pdf, accessed 8 June 2009.

Ofcom (2009) 'Second Public Service Broadcasting Review: Putting Viewers First', 21 January, http://www.ofcom.org.uk/consult/condocs/psb2_phase2/statement, accessed 8 June 2009.

Peacock Report (1986) *Report of the Committee on Financing the BBC*, Cmnd 9824, London.

Schlesinger, P. (2004) 'Do Institutions Matter for Public Service Broadcasting?' Stirling Media Research Institute, University of Stirling, 30 September, http://www.ofcom.org.uk/consult/condocs/psb2/psb2/psbwp/wp2schles.pdf, accessed 17 March 2008.

Tongue, C. and S. Harvey (2004) 'Citizenship, Culture and Public Service Broadcasting: Response to Ofcom Review of PSB, Phase 1', 31 March, http://www.bftv.ac.uk/policy/ofcom040614.htm, accessed 25 January 2008.

Toynbee, P. (2008) 'A Top-sliced Licence Fee will Trigger the BBC's Destruction', *The Guardian*, 22 January, http://www.guardian.co.uk/commentisfree/story/0,,2244725,00.html, accessed 4 February 2008.

Tryhorn, C. (2008) 'Posh TV Programmes Funding Fears', *Media Guardian*, 17 January, http://www.guardian.co.uk/media/2008/jan/17/bbc.television4, accessed March 2008.

VLV (Voice for the Listener and Viewer) (2008) 'Ofcom: Second PSB Review, Phase One – Consultation Response', June, http://www.vlv.org.uk/pages/documents/Ofcoms2PSBReviewP.1TheDigitalOpportunityJune08.doc, accessed 11 June 2009.

3
EU Broadcasting Governance and PSB: Between a Rock and a Hard Place

Maria Michalis

Introduction

Public Service Broadcasting (PSB) has a long tradition in Western Europe. In general, West European countries considered broadcasting a public service, not a competitive industry, and entrusted it to state-owned national institutions. PSB became a feature of the post-war Keynesian welfare order where the interventionist state assumed an extensive role in socio-economic life by directly producing and supplying goods and services.

The concept of PSB has always been elusive. Born and Prosser (2001: 671) have summarised the core normative principles of PSB as '(a) enhancing, developing and serving social, political and cultural citizenship; (b) universality; and (c) quality of services and of output'. Still, there has never existed a single European model of PSB and in practice, throughout history, the diverse national types have adhered to the ideal PSB to highly varying degrees (Humphreys, 1996; Iosifidis, 2007).

The challenges to PSB are many, including technological progress: starting with the advent of cable and satellite distribution in the 1980s, the concomitant growth of transfrontier television, and, more recently, as the pace of convergence continues, digital broadcasting and the internet; market liberalisation which has increased offerings but at the same time fragmented audiences; and the accompanying 'new paradigm' of media policy prioritising economic goals over social and political welfare (Van Cuilenburg and McQuail, 2003). As is clear, these challenges to PSB are not specifically linked to the European Union (EU) but, it is submitted here, need to be understood in the context of more general ongoing developments (Michalis, 2007). These developments concern the broad political and ideological endorsement of market-based solutions, increasingly since the 1980s; the associated shift away from the national Keynesian welfare state towards the internationalising regulatory state that prioritises private consumers over public citizens (Pierre, 1995); and changes in European societies, notably the globalisation process and increasing migration, that have made societies more multicultural and multi-ethnic, upset the close nexus between culture and polity that characterised the ideal nation-state and, finally, challenged notions of a unified national identity and public sphere. The EU is a reflection and a symptom of these wider transformations and its approach towards PSB is a response and at the same time a contributor to the challenges that have arisen. It follows from the above that I perceive the main contestation in the process of European integration to be not about the division of powers between the EU and member states (*more-or-less Europe*), but primarily about the socio-economic character of the European project (*what kind of Europe*).

It is within this framework that this chapter examines EU broadcasting governance and PSB. First, it points to the challenges of trying to reconcile the long established national PSB institutions with the predominantly pro-liberal and pro-competition provisions of the European Treaties. The chapter then analyses how the long-running tension between economic and socio-cultural interpretations of European integration has been resolved in the field of broadcasting concentrating on the main EU policy instrument, the Directive on Television Without Frontiers (TVWF) originally adopted in 1989 and replaced in 2007 by the Audio-Visual Media Services Directive (AVMS). The discussion is also placed within the context of successive reconceptualisations of a European culture and identity. It is argued that although the EU has come to explicitly recognise the importance of socio-cultural policy objectives and citizens' rights, and that is a welcome development, its substantive policy output remains centred on economic and competition considerations.

The balancing act: public service versus competition considerations

A perennial problem in public policy has been the balance between competing values, notably between public service objectives and competition concerns. The founding EC Treaty back in 1957 addressed this issue by providing that its strong competition provisions would apply to public services (so-called services of general economic interest) only in so far as their application would not obstruct the operation of such services (Article 86(2)). In other words, the effective performance of a public interest task prevails over the application of competition rules. Although the conundrum itself is not new, what is interesting is how the actual balance between public services and competition considerations has evolved through the years.

Broadcasting as such did not feature in the founding EC Treaty. There was no need. On the one hand, PSB, the norm at the time, was firmly under national control and regulated as a monopoly. On the other hand, though political aims – notably 'to lay the foundations of an ever-closer union among the peoples of Europe' – were not absent, the main goal of the Treaty was economic: the creation of a common market. The predominantly pro-liberal and pro-competition provisions were, at the time, irrelevant to broadcasting. There was hardly any trade in broadcast services while PSB – closely associated with notions of cultural and national identity, social cohesion and a national public sphere – was a policy domain zealously guarded by national governments.

Paradoxically, even though the EU has no specific competence in broadcasting, it has substantially influenced market developments, principally on the basis of competition rules, where the EU enjoys strong and direct powers. This kind of intervention makes EU broadcasting policy reactive as it responds to the agenda set by opponents of PSB (Jakubowicz, 2004: 294).

Indeed, the very entry of the EU into the field of broadcasting was reactive. In 1974, the European Court of Justice was the first institution to intervene. The Court defined the transmission of television signals as a tradable service and thus established EU jurisdiction over broadcasting but, crucially, accepted limits to competition 'for considerations of public interest'.

As the move from a public service towards a commercial multi-channel television order gathered pace in the late 1980s and early 1990s, commercial interests started to lodge complaints to the EU competition authorities against PSBs. Collins (1994: 146–53) has described the competition decisions of that period as an assault on PSB.

With the gradual liberalisation of public utilities from the 1980s onwards, so-called services of general economic interest have attracted greater attention. In the late 1990s, France was instrumental in initiating an EU-level debate on the likely adverse impact of the internal market and the attendant processes of liberalisation and globalisation on public services. That debate culminated in a new provision in the 1997 Amsterdam Treaty which, though upholding the primacy of competition rules in principle, accepted the importance of public services and placed them, for the first time, among the shared values of the Union (Article 16). This new approach towards public services also needs to be understood as part of the reconceptualisation of European identity in the post-Cold War era away from the traditional national-state model based on ethnic, cultural, linguistic and historical elements, towards a post-national deterritorialised civic identity premised on rights and citizenship (Michalis, 2007: 170–1). The EU would no longer have just an economic face but needed also a political and social dimension that could make it more acceptable to the public at large. A few years later, the Charter of Fundamental Rights, endorsed by European leaders in 2000, established the citizens' right of access to public services, and integrated Article 10 of the European Convention of Human Rights on freedom of expression and information. Still, there is no consensus to grant the EU greater competence over public services (EC, 2007a).

These broader developments concerning public services have occurred in tandem with specific developments in the area of PSB. The aims here have been to demarcate the limits of EU competence and to clarify the application of state aid rules, rules to which, under intensifying market competition, commercial interests have increasingly resorted ever since the 1990s in an attempt to restrict the activities of PSBs. These developments will be examined in turn.

In response to the rising tide of complaints by commercial market players, and as part of the concurrent general endorsement of public services by the EU mentioned above, the 1997 Amsterdam Protocol on PSB established that its remit, organisation and funding remain the responsibility of member states thereby delimiting the actions of the EU in this area and the potential adverse effects of competition rules. Although the Amsterdam Protocol is significant, developments since its adoption somewhat qualify the protection it can still offer for PSB. First, at the time of the Amsterdam Protocol, PSB was still very much about television and radio. At the beginning of the twenty-first century, however, commercial operators are increasingly concerned about the scope of PSB's activities in the emerging multi-platform digital communications landscape. Second, the Amsterdam Protocol was debated and subsequently adopted on the understanding that as long as public funding did not exceed the actual cost of performing the public service obligations (proportionality) such financing fell outside the EU state aid regime and the associated strong competence of the Commission (Wagner, 1999). However, the understanding of what constitutes state aid has since changed with the Altmark ruling in 2003 (see Wheeler's chapter in this volume).

In 2001, in an attempt to end the case-by-case treatment of commercial players' complaints and enhance legal certainty, the Commission adopted guidelines clarifying the application of state aid rules to PSB. In the so-called Broadcasting Communication, the Commission abandoned its earlier attempt to restrict the scope of the public service remit to certain programme types that commercial entities cannot provide, and instead, it recognises that, in line with the Amsterdam Protocol, the definition of the public service remit rests with member states, and that PSB 'is directly related to the democratic, social and cultural needs of each society and to the need to preserve media pluralism' (EC, 2001: para. 11).[1] With regard to the remit and funding of PSB, the Commission can only check for manifest error and possible abusive practices. The Broadcasting Communication sets out three criteria (definition, entrustment and monitoring, and proportionality) that, if fulfilled, make state aid admissible.

Originally, commercial operators questioned the mixed funding scheme of PSBs – that is, combining state funds and commercial revenues. But in the late 1990s, commercial players started questioning the expansion of PSBs into the conventional broadcasting field (thematic channels), while more recently, an issue is their expansion beyond conventional broadcasting, notably in the online environment (Michalis, 2007: 232–8; Moe, 2008).

Although the EU has so far supported PSBs in its state aid decisions, the Commission's reasoning has changed over the years. Whereas in older cases concerning the expansion of PSBs into the traditional broadcasting sector, the Commission did not consider market failure as an important argument in its analysis, in recent cases concerning their expansion into non-traditional broadcasting fields (internet and mobile media), the Commission, breaking with past practice, has viewed market failure as a determining criterion (see EC, 2003).

The competition and state aid cases to date show that the Commission has managed to accommodate public interest considerations in its decisions. However, recent developments in the area of state aid risk squeezing public policy into the rigid structure of competition rules and reducing the assessment of the public service remit to a mechanistic process. In turn, the threat is that member states' discretion to define the public service mission will be reduced while PSB will be boxed into a few areas, prevented from expanding beyond traditional broadcasting and modernising in line with technological developments and audience preferences.

Television without frontiers: cultural unity, cultural diversity and market liberalisation

In the early 1980s, when it was becoming clear that transfrontier satellite broadcasting was possible, the idea of a single broadcasting market gained growing acceptance in the EU, but for conflicting reasons. On the one side, there were those who viewed it as a means to promote cultural and political integration, and in turn as vital for the sustainability of economic integration, which, coupled with temporary external trade protection, would also serve as a bulwark against the potential avalanche of cheap US programming that could easily fill up the extra channels. On the other side, there were those who saw it as a lever to promote market liberalisation that would nurture European champions. Under the first vision, the single broadcasting market required not just negative integration – market-making intervention concerning the removal of market barriers – but also positive integration, that is, market-correcting and shaping measures. In contrast, under the second scenario, the single broadcasting market related only to negative integration. In effect, these two contested visions refer to the balance between socio-cultural and neoliberal pre-competitive concerns, or what Collins (1994) has called *dirigiste* and liberal and more recently (Collins, 2008) communalist and associative EU policy visions.

Originally the EU, prompted by the European Parliament, attempted to foster a single audio-visual space by dismantling national cultural barriers in the name of a dormant European culture and the awakening of shared feelings of European identification (see for instance, European Parliament, 1982). The European Broadcasting Union (EBU)-sponsored transfrontier television ventures in the first half of the 1980s, effectively an attempt at European PSB, were expected to promote cultural unity (Collins, 1998). The belief was that cultural unity – implying a strong and single European identity as well as the dissemination of information perceived as essential for active citizenship and shared political responsibility – could be forged from above. This vision was misconceived for three reasons. First, it supposed that the EU needs a strong European identity, one that builds on the traditional national identity forging tools, and overlooked the fact that the congruity between political and cultural community has been eroded at the national level. The processes of globalisation and European integration have challenged the relative congruence between bounded territory, governance and identity while the historical conceptualisation of national identity based on ethno-cultural affinities has come under question the more multi-ethnic and multicultural European societies have become (Laffan, 1996). Second, it assumed powerful media effects, i.e. that a passive audience watching the same programme would automatically forge a common cultural identification. Finally, it underestimated the persistence of cultural and linguistic differences across the EU.

By the mid-1980s, the second vision of the single broadcasting market was gaining ground. Cultural concerns were quickly overshadowed by the more pressing need of breaking down market barriers to facilitate transborder broadcasting. The liberalisation of domestic television markets was a national, not EU, decision. The TVWF Directive was adopted in 1989 in response to the challenges brought up by satellite television (EU, 1989). Its main aim was to facilitate transfrontier television and thereby foster a single market in broadcasting. The Directive provided for the free movement of television services across the EU, ensuring that member states allow reception of broadcasts from other member states, on the basis of the country of origin principle, whereby a broadcaster must comply with the rules of the member state of establishment. The Directive set out basic common standards throughout the EU relating to advertising and sponsorship, and the protection of minors. In effect, the TVWF Directive extended the two fundamental principles of the single market – harmonisation of minimum requirements and mutual recognition among national rules – to television services.

Although the TVWF Directive was a victory for liberal economic forces, cultural considerations were not absent. Indeed, the most controversial provision related to the so-called quotas requiring broadcasters to devote the majority of their programming to European works.[2] This provision, on the one hand, aimed at increasing the circulation of European programmes within the EU and thus reducing programmes from outside the EU (notably the USA), and on the other, the expectation was that if viewers from one member state were exposed to more programmes from other member states, that would contribute to the fostering of a European cultural identity.

The debate on quotas marked a shift of emphasis away from the earlier preoccupation with 'cultural unity' to 'cultural diversity' internally and 'cultural exceptionalism' externally. In 1993 the political standoff between the EU and the USA over the liberalisation of cultural goods and services risked derailing the entire GATT Uruguay Round of trade negotiations. In effect, under its new conceptualisation, European culture and cultural anti-Americanisation were two sides of the same coin (Schlesinger, 1997).

The quota provision, effectively a cultural policy tool, stands out from the overwhelmingly liberalising provisions of the Directive, but various elements minimise its significance. First of all, the provision requires member states to fulfil the quotas 'where practicable and by appropriate means' making it then a symbolic rather than substantive provision. Second, it represents a political agreement and is thus not legally binding. The Commission monitors its implementation but there are no sanctions for non-compliance. Third, although the Commission reports a generally satisfactory picture of national compliance, especially by PSBs, and in 2007 European works represented 74 per cent of viewing time, given the limited intra-European circulation of television programmes, the quotas tend to be fulfilled by domestic content (Attentional Ltd. et al., 2009). Finally, there is no evidence that in the absence of the quotas the trade deficit with the USA would have been larger and that the measures designed to promote intra-European circulation of programmes have also promoted exports (David Graham and Associates, 2005: section 8.5).

When the Commission launched a review of the TVWF Directive in 1994, attempts, led by France, to make the quota provisions binding and expand them to new services were defeated, thereby leaving the pro-liberal stance of the Directive intact. In the end, the quota provisions remained unaltered. The revision of the TVWF Directive completed in 1997 served to increase legal certainty and improve implementation by clarifying the country of origin principle in order to prevent a race-to-the-bottom whereby broadcasters would choose to operate in the member state with the most liberal regulatory framework. Other elements of this revision included the so-called events of major importance to society (such as the Olympic Games and the Football World Cup) that need to be available on free-to-air channels, and stronger protection of minors.

In the 1990s, the EU also tried to address media pluralism, and media ownership and concentration. Lacking the legitimacy to address them in their own right, the proposals put forward were from the perspective of the internal market. Both national governments and the industry strongly opposed supranational intervention, and still today the EU regulatory framework does not cover these issues. In 2007, the Commission launched a debate on media pluralism, viewing it as essential for preserving the right to information and freedom of expression that underpin the democratic process (EC, 2007b). But the aim of this exercise is the development of non-binding EU-wide indicators for pluralism and not the expansion of the EU's regulatory powers. The 2007 Lisbon Treaty, in the process of national ratification, makes clear that the Union has only a supporting and complementary role in respect of culture (EU, 2007a). Content and culture will thus remain the primary responsibility of member states.

From TVWF to AVMS: technological convergence and public policy objectives

The second revision of the TVWF Directive launched in 2003 clearly illustrates the move that began in the early 1990s away from a European identity modelled on the experience of the national state towards a civic European identity centred on democratic values and human rights, stripped of all the potentially divisive territorially defined ethno-cultural characteristics. The civic conceptualisation of Europeanness that formally began with the Maastricht Treaty in 1992, culminated in the Lisbon Treaty in 2007 which defines European identity in terms of values and principles, including respect for human rights, freedom, democracy, equality, and the rule of law (Article I-2), and more specifically with regard to public services, a high level of quality, affordability, the promotion of universal access and of user rights (Protocol 26). What unites Europeans is not a single culture but shared values.

Indicative of this civic conceptualisation of European identity is the fact that the most contentious issue at the time of the AVMS Directive was not the quotas provisions, but rather its scope, and the question of jurisdiction over transborder audio-visual media services, especially in respect of the fulfilment of public policy objectives. These will be examined in turn.

The AVMS Directive, adopted in December 2007, updates the TVWF Directive to reflect market developments in an increasingly technologically convergent media environment (EU, 2007b). Various factors necessitated the revision. First, it no longer made sense to have rules for broadcast television only since progressively the same content was available online and was largely unregulated. Moreover, viewers were regularly accessing content in both linear and on-demand formats through new technological platforms outside the scope of the TVWF Directive. Finally, the development of new advertising techniques (such as split screen and interactive advertising) and the migration of advertising from conventional to online media putting pressure on traditional advertising revenues increased calls for the relaxation of the prescriptive rules in the TVWF Directive.

The most significant change introduced by the revised Directive is to expand the scope of EU regulation to some on-demand services. The controversy here is about the balance between liberalisation and regulation, and about the impact of technological convergence on regulation. Broadly speaking, PSBs, their European association EBU, and viewers' and listeners' groups supported the expansion of the scope of the Directive to cover all audio-visual media services. In contrast, commercial broadcasters, publishers, telecommunications and internet operators, and new media interests, strongly supported by Britain, were not only against the widening of the scope of the Directive, fearing it could result in prescriptive television-type rules being applied to new media and internet content, but also in favour of greater relaxation of the existing rules. This tension is clear in the final text which aims to avoid widening the scope too much.

The regulatory framework set out in the new Directive is technology-neutral. It covers audio-visual media services provided by any means of an electronic communications network. The starting point is the type of content, not the delivery platform. On the basis of user choice and control, and impact on society, the Directive distinguishes audio-visual media services between 'television broadcasts' and 'on-demand services'. The first category covers linear services provided for simultaneous transmission according to a programme schedule. On-demand services are non-linear and allow the viewer to choose what and when to watch from a catalogue of programmes selected by a media service provider.

Effectively, an on-demand service falls under the Directive, and is thus regulated, if its principal purpose is the provision of programmes which compete for the same audience as television broadcasting. The Directive explicitly excludes the electronic press, audio transmissions and radio services, as well as many on-demand services even if they include audio-visual material, such as services which are 'primarily non-economic' and not in competition with television broadcasting, private websites (blogs) and communications (emails), user-generated content websites, and services where the audio-visual content is incidental to the service such as gambling sites, online games and search engines.

The Directive puts forward a three-tier approach to regulation. A minimum set of mostly negative rules – including the separation of advertising from editorial content, protection of minors, prohibition of incitement to hatred, ban on advertising of tobacco products and prescription medicines, and controls on alcohol advertising – apply to all audio-visual media services. In addition to this basic set of requirements, the Directive places a heavier regulatory burden on scheduled services and a lighter one on the nascent on-demand sector. Scheduled services are subject to more controls (positive regulatory requirements) similar to those contained in the TVWF Directive, including free-at-the-point-of-use access to major events, advertising quotas, quotas of European and independently produced programming, and a new right of access to short news reports of events of high importance to society (Article 3(k)).

The AVMS Directive retains the quotas for European works and – since these quotas, specifying the amount of airtime, cannot be applied to non-scheduled services – adapts the provision to cover on-demand services. Member states have to ensure that on-demand services within their jurisdiction promote, where practicable and by appropriate means, the production of and access to European works, by, for instance considering the share and/or prominence of European works in the catalogue offered by the on-demand service (Article 3(i)(1)). The flexibility of the quotas clause has been retained, making its expansion to the on-demand sector a symbolic, rather than substantive, gesture.

Although the revised Directive retains the country of origin principle and extends it to on-demand services, it somewhat weakens it since it allows member states in specific circumstances to restrict reception of services from a provider established in another member state. It is ironic that the inclusion of online services in the revised Directive that was aimed at strengthening the country of origin principle following the 2000 E-Commerce Directive (EU, 2000), which allows exceptions for public interest concerns, served to dilute it further. Similarly to the TVWF Directive, under the new Directive a member state may require audio-visual media service providers within its jurisdiction to comply with more detailed or stricter rules than those set out in the Directive. However, under the country of origin principle, service providers may avoid such heavier regulatory burden by choosing to operate from member states with a lighter regulatory framework. In view of strong concerns by many member states and in order to prevent the risk of a deregulatory race-to-the-bottom, the Directive exceptionally allows a member state to restrict the transmission of audio-visual media services from another territory where the offending provider has infringed the stricter rules of the targeted member state.

In the case of television broadcasts these exceptional circumstances concern notably the protection of minors, and services that incite hatred. The offending broadcaster has to 'manifestly, seriously and gravely' infringe the stricter rules of the targeted member state (Article 2(a)(2)). The grounds for on-demand services are wider and include public policy, in particular in relation to criminal offences, protection of minors and incitement to hatred; the protection of public health; public security; and the protection of consumers and investors (Article 2(a)(4)).

Overall, the AVMS Directive follows the neoliberal stance of the original TVWF Directive, but again, it is not just about negative integration. Instead of the questionable issue, in view of the results achieved and within the context of progressive international liberalisation, of content quotas and a European culture, this time public policy considerations were the point of contestation. EU competence over culture and content has not been strengthened, but the expansion of the minimal harmonisation rules of the original TVWF Directive beyond traditional television services confirms the importance of public service media while, arguably, the compromise over jurisdiction has strengthened member states' regulatory control over the domestic audio-visual environment, potentially at the expense of the single market.

Conclusion

The starting point of this chapter has been the understanding of the European integration process as a field of contestation which has less to do with the territorial division of powers between member state and EU institutions, and more with the character of the European project. This clash between economic and competition concerns (negative integration) on the one hand, and socio-cultural considerations (positive integration) on the other, is clearly reflected in the field of broadcasting, as this chapter analysed.

EU governance of broadcasting has not been static; rather the balance of interests and values has varied over the years, from cultural and democratic goals in the early 1980s, to economic and industrial objectives increasingly since the mid-1980s. The broad phases of broadcasting governance have been associated with differing visions of a European culture and identity. The original search for a single European identity and the early vision of transfrontier television as the ideal medium to forge cultural unity gave way in the second half of the 1980s to the pursuit of cultural diversity via internal market liberalisation (TVWF Directive) and trade protection (quotas). And since the 1990s, European identity has been reconceptualised once again, this time to emphasise political and civic rights. It is indicative, for instance, that during the drafting of the AVMS Directive, the quotas provisions were not the thorny issue they were previously.

The EU approach towards broadcasting has been more about negative, rather than positive integration. Competition and economic objectives have been prioritised at the expense of socio-cultural objectives. And yet, paradoxically, although the EU broadcasting regulatory approach is less Europeanised, not least because of strong member state opposition, compared to, for instance, the adjacent sector of telecommunications, and is incomplete in the sense that it does not address arguably the most important issues of broadcast regulation, content and ownership, the EU has substantially shaped the broadcasting scene in Europe through the creation of an audio-visual internal market and through its strong competition powers where the EU enjoys autonomy for action.

Having said that, increasingly from the 1990s onwards in the post-Cold War context when it was becoming clear that the EU had to develop a socio-cultural side to complement and make more sustainable its economic side, the EU has explicitly recognised the significance of socio-cultural aims, public services and citizens' rights. But the point is that the Union's substantive output remains centred on economic and competition considerations. In effect, EU broadcasting policy has progressed more through competition decisions and Court judgements than legislation. Stressing the social, cultural and democratic functions of broadcasting, member states have fought hard to keep their competence in politically sensitive areas – such as safeguards for media pluralism and ownership – areas which, as a result, have been deliberately left out of the EU regulatory tentacles. The irony is that the minimal EU media policy framework (rock) that aimed at preserving the media as a cultural and nationally regulated sphere, has laid PSB open to competition law attacks (hard place).

The Commission has so far supported PSBs in its competition decisions, but it increasingly considers the presence or not of private players a determining criterion ('market failure' argument). The private commercial sector is likely to continue to resort to the EU's strong competition powers to try and delimit PSBs' activities, especially as digital convergence has intensified competition in digital media markets. The precarious balance between economic and socio-cultural objectives may change yet again.

Notes

1. In January 2008, in response to market and technological developments, the Commission launched the review of the Broadcasting Communication with the aim to adopt a new Communication by the end of 2009.
2. Another quota required broadcasters to reserve 10 per cent of their transmission time or budget to independent productions.

References

Attentional Ltd. et al. (2009) 'Study on the Application of Measures Concerning the Promotion of the Distribution and Production of European Works in Audiovisual Media Services', Report for the European Commission, http://ec.europa.eu/avpolicy/docs/library/studies/art4_5/final_report.pdf, accessed 4 June 2009.

Born, G. and T. Prosser (2001) 'Culture and Consumerism: Citizenship, Public Service Broadcasting and the BBC's Fair Trading Obligations', *The Modern Law Review*, 64(5): 657–87.

Collins, R. (1994) *Broadcasting and Audio-Visual Policy in the Single European Market* (London: John Libbey).

Collins, R. (1998) *From Satellite to Single Market: New Communication Technology and European Public Service Television* (London: Routledge).

Collins, R. (2008) 'Misrecognitions: Associative and Communalist Visions in EU Media Policy and Regulation', in I. Bondebjerg and P. Madsen (eds), *Media, Democracy and European Culture* (Bristol, UK: Intellect), pp. 287–306.

David Graham and Associates (2005) *Impact Study of Measures (Community and National) Concerning the Promotion of Distribution and Production of TV Programmes Provided for under Article 25(a) of the TV Without Frontiers Directive*, Report prepared for DG Information Society, http://ec.europa.eu/avpolicy/docs/library/studies/finalised/4-5/27_03_finalrep.pdf, accessed 25 May 2009.

European Commission (EC) (2001) 'Communication from the Commission on the Application of State Aid Rules to Public Service Broadcasting', *Official Journal*, C320/5, 15 November.

European Commission (EC) (2003) 'Decision of 1 October 2003, State Aid No. N 37/2003 – United Kingdom BBC Digital Curriculum' (Brussels: European Commission).

European Commission (EC) (2007a) 'Communication on Services of General Interest, Including Social Services of General Interest: a New European Commitment', COM(2007) 735, 20 November (Brussels: European Commission).

European Commission (EC) (2007b) 'Media Pluralism in the Member State of the European Union', SEC(2007) 32, 16 January (Brussels: European Commission).

European Court of Justice (1974) 'Judgment of 30 April 1974 in Case 155/73. Giuseppe Sacchi. Tribunale civile e penale di Biella – Italy', *European Court Reports* (1974): 409.

European Parliament (1982) *Report on Radio and Television Broadcasting in the European Community on Behalf of the Committee on Youth, Culture, Education, Information and Sport* (The Hahn Report), PE Doc 1-1013/81, 23 February (Brussels: European Parliament).

European Union (EU) (1989) 'Directive 89/552/EEC of 3 October 1989 on the Coordination of Certain Provisions Laid Down by Law, Regulation or Administrative Action in Member States Concerning the Pursuit of Television Broadcasting Activities', *Official Journal*, L298/23, 17 October.

European Union (EU) (2000) 'Directive 2000/31/EC of 8 June 2000 on Certain Legal Aspects of Information Society Services, in Particular Electronic Commerce, in the Internal Market', *Official Journal*, L178/1, 17 July.

European Union (EU) (2007a) 'Treaty of Lisbon', *Official Journal*, C306, 17 December.

European Union (EU) (2007b) 'Directive 2007/65/EC amending Council Directive 89/552/EEC on the Coordination of Certain Provisions Laid down by Law, Regulation or Administrative Action in Member States Concerning the Pursuit of Television Broadcasting Activities', *Official Journal*, L332/27, 18 December.

Humphreys, P. (1996) *Mass Media and Media Policy in Western Europe* (Manchester: Manchester University Press).

Iosifidis, P. (2007) *Public Television in the Digital Era: Technological Challenges and New Strategies for Europe* (Basingstoke: Palgrave Macmillan).

Jakubowicz, K. (2004) 'A Square Peg in a Round Hole: the EU's Policy on Public Service Broadcasting', in I. Bondebjerg and P. Golding (eds), *European Culture and the Media* (Bristol, UK: Intellect Books), pp. 277–301.

Laffan, B. (1996) 'The Politics of Identity and Political Order in Europe', *Journal of Common Market Studies*, 34(1): 81–102.

Michalis, M. (2007) *Governing European Communications* (Lanham, MD: Lexington).

Moe, H. (2008) 'Between Supra-National Competition and National Culture? Emerging EU Policy and Public Broadcasters' Online Services', in I. Bondebjerg and P. Madsen (eds), *Media, Democracy and European Culture* (Bristol, UK: Intellect), pp. 307–24.

Pierre, J. (1995) 'The Marketization of the State: Citizens, Consumers, and the Emergence of the Public Market', in G. B. Peters and D. J. Savoie (eds), *Governance in a Changing Environment* (Montreal and Kingston: McGill-Queen's University Press), pp. 55–81.

Schlesinger, P. (1997) 'From Cultural Defence to Political Culture: Media, Politics and Collective Identity in the European Union', *Media, Culture and Society*, 19(3): 369–91.

Van Cuilenburg, J. and D. McQuail (2003) 'Media Policy Paradigm Shifts: Towards a New Communications Policy Paradigm', *European Journal of Communication*, 18(2): 181–207.

Wagner, M. (1999) 'Liberalization and Public Service Broadcasting. Competition Regulation, State Aid and the Impact of Liberalization', http://www.ebu.ch/CMSimages/en/leg_p_psb_liberalization_mw_tcm6-4356.pdf, accessed 8 June 2009.

4

The European Union's Competition Directorate: State Aids and Public Service Broadcasting

Mark Wheeler

Introduction

This chapter will discuss how the European Commission (EC) has regulated media markets with reference to Public Service Broadcasters (PSBs) receiving state aids. The EU Competition Directorate has licensed PSBs in accordance with the protocol of the 1997 Amsterdam Treaty which enabled them to receive public subsidies as long as they did not distort national media markets. The Commission's state aid measures were further established by the European Court of Justice's (ECJ) 2003 Altmark judgment which defined the compensation levels that enterprises could receive in exchange for carrying out public service obligations.

This analysis will consider state aid cases concerning European PSBs to identify how these measures have highlighted questions about funding and compensation. It will consider how state aids have been applied post-Altmark to PSBs in Denmark, the Netherlands, Germany and Ireland. The review will also show how these measures relate to questions concerning the competitive fairness of the PSBs' entrance into the information markets and their control over sports rights. The questions surrounding public subsidies and new technologies became more conspicuous during the Directorate's 2008 investigation of the UK government's attempt to allow Channel 4 (C4) to utilise £14 million of subsidies drawn from the British Broadcasting Corporation's (BBC) licence fee monies for digital switchover. This led to a standoff between the Directorate and the UK government, with the EC Competition Commissioner Neelie Kroes claiming the decision breached state aid rules. Moreover, the EC is in the process of defining a Broadcasting Communication concerning the operation of state aids by PSBs in new media markets.[1]

Therefore, this chapter will discuss how the EU's employment of neoliberal principles has conflicted with normative public objectives to deliver communications services. It contends that state aids are problematic as they cannot encompass the wider social and democratic implications of broadcasting because 'competition law . . . [cannot] . . . grasp more complex operations of cultural ... power which ... media pluralism has traditionally sought to address' (Hardy, 2001: 15). This analysis of the Competition Directorate will consider its role with reference to 'the notion [that] media diversity is crucial to democratic systems ... and should be at the core of any European action' (Iosifidis, 2007b: 520).

The European Union and the audio-visual sector

The EU regulations governing European audio-visual services were developed in Article 151 of the European Treaty (EU, 1997). The EC's competence for audio-visual services was enlarged as technological reforms enabled operators to develop at a pan-European level. There has been a positive harmonisation between television markets through the revised versions of the Television Without Frontiers (TWF) Directive and the 2006 Audio-Visual Service Directive which replaced TWF.

Simultaneously, the EU's audio-visual policies have preserved the cultural priorities that are associated with diverse television services in democratic societies. In developing a regulatory approach, the EU's response signalled a conflict between the economic priorities of industrial competitiveness and the desire to maintain the principles of European cultural identity: 'Broadcasting ... has ... been a notable site where one of the "grand narratives" of the Community has been played out, the battle between the interventionists and free marketers' (Collins, 1994: 23).

These tensions have been exacerbated by the globalisation of communications services which have brought new entrants into the European television marketplace. Thus, one of the constant themes underpinning the EU's policy responses has been liberalisation of the rules governing Europe's television industries with a marginal consideration of the maintenance of public service objectives.

While the Commission has been concerned with content regulation, it has utilised economic or structural forms of regulation to intervene over cultural matters. Consequently, the EU has been concerned with the issues surrounding media concentration to ensure fair competition. Within this context, the EC Competition Directorate has sought to encourage market efficiency in the audio-visual sector to support growth and technical innovation.

The Competition Directorate and the audio-visual sector

The Directorate guarantees the unity of the internal market so that companies can compete on a level playing field in all member states. It seeks to avoid the monopolisation of markets by preventing firms from sharing markets via protective agreements which would enable them to maximise their profits and impede development. Competition rulings are designed to correct market failures by making rulings concerning: mergers or concentrations; state aids wherein public subsidies unfairly distort the market; and abuses of dominant positions. For instance, they should stem any unfair monopolisation of a market by a public or private enterprise.

In applying these measures to the audio-visual sector, the Competition Directorate has become an active player in intervening in the European television markets (Wheeler, 2001: 3). However, the Directorate's neoliberal approach has led to questions about its ability to enhance pluralism. Competition policy has ignored the 'cultural' diversity of content as it cannot recognise citizens' rights and identities. This is because the Directorate does not conceive communications as anything more than a private exchange of goods between suppliers and customers. Therefore, state aids remain 'the area where there is ... the greatest potential for conflict between the policies adopted by national governments and the way in which [the Competition Directorate] might interpret the competition provisions of the EC treaties' (Levy, 1999: 97).

State aid with regard to PSBs

The state aid mechanism is activated if a member state has distorted competitive trade by favouring certain undertakings that are incompatible with the common market. However, exemptions exist, as state aids which have a social character are compatible with the internal market. With regard to the audio-visual sector, the Commission has sought to prevent the implementation of anti-competitive agreements and the abuse of dominant market positions by public service providers by restricting the unfair provision of services or any over-compensation of funds.

The 1997 Protocol of Amsterdam was designed to rectify any market distortion created by subsidies on the competitive balance between public and commercial broadcasters. The commercial channels in Spain, France, Germany, Italy and Denmark had complained about their PSBs' dual forms of funding (licence fees and advertising) which they claimed gave the PSBs an unfair advantage. The Protocol rectified any unfairness between competitive gain and the maintenance of pluralistic services through the public service tradition. It stipulated that national governments were free to determine the method of PSB funding as long as it did not distort competitive trading for the common interest (Papathanassopoulos, 2002: 72).

In 1998, the Competition Directorate placed stricter limits on PSBs by prohibiting state aids when subsidies, alongside advertising revenues, exceeded the costs of meeting public service obligations (ibid.). Despite the Directorate's attempts to introduce draft guidelines for state aids, the Commission's view remained that complaints against PSBs should be considered on a case-by-case basis. Consequently, in February 1999, the Commission opened formal state aid procedures regarding PSBs within Italy, France and Spain (who receive revenues from both state subsidy and advertising) and found their collection of advertising revenues did not unfairly distort the national markets.

The launch of digital services by PSBs also led to complaints by private rivals. In 1998, the Directorate ruled in favour of two German thematic channels Kinderkanal and Phoenix, run by Arbeitsgemeinschaft der Rundfunkstalern Deutschlands (ARD) and the Zweites Deutsches Fernsehen (ZDF) public operations (Aid no: NN70/98. European Union 1999a) and supported Portuguese public broadcaster Radio Televisao Portuguesa (RTP) when it was challenged by the private company Sociedade Independente de Communicacao (Davis, 1998: 94). Further, in 1999, the EC rejected, on the grounds of cultural exemption, a complaint from BSkyB that licence fee funding of the supply of BBC News 24 to cable television viewers was an abuse of European laws on state aids (European Union, 1999b).

On 15 November 2001, the Directorate published its 'Communication on the Application of State Aid Rules to Public Service Broadcasting' to further clarify its approach on state aid rules governing PSBs (EU, 2001). These included:

- Member states can define the extent of the public service and how it is financed and organised.
- The Commission called for transparency to assess the proportionality of state funding and possible abusive practices.
- Member states were asked whenever such transparency is lacking to establish a precise definition of the public service remit, to formally entrust it to one or more operators through an official act and to have an appropriate authority monitor its fulfilment.
- The Commission would only intervene in cases where there is a distortion of competition arising from the aid which cannot be justified with the need to perform the public service (Wheeler, 2001: 6).

As the commercial organisations increased the complaints claiming that PSBs enjoyed an unfair competitive advantage through their greater capacity to invest in programming and services, the EC argued that public financing must be proportional to the range of programming and services to be included in the public service remit. Moreover, the commercial broadcasters' complaints received greater receptiveness due to a legal judgment which had major implications for all types of enterprises that received compensation from public subsidies.

The European Court of Justice: the 2003 Altmark judgment

The Altmark judgment referred to a state aid case against a public transport bus service in Stendal, a rural district in Germany. In 1994, the district council awarded the Altmark bus company with franchises that were accompanied by subsidies to offset the costs incurred due to its public service mission. However, a competing firm NGVA claimed that such subsidies contravened the EU's rules governing state aids. Ultimately, the case was sent from the German Supreme Administrative Court to the ECJ.

The ECJ judgment (case number C-280/00) provided limits concerning the levels of compensation that could be awarded by member states to firms in exchange for public service obligations. The Court held that four conditions should be satisfied to ensure such compensation did not confer an unfair competitive advantage:

- The beneficiary has been entrusted with clearly defined public service obligations.
- The compensation must be calculated in advance in an objective and transparent manner.
- The compensation does not exceed the costs incurred in discharging the public service obligations, taking into account that the beneficiary is entitled to make a reasonable profit.
- The undertaking selected to discharge the public service obligation is chosen pursuant to a public procurement procedure or the level of compensation is determined on what it would cost a well-run undertaking to discharge these obligations (ECJ, 2003).

The judgment demonstrated to member states that state aid rulings would not apply when appropriate competitive tenders and levels of cost were incurred for enterprises carrying out public service obligations. This meant that national governments could organise their public services without having to submit their financing mechanisms for prior Commission scrutiny. However, the ECJ included stringent efficiency tests to ensure that member states would not favour certain undertakings under the guise of compensating them for the costs incurred in discharging their public service obligations. In effect, the Altmark solution effected a quasi-EU regulation on member states.

The EU's application of state aid concerning the compensation of PSB funds

The Altmark judgment was unpopular with European PSBs as it meant they had to defend the legality of their funding regimes. Moreover, the Altmark judgment enabled the Competition Directorate to take a tougher stance when it considered the 'fairness' of the levels of public or commercial compensation paid to PSBs in subsequent EU investigations in Denmark, the Netherlands, Germany and Ireland.

In May 2004, the Commission ordered TV2/Danmark to return the excess compensation of 84.4 million euros plus interest it had received from the Danish government during 1995–2002. The Directorate calculated the Danish state's financing of the PSB had contravened its rulings that state aid might only be received if the financing was proportionate to the public service's net cost. In determining TV2's net cost, the Commission took into account the advertising revenues it had generated from the public service programmes and concluded that TV2 had received too much money in subsidies.

Moreover, the findings revealed that the Danish state had unfairly distorted the national television market by reinvesting annual amounts of excess compensation into the public broadcaster. TV2 had benefited from state measures including interest-free and instalment-based loans, guarantees for operating loans, a corporate tax exemption and ad hoc capital transfers; and had received access to a nationally available transmission frequency on favourable terms. These anti-competitive advantages were enhanced when excess forms of compensation favoured public broadcasters over commercial players, thereby depressing prices (EU, 2004b).

Simultaneously, on 3 February 2004, the Directorate launched a state aid probe into Holland's eight PSBs and their umbrella organisation, the Netherlands Broadcasting Corporation (NOS), to consider whether the Dutch state had provided them with excess funds of 110 million euros since 1992. This investigation accorded with the Commission's preliminary conclusion that the Dutch PSBs received subsidies and ad hoc payments over and above the funds necessary to finance their output.

Most importantly, they had received additional monies for the provision of commercial services which were deemed to be outside of the purview of their public service remit. Furthermore, the Directorate investigated possible forms of 'cross-subsidisation' in which the PSBs' activities in advertising markets and their acquisition of sports transmission rights distorted normal types of market behaviour (EU, 2004a). On 22 June 2006, the Competition Directorate served notice on the Dutch authorities that they were responsible for the recovery of 76.3 million euros plus interest from NOS due to the excess ad hoc funding which had been granted from 1994 to 2005 (EU, 2006).

Within Germany, a state aid investigation was triggered in March 2005 when complaints were brought against the ARD and ZDF concerning their lack of a defined public service remit and an alleged degree of overcompensation drawn from cross-subsidised advertising revenues. In concluding its investigation, the Competition Directorate informed the German government that its PSB financing regime was incompatible with state aid rules. Subsequently, the EU and the German authorities agreed on a two-year programme to implement measures ensuring the proportionality of the levels of licence fee income and advertising funding received by the ARD and ZDF (Dias and Antoniadis, 2007).

By 2007, the EC received commitments from the German authorities including safeguards to separate accounts for public services and for commercial activities, limits on the amount of public subsidies the PSBs could receive in exchange for public service programming costs and transparent definitions ensuring commercial subsidiaries would not benefit from undue public funding. Consequently, the EC concluded that these measures had ensured the German PSBs had conformed to state aid measures in relation to the receipt of their public compensation and the clear division of core and peripheral commercial activities (ibid.).

An investigation of the Irish PSBs Radio Teilifis Eireann (RTE) and Teilifis na Gaeilge (TG4) licence fee funding began in 1999. A commercial operator Television Network Limited (TV3) complained that RTE and TG4 were not properly defined by clear public service obligations and that RTE's use of public funds lacked transparency and proportionality. TV3 also argued that market distortions had occurred due to the over-compensation that RTE had received along with excessive advertising revenues. Subsequently, the Commission declared the existing financial regime did not provide a sufficiently precise public service definition with regard to RTE's and TG4's activities. Further, as the scope of these activities broadened to include commercial activities, the failure to effect an independent control mechanism had led to the PSBs enjoying an unfair competitive advantage over their rivals (EU, 2008b).

In October 2007, the EC upheld the complaint and the Directorate provided a set of remedies to ensure the PSBs complied with the state aid measures. First, Ireland should clarify the public service remit of all its activities outside of broadcasting and this definition should indicate where RTE's activities were of a public or commercial nature. Second, the scope of RTE's public service objectives had to be legally defined and subject to an independent form of regulation. Third, all future forms of funding should be proportionate to the net public service requirements and those commercial activities that existed outside of these commitments should have funds deducted. Fourth, the PSBs should provide separate public and commercial accounts that were consistent with cost allocation rules. Finally, the PSBs' commercial activities had to be carried out in line with market practices and the fulfilment of these measures should be subject to independent external control. Subsequently, in January 2008, the Irish authorities submitted a summary of their commitments, including the creation of a new regulator for the broadcasting sector, and the Directorate accepted that these recommendations accorded with the state aid measures (ibid.).

State aid measures: PSBs, new media enterprises and the financing of sports rights

Simultaneously, the EC has employed state aid measures in relation to the PSBs' entrance into new media markets and their use of public subsidies to act as Public Service Media (PSM). In Holland and Germany, the

Commission made it apparent that while online information services may be included as public service obligations, online activities such as e-commerce and mobile telephone services will not be considered as a 'service of general economic interest' due to the potential profitability and private nature of these communications transactions.

Moreover, with regard to the financing of new media activities, the Commission has commented that clearly defined forms of public commitments have to be applied by PSBs to show how such a service will benefit the public without unfairly distorting the marketplace. Most especially, any proposal for a new media service should demonstrate how it will pursue social, cultural and democratic obligations for the common good. For instance, BBC online has been seen to provide public access to a range of news and information resources. These proposals have to be endorsed by public authorities, although in such a manner that indicates they will not contravene the requirements of state independence and programming autonomy.

These principles have defined the recommendations within the Competition Commission's 2008 Broadcasting Communication concerning state aid in the new media environment (EU, 2008c). Most especially, commercial operators are concerned that PSBs may enjoy unfair competitive advantages through their employment of public subsidies to enhance their online activities. Conversely, some national governments have argued that a PSB presence is required in the online realm due to the problems associated with market failure and the need to protect democratic rights. In particular, the European Broadcasting Union claims that such rules may undermine the principles of subsidiarity by reducing 'the scope of Member States to enable PSBs to make a significant input into the information society' (EurActive, 2009).

In addition, the Commission assessed allegations from complainants in Germany and Ireland about the PSBs' behaviour when acquiring sports rights. The EC contends that all sports, including high-quality premium sports, are essential components of a balanced and varied programming diet. Consequently, PSBs can acquire sports rights as long as there are not 'too many sports' on public channels and if they do not unduly distort the rights markets. Therefore, difficulties occur when certain PSB acquisition practices such as the purchasing of exclusive rights might lead to the possible undermining of a pay-TV company's competitive opportunities. In Germany, the Directorate decided that while the PSBs had secured a significant proportion of sports rights with particular appeal to German audiences, they had not stopped commercial operators from acquiring rights to equally attractive events. Further, exclusivity did not preclude the PSBs from gaining rights to sports events, but the EC did require that unused sports rights should be offered or sub-licensed to third parties.

Effectively, the EC demanded that a wide range of good governance practices must dictate the employment of public subsidies in the areas concerning new media enterprises and television sports rights (Dias and Antoniadis,

2007: 69). However, the difficulties between the Competition Directorate and the PSBs in relation to state aid measures have proved problematic with regard to analogue switch-off/digital switchover.

State aids in relation to digital switchover

The EC's State Aid Action Plan comments that member states may employ state aid to overcome market failures in the transference from analogue to digital services to ensure social cohesion. Most especially, there is a significant danger that only a limited section of the population will benefit from the advantages of digital television. These problems are most acute in the case of terrestrial television services due to spectrum scarcity. Also the parallel 'simulcasting' of analogue and digital transmissions is prohibitively expensive. Further, across member states, terrestrial networks have been employed to fulfil universal coverage obligations. This means that a high percentage of the population should be covered by digital transmissions before any government can contemplate analogue switch-off.

Therefore, the Commission acknowledges that a cohesive switchover can be undermined by difficulties related to the coordination of technological reforms, the dangers that incumbent broadcasters will gain a competitive advantage by delaying switchover and that problems associated with audience uncertainty will undermine universal service obligations. Yet, the Competition Directorate still requires that member states abide by state aid instruments to address switchover, that the level of subsidy remains at an absolute minimum and that it does not distort competition.

Thus, the EC requires the state aid scheme for digital switchover to be proportionate to the public service obligations. The Directorate has sought to examine what effect market failures may have on the switchover process by assessing whether these perceived failures prevent the market from achieving economic efficiency. It is only when these conditions are met that state aid schemes can be considered for approval under Article 87(3)(c) of the EU Treaty (Schoser and Santamato, 2006: 26).

Channel 4's rights to licence fee subsidies concerning digital switchover

In 2008, the Competition Directorate began its investigation of the UK government's decision to allow C4 to utilise £14 million of subsidies drawn from the BBC's licence fee monies to meet the capital costs associated with digital switchover. The recommendation had been a major victory for C4 who had argued that it could face a £150 million per annum shortfall as digital media costs would squeeze its advertising revenue.

Although the UK government had notified the EC of its plans in October 2007, the investigation was triggered by a complaint from Independent

Television (ITV). ITV argued that as C4 remained a public corporation entrusted with a public service remit but is run on a commercial basis (its revenue is received via advertising, sponsorship and subscription) its access to licence fee monies would unfairly distort the British television market. Further, the complainant objected to any possible financial assistance for C4 on the grounds that it had ample cash reserves to meet the costs associated with digital switchover.

Consequently, the Directorate contended that, given the inadequate information it had received from the UK government, it could not fairly assess whether C4 should receive this form of state aid as it was unclear whether the channel's net public service costs were proportionate to its obligations (Kroes, 2008). Further, the EC argued:

- Although switchover may affect C4's profitability, it will not affect its viability and therefore did not constitute a valid reason for state funding.
- A decline in profitability would not mean that C4 could not deliver its public service remit.
- C4's future plans including investments in new media services, the launch of non-PSB channels and the development of video-on-demand services undermined its case for financial support.
- C4's non-public service commercial activities should not benefit from any form of state aid.
- C4's reserves of £145 million covered its transmission costs (Brown, 2008).

Thus, Kroes wrote to the British Foreign Secretary David Miliband commenting that the UK government's plans were in breach of the EC's state aid rules: 'The Commission doubts whether the notified measure is compatible with the common market' (Kroes, 2008). Conversely, C4 attempted to minimise the impact of the interim decision by arguing that the digital switchover assistance from the BBC should be counted as 'special funding' rather than an unacceptable form of state subsidy (Brown, 2008).

On 26 November 2008, the UK Secretary of State for the Department of Culture, Media and Sport (DCMS) Andy Burnham decided to withdraw the £14 million state aid for C4's digital transmission costs. Instead, any potential shortfalls in C4's funding arrangements were to be considered under Ofcom's analysis of PSB funding entitled *Putting Viewers First* (Ofcom, 2009) and within the Communications Minister Lord Carter's interim report *Digital Britain* (DCMS/DBERR, 2009). Officially, C4 welcomed the change in policy claiming that a more wide-ranging review would help secure the channel's long-term health (Andrews and Mason, 2008). However, the questions which have emerged from these reviews associated with either the potential merger between C4 with Five or with BBC Worldwide to ensure the commercial profitability of the channel in a more competitive advertising marketplace, will continue to require the EC's oversight in relation to competition rules. And these issues will return attention to the fundamental division between the neoliberal values of the EU and public enterprises in total or partial receipt of state subsidies (Sweney, 2009).

Conclusion

The EU has sought to enhance expansion of television services through the principles of liberalisation and harmonisation (Wheeler, 2004). Simultaneously, the Commission has attempted to redress any undesirable outcomes of an unfettered marketplace. Therefore, an inherent tension in the EU policy process has been evident as neoliberal values have come into collision with traditional PSB regimes. These divisions have been played out in the Competition Directorate's issuing of state aid procedures concerning the distortion of markets by PSBs through their receipt of public subsidies.

The EC has become involved with several state aid cases concerning the funding and licensing of the PSBs as defined by the protocol of the Amsterdam Treaty and the ECJ's Altmark solution concerning proportionality and over-compensation. Most especially in the post-Altmark era, the Competition Directorate has hardened its stance against PSBs and their anti-competitive effect in terms of market distortion. In the Danish and Dutch cases the Directorate required the repayment of monies received by PSBs which have detrimentally affected opportunities for commercial competitors. In the German and Irish cases the PSBs were subjected to substantial commitments requiring them to define public service obligations, submitted themselves to independent forms of economic regulation and were required to be responsive to proportionate financial regimes.

The application of state aid measures has defined the PSBs' activities in online environments, their opportunities to purchase sports rights and the costs for digital switchover. While the EU recognises that the transition process between analogue switch-off to digital switchover can lead to market failures which have major implications for the principles of social cohesion including the universal access to broadcasting services it has determinedly pursued the principles of market efficiency. The controversy which surrounded the UK government's proposal to allow C4 to access BBC licence fee monies to cover digital switchover costs indicated how the criteria of market distortion and unfair competitive practice have continued to define the EC's approach. The fallout from this case had important ramifications for the costs of UK digital switchover and for C4's corporate future.

Therefore, Competition Commissioner Kroes has applied state aid notifications with a greater ideological fervour than her predecessors. The use of competition policy with regard to state aid measures has meant that PSBs have to clearly define their public worth and has hampered their opportunities to engage in commercial activities which are seen to create market distortions.

This rigid employment of state aid, however, fails to take into account the social, cultural and democratic functions of PSBs and the need to protect these areas of opportunity within a communications market which has become increasingly defined by competitive commercial services. Competition policy with regard to state aid fails to conceive information and communications rights as a public good. This is an issue of vital concern, since communication must be considered as having a significant *social* worth as well as being understood as an economic commodity. In effect, the EU's neoliberal competition policies may enhance market opportunities, but they fail to recognise the cultural complexities of an audio-visual and communications public sphere in which a diverse range of voices is required to encourage representation and aid participation for European citizens. These concerns return attention to the dichotomy between the Commission's interventionists and liberalisers, and suggest that in regard to supranational competition policy it has been the latter who have won the day in establishing EU rulings governing the audio-visual sector.

Note

1. The consultation process concerning the draft 2008 Broadcasting Communication regarding State Aid was ongoing when this chapter was written. The outcome of this process will have significant implications for the future concerning the scope and ability of PSBs to access the online realm.

References

Andrew, A. and R. Mason (2008) 'Government Withdraws Proposal to give Channel 4 State Aid of £14 million', The Daily Telegraph, 26 November, home page www.telegraph.co.uk, accessed 23 January 2009.

Brown, M. (2008) 'Channel 4: EU Minded to Block BBC Aid for Digital Switchover', *The Guardian*, 4 June, home page www.guardian.co.uk, accessed 12 September 2008.

Collins, R. (1994) *Broadcasting and Audio-Visual Policy in the European Single Market* (London: John Libbey).

Davis, W. (1998) *The European Television Industry in the 21st Century*, Financial Times Specialist Report (Media and Telecoms: FT Business Ltd.).

DCMS/DBERR (Department of Culture, Media and Society and Department for Business Enterprise and Regulatory Reform) (2009) *Digital Britain: the Interim Report*, CM7548 (London: Crown Copyright).

Dias, P. and A. Antoniadis (2007) *Increased Transparency and Efficiency in Public Service Broadcasting – Recent Cases in Spain and Germany: Competition Policy Newsletter*, Number 2 (Brussels: European Commission), pp. 67–9.

EurActive.com (2009) 'Member States Seek to Water Down Media Rules', EurActive.com, 24 February, home page http://www.euractive.com, accessed 28 February 2009.

European Court of Justice (2003) Judgment of the Court (Regulation (EEC) No. 1191/69 – Operation of urban, suburban and regional scheduled transport services – Public subsidies – Concept of State Aid – Compensation for discharging public service obligations. In Case C-280/00, REFERENCE to the Court under Article 234 EC by the Bundesverwaltungsgericht (Germany) for a preliminary ruling in the proceedings pending before that court between Altmark Trans GmbH, Regierungspräsidium

Magdeburg and Nahverkehrsgesellschaft Altmark GmbH, third party: Oberbundesanwalt beim Bundesverwaltungsgericht (Luxembourg City: European Court of Justice), 24 July.

European Union (1992) Treaty on European Union (92/C 191/01), signed at Maastricht, 7 February 1992 (Brussels: European Union).

European Union (1997) 'Protocol on the System of Public Broadcasting in the Member States', Protocols to the Treaty of Amsterdam amending the Treaty on European Union, signed at Amsterdam, 2 October (Brussels: European Union).

European Union (1999a) 'Commission Decision on State Aid Financing of Kinderkanal and Phoenix Specialist Channels' (Brussels: European Commission).

European Union (1999b) 'Commission Decision on State Aid Financing of a 24 Hour News Channel out of a Licence Fee by the BBC', SG(99) D/10201, 29 September (Brussels: European Commission).

European Union (2001) 'Communication from the Commission on the Application of State Aid Rules to Public Service Broadcasters', 15 November (Brussels: European Commission).

European Union (2004a) 'Press Release: Commission Launches Aid Probe into Dutch Public Service Broadcasters', 3 February (Brussels: European Commission).

European Union (2004b) 'Press Release: Commission Order Danish Public Broadcaster TV2 to Pay Back Excess Compensation for Public Service Tasks', 19 May (Brussels: European Commission).

European Union (2005) 'EU Competition Policy in the Media Sector (compiled 2005)' (Brussels: European Commission), http://europa.eu.int/comm/competition/publications/studies/ecompilation_2005.pdf, accessed 7 November 2008.

European Union (2006) 'Press Release – State Aid: Commission Orders Dutch Public Service Broadcaster NOS to Pay Back €76.3 Million Excess Ad Hoc Funding', 22 June (Brussels: European Commission), http://europa.eu/rapid/pressReleasesAction.do?reference=IP/06/822, accessed 12 September 2008.

European Union (2008a) 'State Aid: Commission Opens an Inquiry into UK State Financing of Capital Cost for Digital Switchover of Channel 4' (Brussels: European Commission), http://www.ebu.ch/CMSimages/en/BRUDOC_INFO_EN_421_tcm6-60883.pdf, accessed 12 September 2008.

European Union (2008b) 'State Aid E 4/2005 – Ireland State Financing of Radio Teilifis Eireann (RTE) and Teilifis na Gaeilge (TG4)', 27 November (Brussels: European Commission).

European Union (2008c) 'Draft Communication on the Application of State Aid Rules to Public Service Broadcasting', 4 November (Brussels: European Commission).

Hardy, J. (2001). 'Border Crossings: Convergence, Cross-Media Promotion and Commercial Speech in UK Communications Policy', paper for 51st Political Studies Association Conference, University of Manchester.

Humphreys, P. J. (2007) 'The EU, Communications Liberalisation and the Future of Public Service', in K. Sarikakis (ed.), *Media and Cultural Policy in the European Union*, European Studies: an Interdisciplinary series in European Culture, History and Politics, 24 (Amsterdam and New York: Rodopi), pp. 91–112.

Iosifidis, P. (2007a) *Public Television in the Digital Age: Technological Challenges and New Strategies for Europe* (Basingstoke: Palgrave Macmillan).

Iosifidis, P. (2007b) 'Review of Katherine Sarikakis (ed.), *Media and Cultural Policy in the European Union*', *European Journal of Communication*, 22(4): 517–20.

Iosifidis, P., J. Steemers and M. Wheeler (2005) *European Television Industries* (London: British Film Institute).

Kroes, N (2008) 'Letter to the Right Honourable David Miliband, Secretary of State for Foreign and Commonwealth Affairs (Public Version): Subject State Aid NoC 13/2008 – United Kingdom Aid to Channel 4 Linked to Digital Switchover', 2 April (Brussels: European Commission).

Levy, D. A. L. (1999) *Europe's Digital Revolution: Broadcasting Regulation, the EU and the Nation* State (London: Routledge).

Ofcom (2009) *Ofcom's Second Public Service Review: Putting Viewers First*, 21 January (London: Office of Communications).

Papathanassopoulos, S. (2002) *European Television in the Digital Age* (Oxford: Polity).

Schoser, C. and S. Santamato (2006) *The Commission's State Aid Policy on Digital Switchover: Competition Newsletter*, Number 1 (Brussels: European Commission), pp. 23–6.

Sweney, M. (2009) 'BBC Licence Fee "Digital Surplus" Could Help Fund New PSB Body, says Ofcom', *The Guardian*, 21 January, home page www.guardian.co.uk, accessed 23 January 2009.

Wheeler, M. (2001) *EU Competition Issues in the Telecommunications and Audiovisual Industries* (London: British Screen Advisory Council).

Wheeler, M. (2004) 'Supranational Regulation: Television and the European Union', *European Journal of Communication*, 19(3): 349–69.

5
PSB and the European Public Sphere

Barbara Thomass

Introduction

This chapter focuses on the relations between Public Service Broadcasters (with particular reference to the German PSB case) and the emerging European public sphere and analyses the media actions that are necessary to allow for the development of a European public sphere or European public spheres (it will be explained why the plural term is essential). The starting point of the contribution is the following hypothesis: normative theories about the public sphere and the concept of PSB have considerable intersections and can be used for an enhancement of European public spheres. They are not only assumed because of the commonly used word *public*, which is more than a play of words, but refer to a shared content – both concepts comprise the idea of being related and accountable to a public interest. The contribution therefore deals with the following issues: What is a European public sphere? How can it emerge and develop? What is the role of the media in it? Why can PSB play a crucial role in this respect? How does it live up to this role? Which further desiderata result for its performance?

What is a European public sphere?

When trying to identify the complex concept of a European public sphere, the following question should be addressed: Do we consider the public sphere in a normative or an empirical way? Concerning the public sphere which can be described empirically, it is important to state that this is not a holistic system, but a diversity of different public spheres, which are entangled with each other and can be differentiated according to the space they open, the relevant media, the participating audiences and so on. However, when the concept of a European public sphere emerges, reference is mostly made to the political sphere with its normative content – the emphatically desirable construct – as it was once analysed by Habermas. With respect to this connotation we use the notion in the singular form and distinguish two forms of political public sphere: the *liberal-representative* and the *deliberative* model (Gerhards, 1997, 2002).

The liberal-representative idea goes back to John Locke (1689), John Stuart Mill (1861), Joseph Schumpeter (1942), Anthony Downs (1957) and Bruce Ackerman (1989), with the latter in particular discussing the dimensions of the concept within the work of Jürgen Habermas. The liberal-representative model has as a starting point the following preconditions: It is obligatory for a democratic system that political decisions respect citizens' interests as well as the processes by which the will of the citizens is formed. This typically happens at election times; for citizens to be able to make informed decisions they need to be aware of both competing actors of power, and the actions and laws that have been put into practice since the last elections. Thus Robert Dahl (1989: 111) claims the possibility of informing oneself as a criterion of the democratic process in the sense of 'enlightened understanding'.

Under this model the public sphere is created when the citizenship is well-informed, can exercise its free will and can control the political elite. The public sphere is thus the system of citizen observation. The members of the political elite – on their side – are aware that the citizens observe them, take into consideration the attitudes of the citizens, and therefore orientate their actions towards citizens' expectations. This process involves a high degree of transparency and there is assurance that the elite responds to the will of the citizens. Of course in contemporary representative democracies this process is mediated and the public sphere is basically a sphere created by the media, because an immediate contact of people and the elite is no longer possible.

On the other hand, the deliberative model (Peters, 1994) is more demanding as far as the role of the public sphere is concerned. Based on the following three dimensions we find a strong aspiration to define the *quality* of the public sphere (Gerhards, 1997):

- Who is speaking in the public sphere? Joshua Cohen (1989), Jürgen Habermas (1996) and Amy Gutmann and Dennis Thompson (1996), all authors defending the deliberative model, make a plea for a discourse in the public sphere which should be the opposite of a model where only transparency is demanded, and the implementation of decisions is a question of power. The autochthonous, that is to say desirable, public sphere is a sphere where the civil society is participating, for instance through groups of citizens and NGOs, which are immediately bound to the interests and experiences of the citizens.
- What is communication like? The character of the discourse is a central element in the deliberative model. In a real discourse statements are founded by arguments; it is a well-conducted exchange of information and reasoning between the acting groups and political parties. The participants

are performing a dialogue, for instance paying attention and responding to the arguments of others, therefore guaranteeing the rationality of the debate.

- How is the debate characterised? While in the liberal-representative model the outcome of a debate is normally the wish of the majority, under the deliberative model any decision is a consensus which is scrutinised in a discourse; at the least it is a majority position which has undergone such a discourse, and which is not only implemented by administrative power. Here, the public sphere is the system where communication about the common good takes place and the citizen is the central actor without whom this model cannot work. The citizen acts within groups, NGOs, and at the same time as an individual. He or she is interested in contributing to the debate, listening to other views and wants this dialogue to be recognised by those who are taking the decisions. The citizenship wishes any decisions to be informed by those arguments which had been exchanged before, so that all interests are considered.

To sum up, the public sphere in its ideal form operates under the liberal-representative theory in the following way: Elected representatives compete for the approval of the citizens by articulating opinions. The public sphere is realised with the creation of transparency about this competition and the political proposals that are discussed. Providing there is enough transparency, the implementation of decisions of those in power is legitimate. In contrast, the deliberative model supports the inclusion of groups in civil society, that is to say, as many citizens as possible in the public sphere, seeks arguments and dialogue, and the process of deliberation usually ends in a consensus. The deliberative model presupposes the inclusion and participation of all those who are affected by decisions (Eder et al., 1998: 325). Thus, the public sphere might initiate a collective process of learning among citizens, which will inform future debates too.

Assuming the deliberative model as the 'more' normative concept as a basis, the following requirements for communication in the public sphere can be deduced:

- Equality and reciprocity concerning the involvement of those who are part of the public sphere.
- Openness and due capacity for issues and contributions and high competence of all participating in the public sphere.
- Discursive structure, using arguments, dialogue and critique.

These far-reaching demands are useful for assessing the contribution of the media to the creation of a public sphere. In an attempt to apply these requirements to the political reality of Europe, the European public sphere needs the following (and here the often bemoaned 'democratic deficit' of the EU is brought into account):

- The democratisation of the political processes in order to allow for identification in the sense of interest and participation.
- The representation and analysis of European politics in the media which have to live up to their task of creating a public sphere at a European level, and this in manifold ways.

The emergence of a European public sphere

Here we address the question of whether a European public sphere means: either that there are more and more European issues in the national media, or that there is a combination and amalgamation of national public spheres on the European arena. Gerhards (2002: 142) conceives these two possibilities:

- A supranational European public sphere develops as a unified media system, over which contents and information in different countries of the European Union are perceived. After the failed attempts to set up top-down successful media which are especially perceived as European media – for example, Eurikon or Europe TV which were launched with a European Parliament initiative in the 1980s (Kleinsteuber and Thomass, 1999) – this way is no longer promising as there are currently *no* successful Europe-wide media which could cater for a *political* public sphere.
- Europeanisation of the national public spheres develops in the sense that more and more European topics are dealt with in the national media and the media agendas of the European countries are becoming more homogeneous. The national media based in various European countries exchange views and knowledge, so that the audience can learn more about other European countries and their cultures, values and perspectives (van der Steeg, 2000).

A precondition for the latter means for the emergence of a European public sphere is that the nexus of political actors, the public and the media has an interest in European topics and is ready for interaction (Gerhards, 1997: 17). This triad between politicians, audiences and journalists is a complex one, as it requires that politicians care for issues that are relevant to the citizens (as for example addressing an economic slowdown), journalists do pick up those relevant issues (and not only the often critical examples of European bureaucracy), and audiences feel that it is worth paying attention to European topics. It is necessary that there is an incentive for the media to report on Europe – and that there is a responsiveness of the players involved, given that about 80 per cent of national laws are influenced by the EU. However, the fact that audiences showed little interest in the European Parliament elections of 2009 (a record 43 per cent low turnout) and the rather weak media coverage of the electoral campaign confirm that at least two elements of the aforementioned triad, each for different reasons, do not work well.

The role of the media in an emerging European public sphere

For a European public sphere to develop and function, it needs to have informed citizens. Citizens today are informed via mediated institutions, but are the media bringing to the fore issues and topics of wide European interest? If the media are expected to have an interest in European topics, we must ask under which conditions can this interest develop? Typically the media show interest in European themes once a certain degree of attention for European issues is already evident. This is particularly the case when it comes to the commercial broadcast media, because attention, that is viewership, results in more commercial income and profits. While coverage of national matters guarantees high ratings, the market for European topics is more difficult to develop, and this is especially evident in smaller linguistic areas.[1]

To strengthen European reporting, appropriate resources have to be made available for it (competent correspondents and editors at home, who are able to address the topic for the national and local audiences, enough time slots in the programming, and so on) – even if this is not translated directly into attention and audience shares (which means revenues). European reporting thus requires investments in advance. Referring to the market logic of supply and demand, European reporting is a supply without a relevant demand for the time being, while the commercialised media market mainly responds to demand. It works as a buyer's market. The resources for a new product – in this case, a new field of reporting – are usually only made available if this is worthwhile in itself, in other words if there is a market for those new topics. This lack of interest in catering for issues which are not asked for by the audiences but which are relevant for the political public sphere can be considered as a *structural deficit*. It is built into the structure and philosophy of the commercial media which adopt the market logic and ignore the social benefits of the creation of a political public sphere. So the enhanced representation of European reporting – even if this does not result in high ratings and therefore revenues – can hardly be realised within the framework of the logic of commercial media. Can PSB fill this gap?

The performance of PSB is dependent both on the existing regulatory framework, which represents the legal-political dimension, and on market competition, which represents the economic dimension. It is true that public institutions do perform within competitive media ecology and also need to pay attention to ratings, but due to their societal contract which obliges them to fulfil fundamental cultural and social goals, PSBs can more easily be associated with the process of Europeanisation of the public sphere than commercial media. This is because PSB is a form of broadcasting regulation which is based on the consensus that broadcasting has to satisfy certain social and cultural needs, which lie beyond the interests of the consumer. The PSB system is to a certain degree protected against the forces of the market (for example by its unique method of funding through the licence fee), and specific obligations are imposed upon it (for example, the condition to provide universal service, or to cater for certain programme genres such as news, children's programmes, regional content and so on). These privileges and obligations are secured by special statutes or charters (Syvertsen, 2001).

Therefore we can assume that a broadcaster who does not act according to commercial logic can contribute better to the creation of a public sphere in the sense of an informed participating public than a broadcaster who responds to commercial interests. PSB has a capacity for that, because (and/or provided that):

- A form of social accountability is given to the public sphere or its representatives which is guaranteed by administrative structures.
- A public financing model is secured, which can be used for the production and offering of defined public services.
- Content regulation is in place, especially with regard to balance, impartiality and consideration of minority interests.
- Universal service is provided.
- PSBs address the audience as *citizens*, not merely as *consumers* (Brants and Siune, 1992: 102).

That implies that the standards by which public broadcasters are explicitly and implicitly led, and the social control which guarantees the maintenance of these standards, provide the opportunity for PSB to contribute to a European public sphere. It could be included into the remit of PSB that they should give their audiences access to European affairs, intensify European coverage and enhance the debate about European topics. It is this author's view that European coverage could be explicitly included in the mission statements of PSBs.

German PSB and the Europeanising of the public sphere

In the case of Germany, Kaelble (2004) reported a change in the presentation of European messages in the media. In the 1950s–1960s European topics were usually presented completely from a national perspective and the reports were concentrated on how the respective national government acted and which national interests were affected. However, in the last two decades European issues in the German media have been presented more and more from a European perspective. One can interpret this change as a sign of a developing European public sphere. However, Europe-related news in the media has clearly not become more frequent and Europe does not take a central place in the media – except at special times like the debate about the European Constitution.

Eilders and Voltmer (2003) undertook a quantitative content analysis of editorials of European political issues in the German quality newspapers in the period 1994–8. They considered the amount of editorials referring to European topics as a degree of Europeanisation of the public sphere and compared it to the amount of issues which are no longer decided at the national but at the EU level. They concluded that the Europeanisation of the public sphere clearly lags behind Europeanising of policy. Given that the majority of national laws of the member states of the EU are influenced by or fabricated within the European Commission, the quantity of Europe-related editorials in national media is a *quantité négligeable*.

Concerning the distinction between commercial and public broadcasters on the issue of coverage of European affairs, Kleinsteuber (2004) attributes the lack of a participating public sphere to the commercially dominated media-scape in Europe. He argues that private media, preoccupied with the commercial logic, fail to ensure adequate coverage levels of public and political communication. However, empirical findings as to whether PSBs contribute to the realisation of a public sphere are contradictory. On the one hand, Lauf and Peter (2004) found little evidence that an increasing commercialisation of the media system – thus a declining significance of the public sector – would lead to less attention for European matters. They analysed the 1999 European elections with regard to the frequency of European Union representatives on TV news bulletins in various European countries before the elections and found a high proportion of coverage of European elections and presence of the members of the European Commission and the European parliamentarians in the TV news in Scandinavian countries, Spain, Italy and Austria, but there were no significant differences between the reporting of PSBs and commercial broadcasters. On the other hand, Kevin (2001) reported that a greater amount of news concerning the EU appeared in the programming schedules of PSBs than in those of their commercial counterparts. In this study media coverage of European news was monitored during two one-week periods in 1999 and 2000 in eight European countries (France, Germany, Ireland, Italy, the Netherlands, Spain, Sweden and the UK) and it focused on four newspapers and the primetime news broadcasts of the main public service and commercial channels in the countries under scrutiny.

At this point, it is worth noting that electoral coverage, with its tendency towards personalisation, might not be a very suitable observation criterion for the analysis of the agenda of European topics, since electoral coverage by commercial channels is typically (but not necessarily) fairly good. It should not be merely top events – such as European election times, appointment of new European Union Commissioners, or the adoption of a European

Table 5.1 Amount of programmes relating to Europe in German-language PSBs, September 2004

Channel	*Number of weekly programmes*	*Number of single programmes*	*Total number of programmes*	*Total time of programmes in minutes*
ORF 2	1	–	4	120
ARD	1	–	4	120
NDR	1	3	7	240
WDR	8	2	34	1080
RB	3	4	16	490
RBB	1	3	7	239
MDR	1	1	5	150
HR	2	–	8	340
Südwest	–	4	4	165
Phoenix	–	3	3	285
3sat	2	8	16	740
BR Alpha	11	9	53	1846
Arte	4	1	17	650
ZDF	5	–	20	300
Total	40	38	198	6765

Source: Author's analysis.

Constitution – that give helpful hints with respect to the comprehensiveness of European reporting, for almost all media, regardless of public or private ownership, cannot afford to avoid these hot topics. Additional Europe-related programmes should be identified in the whole TV schedule in order to assume that a channel really contributes to Europeanising national public spheres. In fact, it is only the comparative analysis of Europe-related programmes and their formats and a long-term analysis of these data that would answer the questions as to whether European PSBs have adopted the idea of a European public sphere, to what degree, and how this has developed over time. This is an empirical research design that would require a pan-European conjointed effort.

As an illustration of such a comprehensive evaluation, the author analysed in 2004 the programmes in German public media that have titles referring recognisably to Europe (Thomass, 2006). One month was selected (September 2004), which was to a large extent free from outstanding European topics. During this time all public service programmes in Germany were considered, including the first channel (ARD), the second channel (ZDF), the third channels, BR Alpha, Arte, 3sat and Phoenix, as well as German-language ORF 2 in Austria. The amount of programmes relating to Europe is shown in Table 5.1. Table 5.2 lists the formats that were used. The programmes can be assigned to content categories as shown in Table 5.3.

Table 5.2 Formats of programmes relating to Europe in German language PSBs, September 2004

Programme formats	*Number of weekly programmes*	*Number of single programmes*	*Total number of programmes*	*Total time of programmes in minutes*
Documentary/feature	21	30	13	4730
Magazine	18	7	80	1825
Others	1	1	5	210
Total	40	38	198	6765

Source: Author's analysis.

Table 5.3 Content of Europe-related programmes in German PSBs, September 2004

Programme genres	*Number of weekly programmes*	*Number of single programmes*	*Total number of programmes*	*Total time of programmes in minutes*
Politics	15	6	66	1900
History	–	11	11	466
Travel	10	15	55	2104
Culture	4	3	19	690
Cooking/eating	7	–	28	840
Nature	3	2	14	555
Others	1	1	5	210
Total	40	38	198	6765

Source: Author's analysis.

The compilation of the programmes shows the approach that the German PSBs take in order to bring the issue of 'Europe' to the public – with different genres and content which go beyond news and current affairs. It is a step towards the direction that the programme should offer a variety of genres and forms on how Europe is put on the agenda. Europeanisation of national public spheres certainly requires enhanced audience interest in Europe as a political entity, but also in other European themes beyond politics and this is why the author decided to include topics like travel, cooking and nature.

The programmes examined for Europe-related content could be considered against the structural and normative constraints within which PSBs function. In this context one needs to take into account the number of staff the German PSBs have at their disposal dedicated to the coverage of European affairs, as well as regulatory issues such as the Interstate Treaty on Broadcasting and the Länder-based laws. These are described below.

The foreign correspondents of the first channel, ARD, work for the pool of the Länder-based corporations and are responsible for the coverage of the country, where they are based, and/or for a neighbouring area. World-wide there are about thirty correspondent offices, twelve of which are based in Europe. In these twelve correspondent offices altogether nineteen correspondents work for TV and a larger number for radio (ARD, 2008: 306).

Returning to the issue of regulation, in the relevant German broadcasting laws – the Interstate Treaty on Broadcasting and the broadcasting laws of the Länder – there are few hints referring to a Europe-orientated remit. This issue is vaguely addressed in the Interstate Treaty, which merely provides that a 'main part of the programme has to be reserved for European productions'. The soft formulation is problematic as evidenced by the insufficient broadcasting coverage of European content.

Likewise, the legal bases covering the remit of PSBs do not set any clear guidelines with regard to a European coverage. For the NDR, a PSB based in Northern Germany, it is stated in the Interstate Treaty that: 'The NDR has the obligation to deliver to the audiences an objective and comprehensive overview of the international, national and Länder events in all substantial areas of life' (§5 NDR Interstate Treaty, author's translation). Also for the WDR, based in the western part of the country as another example, no specific provisions had been taken regarding the Europeanising of the coverage: 'The WDR has to deliver in its programmes a comprehensive overview of the international and national events in all substantial areas of life' (WDR Law §4). Furthermore, the law provides specifications for the programme remit with regard to regional stratifications of the broadcasting area, though not on supra-regional areas.

Desiderata for Europe-related activities of PSBs

So how can PSBs contribute to the creation of a European public sphere? As no comparative data and/or strong evidence are yet available, the following issues represent merely the author's evaluation and could serve as a starting point for a possible contribution of PSB to Europeanising of the public sphere.

- *The opening of broadcasting spaces*. This is in fact already the case, regarding the language areas in Europe, where national broadcasters from one country send their messages to other countries with the same language. This could be expanded systematically within a network of PSBs, for example by transnational agreements concerning the carrying of public service programmes in the cable networks, which would also carry the programmes of PSBs located in small nations for the attention of larger ones ('must carry' regulations).
- *The extension of the national mandates to a European scale*. For such an extension of the broadcasting spaces, which can be managed technically,

it would be necessary to extend the remit of the PSB and this could be reflected in the programming. The EU's policies to constrain the significance of the PSBs run contrary to this perspective at present. Although the Amsterdam Protocol of 1992 made a strong case for the cultural importance of the PSBs, the majority of the EU policies which are informed by the will of economic deregulation and strengthening of competition in the audio-visual markets tend to undermine the broad acceptance of PSB.

- *The extension and/or consolidation of the diversity principle*. Diversity (of opinions, as well as programme formats and genres), belongs to a consensual basic leitmotif of media politics. It could be interpreted in a wider frame reflecting cultural diversity in Europe as a realisation of the European space.
- *The forming of a European audio-visual space that would cater for both the consumer and the citizen*. The norms and regulations as they are promoted by the EU Commissioner for Competition concentrate on the audiences as *consumers* – this has been the case since the adoption of the Television Without Frontiers Directive, and it is supported by several rulings concerning the debate about whether PSB licence fees are state subsidies. Looking at the European audio-visual space not only as a market but as part of the public sphere – thus addressing the *citizen* – would leave more opportunities for PSBs to elaborate their Europe-related programmes.
- *Cooperation transgressing national borders*. In bilateral cases (for example, channel Berlin Brandenburg RBB with Polish television or the cooperation between the Nordic PSBs) such cooperation between the public broadcasters in transnational regions already exists. This could be expanded on a Europe-wide scale.
- *The development of exchanges*. Mutual internships of editors and journalists of the PSBs of other European countries could promote mutual understanding, interest and possibilities of investigation and research beyond existing resources. As shown above, correspondents of the broadcasters only exist in some European capitals. Professional cooperation, for example between journalists and newsrooms, could overcome these limitations by exchanging footage and investigated material as is done already by the European Broadcasting Union.
- *The use of online media by PSBs to address audiences beyond national borders*. National broadcasting areas are becoming obsolete with the emergence of the internet and the public channels have to adjust themselves to this reality. Regarding the Europeanising of the national public sphere this could be used in a productive way. Online services which clearly address European users, for example, would allow this possibility.
- *The development and diversification of Europe-related formats*. The formats specified above are still far from covering the whole possible spectrum. Video bridges, by which national realities can be brought together in one

programme or European entertainment programmes are single examples, which could be developed further.

These proposals run contrary to the deregulatory trend that prevails in the European broadcasting sector, for they place emphasis on the deliberate use of a socially accountable medium, which needs to be optimised for social purposes. In the 1980s, various European parliamentarians wanted to manufacture the European public sphere via a medium that had been planned top-down (Eureka) and had not taken into account the existing broadcast structures (Kleinsteuber and Thomass, 1999). This chapter is a plea for a procedure that considers the existing structures and legal conditions of the broadcasting sphere, takes the perspective of the European public sphere (or public spheres) seriously and accompanies it by media policies.

Note

1. For an explanation of the small market paradigm, which states that media in small linguistic areas have less chance to cater for media topics of minor interest as they cannot use the advantage of economies of scale, see Meier and Trappel (1992).

References

Ackerman, B. (1989) 'Why Dialogue?' *Journal of Philosophy*, 86: 5–22.

ARD (Arbeitsgemeinschaft der Rundfunkanstalten Deutschland) (ed.) (2008) *ARD Jahrbuch 2008* (Frankfurt: ARD).

Brants, K. and K. Siune (1992) 'Public Broadcasting in a State of Flux', in K. Siune and W. Truetzschler (eds), *Dynamics of Media Politics* (London: Sage), pp. 101–15.

Cohen, J. (1989) 'Deliberation and Democratic Legitimacy', in A. Hamlin and P. Petit (eds), *The Good Polity: Normative Analysis of the State* (Oxford: Blackwell), pp. 17–34.

Dahl, R. A. (1989) *Democracy and its Critics* (New Haven and London: Yale University Press).

Eder, K., K.-U. Hellman and H.-J. Trenz (1998) 'Regieren in Europa jenseits öffentlicher Legitimation? Eine Untersuchung zur Rolle von politischer Öffentlichkeit in Europa', in B. Kohler-Koch (ed.*), Regieren in entgrenzten Raeumen. Politische Vierteljahresschrift*, 29 (Opladen: Westdeutscher Verlag), pp. 321–44.

Eilders, C. and K. Voltmer (2003) 'Zwischen Deutschland und Europa. Eine empirische Untersuchung zum Grad von Europäisierung und Europa-Unterstützung der meinungsführenden deutschen Tageszeitungen', *Medien & Kommunikationswissenschaft* 2(51): 250–70.

Gerhards, J. (1997) 'Diskursive versus liberale Öffentlichkeit. Eine empirische Auseinandersetzung mit Jürgen Habermas', *Kölner Zeitschrift für Soziologie und Sozialpsychologie* 49: 1–39.

Gerhards, J. (2002) 'Das Öffentlichkeitsdefizit der EU im Horizont normativer Öffentlichkeitstheorien', in H. Kaelble, M. Kirsch and X. Schmidt-Gernig (eds), *Transnationale Öffentlichkeiten und Identitäten im 20. Jahrhundert* (Frankfurt am Main: Campus), pp. 135–58.

Gutmann, A. and D. Thompson (1996) *Democracy and Disagreement* (Cambridge, MA: Belknap Press of Harvard University Press).

Habermas, J. (1996) 'Drei normative Theorien der Demokratie', in J. Habermas (ed.), *Die Einbeziehung des Anderen. Studien zur politischen Theorie* (Frankfurt am Main: Suhrkamp), pp. 277–92.

Kaelble, H. (2004) *Die Genese einer europäischen Öffentlichkeit. Anzeichen und Defizite der politischen Willensbildung auf europäischer Ebene,* available at http://www.nzz.ch/dossiers/2003/eukonvent/2004.04.24-zf-article9GUQ0.html, accessed 29 October 2004.

Kevin, D. (2001) 'Debates about Europe in the National News Media', in P. Bajomir-Lázár (ed.), *Media and Politics* (Budapest: New Mandate Publishing House), pp. 219–34.

Kleinsteuber, H.-J. (2004) 'Strukturwandel der europäischen Öffentlichkeit? Der Öffentlichkeitsbegriff von Jürgen Habermas und European Public Sphere', in L. M. Hagen (ed.), *Europäische Union und mediale Öffentlichkeit. Theoretische Perspektiven und empirische Befunde zur Rolle der Medien im europäischen Einigungsprozess* (Köln: Herbert von Halem Verlag), pp. 29–46.

Kleinsteuber, H.-J. and B. Thomass (1999) 'Der deutsche Rundfunk auf internationaler Ebene', in D. Schwarzkopf (ed.), *Rundfunkpolitik in Deutschland. Wettbewerb und Öffentlichkeit. Band 2* (München: Deutscher Taschenbuchverlag), pp. 1008–71.

Lauf, E. and J. Peter (2004) 'EU-Repräsentanten in Fernsehnachrichten. Eine Analyse ihrer Präsenz in 13 EU-Mitgliedstaaten vor der Europa-Wahl 1999', in L. M. Hagen (ed.), *Europäische Union und mediale Öffentlichkeit. Theoretische Perspektiven und empirische Befunde zur Rolle der Medien im europäischen Einigungsprozess* (Köln: Herbert von Halem Verlag), pp. 162–77.

Locke, J. (2003 [1689]) *Two Treatises of Government* and *A Letter Concerning Toleration*, ed. I. Shapiro (New Haven and London: Yale University Press).

Meier, W. A. and J. Trappel (1992) 'Small States in the Shade of Giants', in K. Siune and W. Truetzschler (eds), *Dynamics of Media Politics: Broadcasts and Electronic Media in Western Europe* (London: Sage), pp. 129–42.

Mill, J. S. (1972 [1861]) *Utilitarianism, On Liberty and Considerations on Representative Government* (London: Dent).

NDR (1992) *Gesetz zum Staatsvertrag über den Norddeutschen Rundfunk (NDR)*, 10 March 1992 (Interstate Treaty on the North German Broadcasting Corporation).

Peters, B. (1994) 'Der Sinn von Oeffentlichkeit', *Kölner Zeitschrift für Soziologie und Sozialpsychologie*, 34: 42–76.

Rundfunkstaatsvertrag vom 31. August 1991, in der Fassung des fünften Rundfunkänderungsstaatsvertrags, in Kraft seit dem 1. Januar 2001 (Broadcasting Interstate Treaty from 1991, in force since 1 January 2001).

Schumpeter, J. A. (2005 [1942]) *Kapitalismus, Sozialismus und Demokratie* (Stuttgart: UTB).

Syvertsen, T. (2001) 'Challenges to Public Television in the Era of Convergence and Commercialisation', *Television and New Media*, 4(2): 155–75.

Thomass, B. (2006) 'Public Service Broadcasting als Voraussetzung europäischer Öffentlichkeit – Leistungen und Desiderate', in W. R. Langenbuche and M. Latzer (eds), *Europäische Öffentlichkeit und medialer Wandel* (Wiesbaden: VS Verlag für Sozialwissenschaft), pp. 318–29.

van der Steeg, M. (2000) 'An Analysis of the Dutch and Spanish Newspaper Debates on EU Enlargement with Central and Eastern European Countries: Suggestions for a Transnational European Public Sphere', in B. Baerns and J. Raupp (eds), *Information und Kommunikation in Europa. Forschung und Praxis* (Berlin: Vistas), pp. 61–87.

6
Civic Engagement and Elite Decision-Making in Europe: Reconfiguring Public Service News

Farrel Corcoran

Introduction

As the central position of the lobbying industry within the decision-making apparatus of the European Union (EU) slowly becomes clearer, its political function can be seen as well outside the parameters of normal democratic scrutiny. To engage with the process of developing laws and regulations affecting the governance of 500 million citizens in a 13 trillion euro economy, an estimated 15,000 full-time lobbyists are engaged in bringing corporate influence to bear on the work of a range of EU institutions. Organised into about 1000 lobby groups, these include public relations and public affairs companies, the lobbying services of law firms, industry-funded think-tanks and 'European Affairs' offices run by private corporations. The annual turnover of all this intense promotional activity, dedicated to shaping a complex regulatory regime to its own advantage and ameliorating its impact on clients, is close to 1 billion euros. Its success in reaching into the heart of EU politics can be seen vividly in those cases where concrete text amendments drafted by industry lobbyists end up as EU law, when parliamentarians in effect become intermediaries and transfer lobbyists' demands into legislation (Hoederman, 2007). Some of these lobbyists are part of global information management giants that Dinan and Miller (2007: 15) call 'deeply obscure organisations' rarely heard of in public (such as WPP, Omnicom or Interpublic). To some observers, when lobbying reaches a critical mass of persuasive power, it subverts democracy.

In this chapter, we ask whether the media are fulfilling their watchdog role in keeping citizens fully informed about the major decisions that are being made by the EU. What if the watchdogs are sleeping in their kennels? Serious concerns are being raised about the recent emergence of the power of global information management conglomerates, the result of transnational mergers between advertising, marketing, public relations and lobbying firms. Is public service communication, pushed hard by competitive forces to emphasise its entertainment potential over its informing role – and in some cases to merge the two into something dubbed 'infotainment' – up to the task of ensuring that the national public spheres and the machinery of government in Brussels are not so tightly managed by business lobbying that the public interest loses out? The situation in Europe is not unlike what is happening in Washington, DC. Because of a powerful set of structural limitations in the US political system, the media lack the ability to bring the bulk of American citizens into anything resembling a deliberative public sphere (Entman, 2004: 164). In Europe, nothing like a public sphere has yet emerged, though we can see the footprints of an elite sphere, that is, circuits of information exchange in the political system from which the public is excluded, reinforcing particular power elites (Corcoran and Fahy, 2009). In both the US and Europe, news media are crucial in the distribution of social power, as they play an important role in shaping the discourse within which decision-makers operate. The question here is how Public Service Media (PSM) should function to redress the imbalances in access to decision-making that have arisen in the centre of European democracy, where information is managed ever more skilfully by sectional interests. There is an urgent need for PSM to adapt quickly to what is a new layer of democratic deficit in the EU.

We look first at the problematic nature of democracy in the EU, and then suggest a solution, in the form of a reconfiguring of public service communication. We begin by examining common versions of the democratic deficit and the crisis in the democratic legitimacy of the EU as it grapples with the ratification of the Lisbon Treaty (www.opendemocracy.net). There is substance in Ward's (2007: 132) contention that the dominant EU approach to the role of Public Service Broadcasting (PSB) has for many years reduced the challenge of maximising the potential of broadcasting to a cultural level, without also seeing the significance of its informational role in democratising the Union. The question here is how should public service news redefine its mission so that civic engagement with European politics is enhanced?

Democratic deficit

There is a 'standard version' of the democratic deficit in the EU, widely shared by media commentators, academics and many citizens (Weiler et al., 1995: 534). Firstly, there is the domination of policy-making by executive actors (ministers from member states acting together in the Council, and government-appointed officials in the Commission) who do their work well beyond the scrutiny and control of national parliaments, even when these have well-functioning European Affairs committees. In effect, governments can ignore their national parliaments when making decisions in Brussels and the European Parliament is too weak, with only limited powers to delay items on a legislative agenda set by the Council. Parliament now has the power to veto the governments' candidates for the Commission, for instance, but the

Council, not the Parliament, is still the agenda setter in this process (Follesdal and Hix, 2006: 535).

Secondly, elections to the European Parliament are actually 'second order national contests' (Marsh, 1998: 536), treated by media and politicians as mid-term national elections, characterised by protest votes against the performance of parties in national governments. Very little campaign attention is focused on European personalities or the shaping of the EU policy agenda. The Parliament is therefore both institutionally and psychologically 'too distant' from many citizens, who are unable to identify with the EU, and frustration with this is periodically vented in a treaty referendum.

The ultimate result of this alienation from EU governance is a 'policy drift' away from voters' ideal policy preferences, as their governments, freed from the national constraints of parliament, courts and interest groups, promote policies in Brussels that they cannot pursue domestically. Crucially, this includes the long-term shaping of a neoliberal regulatory framework. Since the Parliament is not the dominant institution in European governance, multinational corporations have a greater incentive to organise lobbying at the Commission level than do diffuse interests such as trade unions or consumer associations. The result is that policy outcomes are skewed more towards the owners of capital than would be the case at the national parliamentary level, where policy compromises are much more likely (Follesdal and Hix, 2006: 537).

This standard way of analysing the democratic deficit in Europe no longer tells the full story. It needs to be expanded to take into account the surge in lobbying power that typifies the present phase in the development of the EU. The concepts of 'access' and 'voice' are useful here in addressing what are obviously communicative aspects of the democratic deficit. 'Voice' strategies refer to public actions, including protest politics and media campaigns organised at specific venues where political bargaining takes place – parliamentary committees, advisory bodies and technical committees – in order to influence media agendas and indirectly the political system. This can be done by creating a large public presence, for instance when Greenpeace holds a press conference while the European Parliament debates genetically modified organisms, or by targeting smaller, more specialised constituencies, as when an op-ed piece in the *Financial Times* is penned by the Chair of the European Roundtable of Industrialists, aimed at business and financial elites (Beyers, 2004). By contrast, 'access' strategies transmit information directly from interest associations to policy-makers in closed settings. Much of this information is operational or technical, deployed to affect decision-making agendas or frame issues in particular ways, or to signal support for or opposition to legislative or regulatory initiatives. Beyers (2004) makes the further distinction between 'diffuse' and 'specific' interest associations. Some lobbying groups defend diffuse or fragmented interests, linked to broad and general segments of society sensitive to issues of public concern, grounded in personal values. Others defend very specific interests linked to clearly defined business sectors with a restricted membership structure, equipped with the expertise to collect specialised information and focus it precisely with the bureaucracy: employers' or trade associations, organisations representing agriculture, banking, telecommunications, the chemical industry and so on.

Greer et al. (2008) remind us that there are also differences in the way different countries engage with policy-making in the EU, due to a combination of weaknesses in civil society and years of experience as members of the Union. They find that the structure of interest groups in Southern Europe results in lower participation in EU interest group politics compared to Northern Europe. The same is true of post-communist states, which are less successful in lobbying in Brussels, though larger states, like Poland, can be heard more than smaller ones. The countries of north-west Europe, on the other hand (especially Britain, the Netherlands and Germany) developed a culture of lobbying earlier than others, gaining first-mover advantage as effective EU lobbyists (Lahnsen, 2003).

Civil society

Can a 'healthy interest group ecology' reinforced by a strong civil society of active citizen organisations lead to more democratic decision-making? This question is increasingly finding its way into debates about reforming the EU. Greater transparency in interest group politics in Brussels should strengthen democracy by ensuring that the 'mobilisation of bias' in policy-making is not structured systematically towards private corporate interests but leaves the way open for the inclusion of civil society. This is all the more necessary in a situation where news about the EU is filtered through the lens of national perspectives, trans-European media are poorly developed, the emergence of a European public sphere is problematic and the Europeanisation of national public spheres is developing so unevenly (Heikkila and Kunelius, 2008; Pfetsch et al., 2008).

In an attempt to address criticisms of the democratic deficit, the EU proposed changes in its relationship with civil society, especially in declaring its strategic objectives for 2000–5 (European Union, 2000). The turn to civil society in EU official thinking, and its championing of the idea of European civil society, 'is an attempt to flow democracy into the system of EU governance while also seeking to prevent the ebbing of whatever democratic legitimisation may have been achieved through representative democracy' (Armstrong, 2002: 106). But there is growing resistance to EU attempts to rediscover civil society by presenting it as a support for representative democracy rather than as a radical experiment in direct democracy at European level, and there is criticism of the slow performance of the Commission in involving civil society in expert groups, widely used to get external advice before policy is decided. In the view of several interest groups monitoring policy-making in

Brussels, little is currently being done to ensure that a diversity of viewpoints is being heard in the consultation process.[1]

The growing pressure for a mandatory register includes the obligation to disclose the names of all lobbyists, more detailed requirements for financial disclosure and measures to prevent the 'corporate capture' of expert groups by lobbyists whose influence on EU decision-making is too powerful. The Alliance for Lobbying Transparency (ALTER-EU), for instance, complains that too often a privileged position is granted to business advice, compared to the input of public interest experts, and that in areas such as biotechnology and coal, expert advice is completely controlled by industry. The names of many lobbyists with a strong influence on policy are not available to the public and there is no way to hold them accountable for the advice they give to expert groups and Commission cabinet officials. The Commission's guidelines in its own voluntary lobbying register are often ignored, especially the need to ensure that 'the risk of vested interests distorting the advice of Expert Groups should be minimised' (ALTER-EU, 2009). For its part, the Commission defends its use of business representatives, claiming the quality of their expertise justifies their presence.

Compromised news

There is great potential for civil society to help close the democratic gap between structures of governance and the public interest if public service news can be reconfigured in an online environment so that the communicative space of 'Europe' can truly function in the service of democracy.

News sources are the first challenge. Because of macro-level changes in media systems in recent years, news sources can take advantage of unprecedented opportunities to become proactive in shaping the output of journalists – a process Louw (2005) labels the 'PR-ization of politics'. Broadcast deregulation produced the conditions for a radical increase in broadcast news outlets, generating ever more time to be filled by newsrooms depleted by diluted revenues, as private broadcasters consolidated their hold on advertising revenues, and in recent times, their claim on top-sliced licence fees too (Iosifidis, 2008). Tighter news budgets lead to more recycling of news content, a reduction in costly types of news (such as investigative journalism or foreign affairs coverage) and a heavier reliance on easily available news sources. Increasingly, these sources include PR companies focused on providing news stories to hard-pressed newsrooms. A great amount of news today is initiated by special interests with the ability to create 'pseudo-events', photo opportunities, press conferences, managed leaks and other information management tools for leveraging clients' interests into the construction of news. There is a multiplier effect as news content is passed back and forth between different media, both online and traditional, as a 'news source cycle' takes hold (Messner and Watson DiStaso, 2008).

In parallel with these developments in news production, the PR industry has expanded greatly. It grew eleven-fold between 1979 and 1998 in Britain (Miller and Dinan, 2000) and the changing architecture of the media system opened the way for PR to gain widespread influence as a dominant news source. There is good empirical evidence of the scale of the problem in Britain (Lewis et al., 2008). While the number of journalists in the national press has remained fairly static, they now produce three times as much copy as they did twenty years ago. This increases their dependence on 'ready-made' news and limits their opportunities for independent journalism. Nearly a fifth of press and broadcast stories are derived wholly or mainly from PR material, some of it taken word for word from press releases. Less than half the stories analysed were entirely free from traceable PR. The main source of PR is the corporate/business world, which is more than three times more successful than charities, NGOs and civic groups in gaining access to the journalists. PR materials are also widely used in the major news agencies, which feed news stories to a whole range of media. A major escalation of this trend in providing what Gandy (1982) calls 'information subsidies' to news organisations was the Bush government's use of video news releases in US television, made by PR companies using 'fake' journalists to tell 'good' stories, delivered free to cash-strapped channels (Barstow and Stein, 2005).

PSBs have not escaped the political economic pressure to compete with the entertainment sector for audiences and to skew news agendas towards entertainment values that increase the salience of celebrities, crime, sports, 'on-the-spot' news and lifestyle features (Franklin, 1997), pulling broadcast news further away from the serious scrutiny of politics and the dissemination of information supporting citizens' need to make informed choices about major social issues. In the case of European politics, a particular set of limitations severely constricts the flow of information that could address citizens with a true plurality of voices and arguments, and thus deepen the democratic legitimacy of the EU. The 'European Quarter' in Brussels is the site where decisions and dominant ideologies are forged that have a significant, extended, material impact on people's lives, where a range of different interests struggle and manoeuvre to influence decision making. The AIM (2009) project has explored the work culture of the large journalistic corps accredited to Brussels and its relationship with the national media, with their firmly entrenched beliefs about the lack of audience interest in EU news. Ethnographic methods were used to explore what happens to the high volume of information that is produced daily in Brussels and how much of it becomes news. Despite the active role played by the Commission in information management, the major finding of AIM is that only a relatively small amount of news actually flows through from Brussels to member states and this is almost always firmly framed within a national perspective: if there is no easily identifiable *national* angle to a story, it does not become news.

The future of public service news

What are the implications of all this for the future of public service news? The question has to be posed in the face of what seems like an ever-widening disparity. On the one hand, we see large areas of EU policy being shaped, not through public debate and the sharing of alternative arguments in a rationally ordered public sphere, but through the presentation of partial information, often in a covert manner, in an elite sphere that is not open to public discussion, filtered to favour goals which, unless publicised and challenged, benefit only corporate interests and damage the democratic process (Williams, 2007). On the other hand, healthy democracy needs public interest news, and this should be measured in terms of the role it plays in illuminating major social issues and helping citizens to make informed choices about public policy. We now have the technological infrastructure for creating a newsgathering environment that is democratic, responsible to the public interest and popular.

The PSB normative paradigm is based on more than wishful thinking. Empirical research demonstrates significant connections between patterns of news coverage and levels of public knowledge. Curran et al. (2009) compare what is reported in news and what the public knows, in four countries with different media systems: the market model (United States), public service (Denmark and Finland) and a dual model (Britain). The comparison shows that public service television devotes more attention to public affairs and international news, fosters higher levels of news consumption and creates greater public knowledge in these areas than the market model. Holtz-Bacha and Norris (2001: 138) suggest that a virtuous circle is created between media habits and political knowledge, in which people who have a sense of civic engagement tend to watch hard news, while repeated exposure to such content increases levels of civic engagement.

The question now is how the infrastructure for public service newsgathering can be made to work better in the public interest. If we accept the near impossibility of creating civic engagement in the full public sphere via mass media (Entman, 2004), the answer must lie in delivering a better news service at least to civil society and those minority audiences that are politically aware. This can be done by using online resources, such as hypertextuality, to provide layers of more in-depth news that go deeper into public issues than what is appropriate in a television newscast designed for a mass audience. It means paying particular attention to new multi-platform strategies for PSB that will be successful even as the audiences created by scheduled terrestrial television tend to wither away. BBC Online, the leading content site in Europe, regarded as the industrial benchmark by many broadcasters, is currently experimenting with public service news online, especially in its iCan portal. This is designed to help people engage with politics by delivering in-depth information on a myriad of civic issues and connecting them to other websites. The expanding technical capabilities of the internet are creating innovative, rich, interactive content very different from the familiar linear narrative of traditional PSB news, with its 'central, centrist news operation of unitary values' (Lee-Wright, 2008: 257). Many of the new virtual communities of interest are already familiar with web models that aggregate news sources (for instance, MyYahoo.com, MyWeb2.0com) or predict news content interest to registered users (findory.com and reddit.com) or offer 'collaborative filtering' of news (Slashdot.org, Digg.com or Newsvine.com), as social networking becomes the key driver attracting users to sites that have moved closer to what was traditional news territory.

PSB portals can be used not just to direct users to proprietary content or affiliated sites, but to open up a wider range of information by connecting people to other independent public service content and civil society sites. This means lending the trusted brand of national PSB, with the interactive 'red button' technology of their iTV systems, to guide viewers, including older, non-literate 'digital immigrants' to new online spaces that can be accessed 'whenever, wherever, however' users want to seek news (Bennett, 2008: 289). The television screen becomes a democratic internet portal that organises access points to a much broader range of content than is accommodated on traditional, high cost, theatrical, linear television newscasts, designed with a mode of address to maximise audience reach. Internet technologies bring not only ubiquity, immediacy and multimediality, but also the advantage of exchanging ideas with journalists and user generated content such as that available at Wikinews, as well as hypertextuality, which enables citizens to pursue their own online pathways without being directed by established news organisations (Trappel, 2008: 316). Extensive application of hyperlinks in public service news sites can guide the user to an array of wide-ranging independent sources, working against the grain of commercial tendencies to use market-infused filtering methods that result in information enclosure. As Moe (2008: 331) argues, this commitment to an open, multipurpose space on the internet, defying commercial enclosure, could actually yield a high level of trustworthiness for PSB and advance user participation and civic engagement.

The rationale behind online media provided by PSB is that the public service remit should not be confined to specific technologies, such as radio or television transmission, but should be regarded as a service catering for majority and minority interests, a benchmark for quality and diversity, now resituated in the 'linked space' of the internet. However, PSB online activities are disputed by private sector competitors who claim that expansion into non-broadcast digital services overstretches the public service remit. Private sector complaints about ARD and ZDF in Germany, charging that their expansion into new media represented unfair competition, were dismissed by the European Commission in 2007 on the principle of technology neutrality. This means that provided rules respecting fair competition for television broadcasting are observed in a digital environment, PSB is entitled to expand its remit into new services (Trappel, 2008). It is unlikely that we have seen the last of complaints advocating a 'monastic' model for PSB, rather than the 'full portfolio' public service communicator model. The Commission requires governments to specify the 'public added value' of new media activities and the Amsterdam Treaty Protocol currently protects the sovereignty of member states in defining the precise remit of their PSB systems. But consultations are already beginning on a new Broadcasting Communication and governments will inevitably be pressed to define more precisely just how the remit is to be adapted to the digital age, so that commercial broadcasters' fears about PSB 'mission creep' can be allayed (Donders and Pauwels, 2008: 207).

Conclusion

The notion of the public sphere that underpins this view of both the decrepit state of European democracy and possible media remedies is closer to Mouffe's (2005) idea of 'agonistic' democracy than the classic Habermasian approach. In an agonistic theory of the public sphere, questions of power are placed at the very centre, hegemonic political projects confront each other squarely and contestation is the norm. The Habermasian public sphere, enshrined as the conceptual foundation for much media research, is presented as a multitude of overlapping and intersecting parts, a 'sounding board for problems', a 'warning system with sensors' spread throughout society where questions of power are not particularly highlighted. When we consider the democratic deficits of the EU, the dominance of elite spheres in Brussels, the covert lobbying inherent in much EU decision-making, the weakening of the public service ideology and economy supporting PSB, and the dominant influence of private sources in shaping news agendas, it is timely to argue that a reinvigorated civil society must be an essential part of an agonistic European public sphere.

This is already happening in the USA. The loose but densely networked global justice movement has produced an array of large-scale transnational protest activities, empowering people, for instance, to oppose the Iraq war (Bennett et al., 2008). Such movements rely heavily on the internet for managing political information and reduce the need for formal, time-consuming brokerage of activist coalitions such as characterised protest movements in the Vietnam era. A similar movement is taking shape in Europe, organised online, to increase the transparency of lobbying in Brussels (ALTER-EU, 2008). New communication strategies should use the trustworthiness of the PSB brand, built up since the middle of the last century, to guide citizens and social movements to information that can open up the world of European decision-making that is too often closed to the public. Much of that information is already within reach by the under-utilised Brussels press corps, who presently find it so difficult to have their stories accepted as news by domestic editorial gate-keepers. Studies of online media audiences indicate that left to themselves, individuals and groups using new media technology choose to follow a public service ethos (Harrison and Wessels, 2005).

A key question remains. Can the culture of PSB tolerate a reassessment of one of its core principles in order to facilitate this kind of information pluralism? Can the core twentieth-century values of objectivity and impartiality, built into the remit of PSB since its inception, be made suitable for the twenty-first century? Pressed by the Commission in future to define even more tightly what that remit will look like in a fully digital world, will governments be willing to take a fresh look at impartiality, which was originally designed to protect multiparty parliamentary systems, even if it means that new online technologies and new virtual communities living in our multicultural societies demand it? The controversial decision of the BBC refusing to broadcast an appeal for funds for the reconstruction of Gaza, after the Israeli assault of January 2009, for fear that it would damage its reputation for impartiality, despite condemnation of the decision by other broadcasters and government ministers, indicates that core PSB instincts are very deeply embedded in some institutions.

But the old values are being interrogated in other places. The question of impartiality in news has been raised by the British regulator Ofcom (2007) in the context of new evidence of disengagement from broadcast news among young people and ethnic minorities, who increasingly feel that news is of little current relevance to them. They perceive bias and exaggeration in what they are being told. New forms of information exchange – blogging, self-posting of videos on the internet, social networking – hold more appeal for disaffected groups who are not in tune with mainstream media. If they are not to be abandoned to an online culture of phatic communication (Miller, 2008), that is, communication used merely to establish social contact rather than to share ideas, then PSM need to find ways to engage with them. This will involve some effort in designing attractive spaces online and guiding people to them. Experimental evidence suggests that websites specifically designed to encourage people to engage in public life can be successful in cultivating civic engagement attitudes (Kurpius, 2008).

The universal requirement for due impartiality may actually impede the expression of genuine diversity of views and a less rigid approach might encourage greater engagement among a younger internet generation that views the gate-keeping role of mainstream media as a sign of elitism. It is becoming obvious that in a fully digital environment, enabling and encouraging access to a wide range of counter-hegemonic discourse to support new social movements in civil society, impartiality will be increasingly difficult to enforce and will come to be seen not as an end in itself but as part of a broader focus on acquiring information, based on trust in a particular source. For those who dream about the emergence of a European public sphere, with a transnational media system capable of generating a simultaneous discussion of European issues across member states within a shared European frame of reference, settling for a PSM system that aims primarily to empower civil society against elite consolidation may have to be the default option for some time to come.

Note

1. These include the Corporate Europe Observatory, Lobby Control, Spin Watch, the EU Civil Society Contact Group, the European Consumers Organisation and the Alliance for Lobbying Transparency (ALTER-EU), a coalition of 140 civil society groups calling for EU lobbying disclosure legislation, dedicated to exposing deceptive lobbying (such as interest associations lobbying against environmental standards, while also presenting their clients in PR campaigns as champions of environmental protection) and ensuring that the whole lobbying process becomes as transparent as possible.

References

AIM (Adequate Information Management in Europe) (2009) www.aim-project.net, accessed 15 February 2009.

ALTER-EU(2009) www.alter-eu.org, accessed 15 February 2009.

Armstrong, K. A. (2002) 'Rediscovering Civil Society: the European Union and the White Paper on Governance', *European Law Journal*, 8(1): 102–32.

Barstow, D. and R. Stein (2005) 'Is it News or Public Relations? Under Bush, Lines are Blurry', *New York Times*, 13 March.

Bennett, J. (2008) 'Interfacing the Nation: Remediating PSB in the Digital Television Age', *Convergence*, 14(3): 277–94.

Bennett, W. L., C. Breunig and T. Givens (2008) 'Communication and Political Mobilisation: Digital Media and the Organisation of Anti-Iraq War Demonstrations in the US', *Political Communication*, 25(2): 269–89.

Beyers, J. (2004) 'Voice and Access: Political Practices of European Interest Associations', *European Union Politics*, 5(2): 211–40.

Corcoran, F. and D. Fahy (2009) 'Exploring the European Elite Sphere: the Role of the *Financial Times*', *Journalism Studies*, 10(1): 100–13.

Curran, J., S. Iyengar, A. B. Lund and I. Salovara-Moring (2009) 'Media System, Public Knowledge and Democracy: a Comparative Study', *European Journal of Communication*, 24(1): 5–21.

Dinan, W. and D. Miller (2007) *Thinker, Faker, Spinner, Spy: Corporate PR and the Assault on Democracy* (London: Pluto Press).

Donders, K. and C. Pauwels (2008) 'Does EU Policy Challenge the Digital Future of PSB?', *Convergence*, 14(3): 295–311.

Entman, R. M. (2004) *Projections of Power: Framing News, Public Opinion and US Foreign Policy* (Chicago: University of Chicago Press).

European Union (2000) *Shaping the New Union* (Brussels: European Commission).

Follesdal, A. and S. Hix (2006) 'Why There is a Democratic Deficit in the EU: a Response to Majone and Moravcsik', *Journal of Common Market Studies*, 44(3): 533–62.

Franklin, B. (1997) *Newzak and News Media* (London: Arnold).

Gandy, O. (1982) *Beyond Agenda Setting: Information Subsidies and Public Policy* (New York: Ablex).

Greer, S. L., E. M. da Fonseca and C. Adolph (2008) 'Mobilising Bias in Europe: Lobbies, Democracy and EU Health Policy-Making', *European Union Politics*, 9(3): 403–33.

Harrison, J. and B. Wessels (2005) 'A New Public Service Communication Environment? PSB Values in Reconfigured Media', *New Media and Society*, 17(6): 834–53.

Heikkila, H. and R. Kunelius (2008) 'Ambivalent Ambassadors and Realistic Reporters', *Journalism*, 9(4): 377–97.

Hoederman, O. (2007) 'Corporate Power in Europe: the Brussels Lobbycracy', in W. Dinan and D. Miller (eds), *Thinker, Faker, Spinner, Spy: Corporate PR and the Assault on Democracy* (London: Pluto Press), pp. 261–77.

Holz-Bacha, C. and P. Norris (2001) 'To Entertain, Inform and Educate: Still the Role of Public Television', *Political Communication*, 18(2): 123–40.

Iosifidis, P. (2008) 'Top Slicing and Plurality in PSB: a European View', *Intermedia*, 36(1): 30–6.

Kurpius, D. (2008) 'Public Life and the Internet: If you Build a Better Website, Will Citizens be Engaged?', *New Media and Society*, 10(2): 179–210.

Lahnsen, C. (2003) 'Moving into the European Orbit: Commercial Consultants in the European Union', *European Union Politics*, 4(2): 191–218.

Lee-Wright, P. (2008) 'Virtual News: BBC News at a Future Media and Technology Crossroads', *Convergence*, 14(3): 249–60.

Lewis, J., A. Williams and B. Franklin (2008) 'A Compromised Fourth Estate? UK News, Journalism, Public Relations and News Sources', *Journalism Studies*, 9(1): 3–21.

Louw, E. (2005) *The Media and Political Process* (London: Sage).

Marsh, M. (1998) 'Testing the Second-Order Election Model after Four European Elections', *British Journal of Political Science*, 28(4): 591–607.

Messner, M. and M. Watson DiStaso (2008) 'The Source Cycle: How Traditional Media and Weblogs Use Each Other as Sources', *Journalism Studies*, 9(3): 447–63.

Miller, D. and W. Dinan (2000) 'The Rise of the PR Industry in Britain: 1979–98', *European Journal of Communication*, 15(1): 5–35.

Miller, V. (2008) 'New Media, Networking and Phatic Culture', *Convergence*, 14(4): 387–400.

Moe, H. (2008) 'Dissemination and Dialogue in the Public Sphere: a Case Study for Public Service Media Online', *Media Culture and Society*, 30(3): 319–36.

Mouffe, C. (2005) *On the Political* (Oxford: Routledge).

Ofcom (2007) *New News, Future News: the Challenge for Television News After Digital Switchover* (London: Office of Communications).

Pfetsch, B., S. Adam and B. Eschner (2008) 'The Contribution of the Press to Europeanisation of Public Debates', *Journalism*, 9(4): 465–92.

Trappel, J. (2008) 'Online Media within the Public Service Realm?', *Convergence*, 14(3): 313–22.

Ward, D. (2007) *The European Union Democratic Deficit and the Public Sphere* (Berlin: IOS Press).

Weiler, J., U. Haltern and F. Mayer (1995) 'European Democracy and its Critique', *West European Politics*, 18(3): 4–39.

Williams, G. (2007) 'Behind the Screens: Corporate Lobbying and EU Audiovisual Policy', in W. Dinan and D. Miller (eds), *Thinker, Faker, Spinner Spy: Corporate PR and the Assault on Democracy* (London: Pluto Press), pp. 196–211.

7

For Culture and Democracy: Political Claims for Cosmopolitan Public Service Media

Katharine Sarikakis

Introduction

The model of Public Service Broadcasting has occupied a special place in the narratives of democracy and citizenship in the Western world. Historically, the PSB system has been seen as a major European institution that embodies all that is considered noble and diachronic of European values, with particular attention to democratic standards of political and civic life and commitment to culture and education as the elements of 'good life' and enlightenment. Moreover, we associate Public Service Broadcasting/Media (PSB/PSM) with the concepts of citizenship and democracy, culture and pluralism.

However, there are two problems with this picture: first, state-controlled broadcasting systems, often under the rubric of 'public', have been associated with bias, nepotism and inefficiency; second, in European Union (EU) policy-making, and in national contexts, there is an assault on PSM. Resistance to PSM across three intersecting fronts forms the 'market argument' against PSM. First, in terms of PSM's normative foundations, the availability of digital media providing a plurality of content is seen to render PSM's role of catering for minorities obsolete. Second, there is resistance to PSM's expansion to new media technologies, such as mobile communications or the internet. Third, private media challenge the PSM system of funding by claiming anti-competitive behaviour and distortion of the free market. These arguments are discussed in this and previous publications on PSB policy extensively (Chakravartty and Sarikakis, 2006; Humphreys, 2007; Pauwels and De Vinck, 2007; Wheeler, this volume). The focus of this chapter is to explore through statements and policy initiatives the setting of international standards for PSM, and the relationship of PSM to culture and democracy. What is at stake if PSM are considered market competitors and are governed strictly by competition criteria? To appreciate the potential consequences of the market regulation of public service communication, it is important to retrace, review and revise the connections of PSM to citizenship and culture.

The main argument here is concerned with the delimitations of PSM as institutions considered predominantly 'national', and with the dominance of a restricted sense of citizenship they come to serve. I suggest that a more open, bolder even, approach to their role will bring PSM closer to the ideals they represent.

Cultural citizenship and the public sphere

Certainly the picture of PSM around Europe – and the world – has many facets. For even the noblest of intentions underwriting the function of PSB/PSM, there is the element of 'cultural governmentality', the moulding of the public, the projection of certain values. Referring to the birth of the BBC, Bailey (2007: 98) notes that the ideological motivation behind the corporation was its underlying aim to govern and 'civilise' a nation: 'if representative democracy and Parliamentary sovereignty were to function in a way that served the interests of the state, it was imperative that the newly enlarged electorate be taught how to exercise their democratic rights and duties not as members of a social class but as social citizens'. Although the BBC's role today is not necessarily to 'civilise' the masses or to bring High Art to the 'underclasses' similar undertones are present in debates around the democratisation of the media for democracies in transition, such as in Eastern Europe or in African states. The normative aspects of PSM systems, as these gradually expand to the utilisation of technologies beyond broadcasting, focus on the *responsibility* to provide educational, informational and cultural resources for the betterment of citizens and society.

The BBC has served as the archetypal form of public service communications since its inception, not only domestically in British society, but also internationally through the wide reach of its programmes and online content. The worldwide leadership of the BBC places it on a pedestal in the stardom of PSBs – a position it occupies alone. It is almost inevitably necessary to refer to the BBC when discussing the role of PSBs in culture and democracy, because of its strong sense of identity and record of success and public appreciation not always enjoyed by other national PSBs. The importance of these elements vis-à-vis the current political, market and technological pressures facing all PSBs lies in the fact that they hold the organisation in a much stronger position to negotiate and determine its future, than others. Even so, the public service ideal has been under pressure to adapt to a format 'friendlier' to the demands of the market and the private media, which favour regulation hostile to financial support for PSM by the state.

As a cultural technological means of governing the citizenry, PSM are organised symbolically and materially around common understandings of national culture and citizenship. Their remit is operationalised within the territory of the nation-state and they are conceptually tied to the (idea of serving the) 'nation'. They provide a space for public debate with the potential to fulfil, at least partially, some of Habermas' conditions for the emergence and function of the public sphere, namely freedom from private interests, plurality of voices, debate over issues of common concern and with the public good in mind, constitutional equality to allow freedom of speech, and freedom from fear of retribution. Both the concepts of citizenship and public sphere have been largely connected to and also depend on the nation-state and the national territory – geographically, politically and culturally. As concepts and conditions tied to the national, they determine the scope and normative development of PSM.

However, three factors present a challenge to the normative foundations and forms of operationalisation of PSM today, even beyond economic and political pressures. For a start, the technological advancement of communication and its global character turns the 'national' into a part of a more complex international and transnational dynamic. Second, not only global trade but also human mobility and migration challenge the ways in which the national and the cultural intersect, critically challenging assumptions of unproblematic associations between national territories and homogeneous cultures. Finally, integrated economies and an increasingly complex system of multi-tiered policy-making at a global level pose new challenges to the exercise of citizenship. Wallerstein's (1974) concept of peripheries and semi-peripheries continues to express not only relations among nations but also inner-national dynamics. Politics and policies are located in complex relations among a core group of nation-states and the global private sector. Moreover, within the centres/cores there is a multiplicity of peripheries and semi-peripheries, socio-economic divisions of class, race and gender, inequalities in access to decision-making processes and restriction in self-determination. A world of peripheral entities constructed as cultural, ethnic, linguistic *minorities* constitute further geographies of gender and generational inequality.

While on the one hand the proliferation of new media allows myriad ideas to be voiced, the citizenry is becoming marginalised and disaffected from the mainstream political process. This also means that the more citizens become disengaged from the formalised political processes, the more pressing becomes the need to redefine and expand our understanding of citizenship and its realisation. To fully encompass the dimension of citizenship in the multimedia era is to expand the normative remits of PSM to embrace not only a set of duties and rights associated with conventional understandings of national citizenship, but also to engage with citizenship's economic, social, political and cultural dimensions (Lister, 2005). These are the fields where the existence of substantive, as opposed to formal citizenship is tested, in other words, these dimensions indicate whether the material and symbolic conditions for the exercise of citizenship are adequate and sufficient. Not only material but also symbolic inequalities in everyday life affect the ability of the public to enjoy social justice. For example, global capital flow creates new and heightens old processes of social exclusion and discrimination; our understanding of citizenship must expand to encompass new patterns of public participation (Stevenson, 1999, 2003). At the same time the informational or knowledge economy has profound effects upon labour relations, the role of the media and the effects of technology (Mosco, 2008). Within this context, the role of the media as facilitators of public spheres, within which new and old forms of injustice can be challenged, is important: the power to 'name', that is, to assign significance to a problem or to render one invisible, defines some of the most important aspects of citizenship. It defines our existence as political and cultural beings, since the power to name can construct claims and even ideas as legitimate or illegitimate.

This cultural dimension of citizenship is itself the crossroads whereby the material and symbolic conditions of making sense of and further constructing the world intersect. These are the conditions within which individuals and societies are expected to develop. In order for this process to materialise, it is necessary that citizens have the means to the terrains of construction and contestation. Arguably, this translates into access to information, knowledge, experience and participation (Murdock, 1999), as citizenship 'is not merely membership of a society' but of civil society (Delanty, 2000: 134), which is defined as the public space between state and society. Therefore the participation of individuals in civil society defines their involvement in the debates surrounding issues of public concern but also issues considered of a 'private' nature, such as sexual identity, domestic violence, religious orientation and others. These 'private' issues become of public importance when they hinder individuals' self-realisation and self-governance. In multicultural, cosmopolitan societies, difference stands out as an important element contesting fantasies of homogeneity. However, 'difference' cannot justify inequality or practices that restrict the freedom of self-governance; indeed, autonomy, choice and human flourishing can only derive from cultural possibilities once egalitarian politics is recognised. Citizenship for Habermas is not rooted in a particular form of life identifiable with a cultural community: although communication should be directly linked with the law, the formal recognition of citizens as political beings, this should – and does – take place within as well as beyond the nation-state.[1] However, migrants' political status in Europe is precarious, leading to fragmentation, alienation and marginalisation.[2] The role of culture as an object and aim of policy, but also as one of the cornerstones of PSM, brings to the fore systems of representation, discursive practices and ways of political participation.

While citizens are becoming fragmented, public spaces are shrinking: what is at stake in the process of globalisation, according to Mattelart and Mattelart (1992), is the commercialisation of public spaces where participation of citizens materialises. Commercialisation, privatisation and concentration of media ownership, or the 'feudalisation' of public spaces, are the dimensions most prominent at the beginning of the twenty-first century. But it is not unimportant that public spaces are occupied by a few companies owned by the strongest Western media in the world.[3] The commercialisation of public spaces, whether by national or transnational capital, undermines the enrichment of social debate. Furthermore, the expansion of private interests into the public space reduces citizens' participation to increasingly confined spaces and their role to that of consumers. The challenges communicative spaces are presented with, deriving from state control and market pressures, have been considered to be the core obstacles to PSM as the main facilitator of a public space. In the following section, I explore European and international policy statements and declarations as to the significance of PSM for culture and democracy and discuss the extent and complexity of vision in view of the changing environment within which PSM are required to operate.

Public service communication in the EU

At the time of writing, state and private broadcasters were making their arguments against or in favour of the European Commission Communication on state aid for broadcasting (European Commission, 2008) that seeks to impose restrictions and conditions on the range of activities non-commercial, non-private broadcasters can undertake. The tone and spirit of the Communication clearly prioritises competition interests as its starting point (see Wheeler, this volume). A decade after the PSB Protocol to the Amsterdam Treaty, the question of the PSM, contrary to a widespread belief at the time, has not been answered once and for all. Debates in Europe regarding the state and future of PSB had entered the EU polity and Europe's mediated public spheres within a climate of intensified liberalisation policies and in particular in the aftermath of the first Television Without Frontiers Directive, in the 1980s. The most dramatic change of these decades preceding the Protocol was the privatisation of state or publicly owned services in Europe, including the airwaves. The 'liberalisation' wave seriously affected the media sector: transnational corporations entered the European market powerfully, dominating entertainment and popular culture programmes in most countries, challenging PSBs' audience shares. A few important changes characterise this structural transformation of the media, namely overall increased media content production and a shift in the focus and quality of content from information to entertainment. These changes came hand in hand with the concentration of the infrastructure of communications ownership and were further accompanied by an intensified marketisation of the public, which was now redefined into audiences, users and consumers (Castells, 1996; Chakravartty and Sarikakis, 2006; Haywood, 1998; Lyon, 1988).

Under these conditions, compounding the pressures deriving from the 'market argument' a re-evaluation of the remit of PSM is now necessary, which must be made in PSM's own terms and those of the societies they seek to serve. For this exercise, self-reflection and an outward reach for transnational alliances are two important strategies. Within this context, the role of PSM to sustain citizenship becomes all the more urgent, but also increasingly complex. In the current climate of violent conflict, whether in the form of terrorism or military intervention, the ripples of encounters across the world unavoidably reach our doorstep. Paradoxically, the challenges under which PSM find themselves economically and politically deprive them of the means to invest in an emancipatory, globally oriented and nationally acted cultural governance.

However, the force of the conviction of the value of PSB in European society and citizens' lives has found its way to the Amsterdam Treaty of 1997 after a dramatic series of events that brought governments and PSBs into alliance against proposals for the restriction of member states' right to support them (Collins, 2008; Moe, 2008). According to the Amsterdam Treaty:

> the system of public broadcasting in the Member States is directly related to the democratic, social and cultural needs of each society and to the need to preserve media pluralism. (Treaty of the European Community, 1997: para j)

The protection of PSB/PSM has been raised again as a problem for the internal market in terms of competition. Evidently, the values that PSM represent as a form of media and their raison d'être are only tolerated as long as they do not interfere with the economistic and market-driven framework of EU policy: PSM support is accepted as long as 'such funding does not affect trading conditions and competition in the Community' (Treaty of the European Community, 1997).

There are some important dimensions to the Protocol: first, there is a symbolic significance attached to the recognition of PSB/PSM and their role in political, cultural and social dimensions of citizenship. Second, the Protocol extends European jurisdiction into the definition of public service organisations in general and PSBs specifically, aiming, albeit unsuccessfully, for exemption from competition rules. Third, the Protocol constitutes the first such legally binding international agreement with far-reaching implications in the normative construction of policy agenda-setting and debate, at an international level but also within national settings in non-European countries. The Protocol also echoes international efforts to defend PSM against autocratic and oppressive political regimes that interfere with the operationalisation of (something like) a (mediated) public sphere; to acknowledge and counterbalance the difficulties imposed by market pressures on PSM to function under competition conditions; and to tackle established media cultures that are hostile to communicative and informational democracy.

The unsuccessful Lisbon Treaty would have put new obligations on the EU to respect essential state functions and protect certain key areas that have a commercial element and which are included in internal market rule – which is now the basis for renewed assault on PSM on the grounds of anti-competition funding. The provision for public services (Services of General Interest: SGEI) would include the public service media. The Treaty would have provided the legal basis now missing for legislation that would safeguard universal public service delivery. As with Article 36 of the Charter of Fundamental Rights, the Protocol[4] to the Reform Treaty would have provided a coherent framework for EU action that would reassert 'a number of operational principles' such as:

- The essential role and the wide discretion of national, regional and local authorities in providing, commissioning and organising SGEIs in such a way as to meet as closely as possible the needs of users.
- Respecting the diversity of services as well as the needs and preferences of users, due to the diversity of situations (geographical, economic, social and cultural).
- A high level of quality, safety and affordability of services of general economic interest.
- Ensuring equal treatment, universal access and upholding user rights (European Commission, 2007: 9–11).

This legal framework would have proven useful for the support of PSM and their classification in clear terms as public services worth the same protection as health and education. Certainly, this missing piece in the fabric of supranational policy in the EU would have also required a set of supporting policies of a cultural and social orientation and a deeper integration of those with policies relating to PSM.

The structural changes of the media industry in Europe brought consequent changes in the 'product' of the sector and shaped its 'end-user', that is, the public. The outcome of these changes was that while private media benefited enormously from the opening up and creation of new markets, PSM remained not only under old regimes of governance either controlled by the state or seriously under-resourced but they also had new challenges and pressures ahead of them. Discussing the effects on the Belgian PSB Burgelman and Perceval (1996) argued that it is 'absurd to discuss the crisis of public service broadcasting in terms of programme quality or public perception' when the problem of lack of political autonomy remains largely unresolved. Funding and resourcing are inherent aspects of this same question of political dependency. The changing, often declining support for PSM following the general collapse of state functions in the public domain represents a major obstacle to an independent and public-attuned shaping of communication services. It is not possible to speak of 'one' PSM system, as the historical experience of nation-states has brought to the fore distinct organisational models and understandings and interpretations of the media.

International standards for PSM

Despite differences in history, structure and success, there are certain core normative values that are nearly universally accepted as the minimum standards for PSM. Two declarations by UNESCO and the Council of Europe were made with a view to the changing media environment, the increased needs for appropriate public spaces for education and information, cultural diversity and social cohesion. They were drafted in the early and mid-1990s to offer a counter-marketisation argument during the deliberations in Europe and the Uruguay Round of the GATT negotiations. These declarations inserted into the debate two important elements: the role of community radio as a public service medium and the need for preservation of a comprehensive public service programme system amidst provision from private media. Clause 5 of the Declaration of Alma Ata in 1992 by UNESCO calls upon states to respect these core values in supporting the development of PSB:

> To encourage the development of journalistically independent public service broadcasting in place of existing State-controlled broadcasting structures, and to promote the development of community radio.
>
> To upgrade educational broadcasting through support for distance education programmes such as English language instruction and formal and non-formal education, literacy programmes, and information programmes on AIDS, the environment, children, and so on. (UNESCO, Declaration of Alma Ata, 1992: Clause 5, section II, p. 8)

Two years later, in 1994, the Council of Europe called upon states to:

> Undertake to guarantee at least one comprehensive wide-range programme service comprising information, education, culture and entertainment is accessible to all members of the public. (Council of Europe, 1994: Resolution No. 1 – Future of Public Service Broadcasting)

This Resolution emphasises the need to maintain a comprehensive PSM system by upholding values of universal access and diverse services. Both statements were made amidst the debates for liberalisation and the emergence of the Information Superhighway with protagonists the USA and the EU.

However, not only Western European PSM but also those located in the 'transitional' democracies of Eastern Europe and other parts of the world, such as certain African and Asian countries, face the challenges of the need for a comprehensive PSM system and the pressures deriving from either state control or market imperatives. Despite hopes for an opening of the media towards more pluralistic systems, the dominance of political elites over state media in Eastern Europe continued undisturbed, often conflating state media with PSM 'as a cover for paternal or authoritarian communication systems' (Williams, 1976: 134 cited in Splichal, 1995: 63). There is little evidence to suggest that a social or public media system will develop in Eastern Europe as most policies have introduced media liberalisation without at the same time articulating a PSM model, either normatively or structurally (Jakubowicz, 1996, 2004; Vartanova and Zassoursky, 1995; Zernetskaya, 1996). The same is true for other Western EU countries that have failed to revisit the structural and normative makings of their PSM. Still, the variety of visions and professional cultures may offer the potential for the enrichment of PSM in the West and add to the sensitivity and awareness necessary for the coverage and representation of world affairs and minority social groups.

International declarations and regional instruments for PSM standards have sought to provide a critical normative framework by emphasising the need for change across societies, whether those under state-controlled media or those with existing PSM. For example, the African Charter on Broadcasting envisions a PSB system and the Asian Media Summit Recommendation to the World Summit on Information Society (WSIS) through their respective paragraphs sought to emphasise the core elements of PSM. The African Charter echoes some of the core issues of audio-visual policy in the EU, such as the clear mandate for PSM, adequate funding and content quotas for independent works (paragraphs 1, 5 and 6). It openly emphasises the demand for change in the public service system of communication:

> All State and government controlled broadcasters should be transformed into public service broadcasters, that are accountable to all strata of the people as represented by an independent board, and that serve the overall public interest, avoiding one-sided reporting and programming in regard to religion, political belief, culture, race and gender. (Winhoek +10 African Charter on Broadcasting 2001: para 1)

The African Charter on Broadcasting brings attention to the requirement for independent governance for PSM through appropriate bodies and their freedom from state interference (para 2); the establishment of an appropriate mechanism for adequate funding that guarantees PSM independence (para 5); and the introduction of content quotas for independent work (para 6). The same values are echoed in the Asia Media Summit (AMS) (2005) Declaration that asks governments to:

- Promote public service broadcasting, and ensure its independence from political and commercial pressures.
- Provide public service broadcasting organisations with adequate funding to enable them to provide high quality services while remaining viable and maintaining their independence.

In the light of the ongoing debates at the WSIS at the time, and in view of the technological and market-focused tone of the debates, the AMS drew attention to the importance of non-profit community broadcasting and the need for the allocation of suitable frequencies. These two sets of standards sought to provide a positive intervention of an international character in two mutually complementing ways: the African Broadcasting Charter (2001) aims at the African domestic relations of state and media, and the AMS (2005) recommendation targeted the workings of the second phase of the WSIS in Tunis. Both statements echoed the main challenges faced by PSM around the world in varying degrees and those of local and transnational communities, such as structural spaces for the provision of community media and resources for the sustenance of a long-term comprehensive PSM agenda. They also pointed to the need to acknowledge, value and facilitate diversity of opinion and culture as part of PSM mission statements. The UNESCO Cultural Diversity Declaration (2002) sought to include these dimensions in the PSM paragraph 12, as an inherent part of policy principles to support cultural diversity and with it cultural citizenship and democratic participation, elements central to the PSM mission. States should undertake to 'promot[e] the role of public radio and television services in the development of audiovisual productions of good quality, in particular by fostering the establishment of cooperative mechanisms to facilitate their distribution' (UNESCO, 2002: para 12).

The transnational, nearly universal values advocated by regional summits and forums of representatives of citizens and PSBs demonstrate not only the common values underlying the demand for public service communications. They point also to the interconnection of human experience in terms of a broader sense of citizenship and public participation.

Conclusion

The PSB system has begun to adapt to the possibilities offered by new media technologies and it is gradually becoming recognised that the system should be allowed to evolve. More specifically, a new PSB/PSM profile is needed to represent diversity of opinions and experiences and provide a platform, a common realm that would not only guarantee and promote universal access but also *educate* and provide citizens with a minimum of skills for them to participate in public affairs and enable them to construct critical arguments. This could be achieved through a reorientation of PSM so that they consider themselves universal not only in reach but also in character: refocusing away from the strictly 'national', PSM would offer through their programming and employment policies a proactive, inquisitive and socially sensitive agenda for connection among nations, cultures and peoples. To do that, the guarantee of availability of resources by the state as well as internal reorganisation would be necessary; the practicalities of this change would be largely determined by the cultural and other diversities of regions and localities. PSM are called upon to provide programming that helps building cultural cohesion, yet offer a forum for the representation of 'minorities' and special groups, succeed in providing a balanced political coverage and educational programmes, fulfil journalistic values of impartiality and objectivity and act as a watchdog of the government. PSM are expected to cater for quality and work for universality. Moreover, they are seen as one of the most important 'commons' alongside independent and community media. Yet, their 'public service' character, in structural, economic and cultural terms, is neither adequately protected nor supported in the range of policies available at a European and international level.

Culture and democracy are not served by the market in any satisfactory or complete manner. Cultural citizenship and the centrality of communication in contemporary societies are inextricably linked to communicative spaces, rights and dialogue. New forms of public participation and the need for new popular imagery emerge in everyday life as societies become more connected in economic, communication and political terms. PSM have the potential to play the noble role of connecting the broader international affairs to those experienced as personal ones, such as those of gender, race or disability and help elaborate political claims for social justice. In line with technological, socio-cultural and political developments a European redefined PSM system would protect and promote an extended form of citizenship that encourages the ethics of care and connection and a continuous critical approach to established practices and dominant values. A new PSM 'contract' with the public should ensure that the cultural possibilities available to communities and individuals are adequate and sufficient to enable participation in the political and cultural processes as mediated by communicative spaces (and the public sphere) and enacted upon by the civil society.

Notes

1. Such is the concept of European citizenship and certain aspects of Commonwealth membership, for example.
2. Migrants may or may not be granted political citizenship and formal rights enjoyed by the majority population; undocumented people seeking asylum are deprived almost of any rights, often even their human rights, while members of specific minorities, such as the Roma, face discrimination on a cultural level so strongly that it renders their formal rights almost ineffective (see European Commission, 2005).
3. In 2005 and 2008 reports on the state of the media in Eastern Europe have raised serious concerns about the level of media ownership concentration and control of audience markets. In 2009, even in the midst of a global financial crisis and despite an overall slight drop in growth, media and communications industries outpace any other industrial sector in the USA, while mobile and digital technologies continue at an exponential growth rate (Seeking Alpha, 2009).
4. Supported by the European Public Service Union representing over 8 million workers in 200 unions from 36 countries, the European Public Health Alliance, the European Trade Union Council for Education, the European Anti-Poverty Network, the European Disability Federation, the European Women's Lobby, Mental Health Europa, and the European Trade Union Confederation.

References

African Broadcasting Charter (2001) Final Report, 'Ten Years On: Assessment, Challenges and Prospects', http://portal.unesco.org/ci/en/files/5628/10343523830african_charter.pdf/african%2Bcharter.pdf, accessed 20 January 2009.

Asian Media Summit (2005) Kuala Lumpur Recommendations to the WSIS in Tunis, 9–11 May, http://portal.unesco.org/ci/en/files/19279/11199412917AMS_Recommendation_2005.pdf/AMS% 2BRecommendation% 2 B2005. pdf, accessed 26 January 2009.

Bailey, M. (2007) 'Rethinking Public Service Broadcasting: the Historical Limits to Publicness', in R. Butsch (ed.), *The Media and the Public Sphere* (Basingstoke: Palgrave Macmillan), pp. 96–108.

Burgelman, J. C. and P. Perceval (1996) 'Belgium: the Politics of Public Broadcasting', in M. Raboy (ed.), *Public Broadcasting for the 21st Century* (Luton: John Libbey Media).

Castells, M. (1996) *The Information Age: Economy, Society and Culture*, vol. I: *The Rise of the Network Society* (Oxford: Wiley-Blackwell).

Chakravartty P. and K. Sarikakis (2006) *Media Policy and Globalization* (Edinburgh: Edinburgh University Press).

Collins, R. (2008) 'Misrecognitions: Associative and Communalist Visions in EU Media Policy and Regulation', in I. Bondebjerg and P. Marsden (eds), *Media Democracy and European Culture* (Bristol, UK: Intellect), pp. 287–306.

Council of Europe (1994) 4th European Ministerial Conference on Mass Media Policy, The Media in a Democratic Society, Resolutions and Political Declaration, Prague, Czech Republic, 7–8 December, http://www.ebu.ch/CMSimages/en/leg_ref_coe_mcm_resolution_psb_07_081294_tcm6-4274.pdf, accessed 7 February 2009.

Council of Europe (2007) Recommendation CM/Rec(2007)3 of the Committee of Ministers to Member States on the Remit of Public Service Media in the Information Society, http://www.ebu.ch/CMSimages/en/leg_ref_coe_recommendation_informationsociety_310107_tcm6-50019.pdf, accessed 12 March 2009.

Delanty, G. (2000) *Citizenship in the Global Age: Culture, Society and Politics* (Buckingham: Open University Press).

European Commission (2005) 'The Situation of the Roma in an Enlarged EU', DG Employment and Social Affairs (Luxembourg: Office for Official Publications of the European Communities).

European Commission (2007) Communication from the Commission to the European Parliament, the Council, the European Economic and Social Committee and the Committee of the Regions accompanying the Communication on 'A Single Market for 21st century Europe' – Services of General Interest, Including Social Services of General Interest: a New European Commitment, http://eur-lex.europa.eu/LexUriServ/LexUriServ.do?uri=COM:2007:0725:FIN:EN:PDF, accessed 17 February 2009.

European Commission (2008) Communication from the Commission on the Application of State Aid Rules to Public Service Broadcasting, http://ec.europa.eu/competition/state_aid/reform/broadcasting_communication_en.pdf, accessed 29 January 2009.

Haywood, T. (1998) 'Global Networks and the Myth of Equality', in B. Loader (ed.), *Cyberspace Divide* (London: Routledge).

Humphreys, P. J. (2007) 'The EU, Communications Liberalisation and the Future of Public Service', in K. Sarikakis (ed.), *Media and Cultural Policy in the European Union*, European Studies: An Interdisciplinary Series in European Culture, History and Politics, 24 (Amsterdam and New York: Rodopi), pp. 91–112.

Jakubowicz, K. (1996) 'Poland: Prospects for Public and Civic Broadcasting', in M. Raboy (ed.), *Public Broadcasting for the 21st Century* (Luton: John Libbey Media).
Jakubowicz, K. (2004) 'Ideas in Our Heads: Introduction of PSB as Part of Media System Change in Central and Eastern Europe', *European Journal of Communication*, 19(1): 53–74.
Lister, M. R. A. (2005) *Citizenship: Feminist Perspectives* (Basingstoke: Palgrave Macmillan).
Lyon, D. (1988) *The Information Society: Issues and Illusions* (Cambridge: Polity).
Mattelart, A. and M. Mattelart (1992) *Rethinking Media Theory* (Minneapolis: University of Minnesota Press).
Moe, H. (2008) 'Between Supranational Competition and National Culture? Emerging EU Policy and Public Broadcasters' Online Services', in I. Bondebjerg and P. Marsden (eds), *Media Democracy and European Culture* (Bristol, UK: Intellect), pp. 307–24.
Mosco, V. (2008) 'Creative Differences: the Problems of Outsourcing Knowledge and Media Labor', in J. Wasko and M. Erickson (eds), *Cross-Border Cultural Production: Economic Runaway or Globalization?* (Youngstown, NY: Cambria Press), pp. 85–116.
Murdock, G. (1999) 'Rights and Representations: Public Discourse and Cultural Citizenship', in J. Gripsrud (ed.), *Television and Common Knowledge* (London: Routledge).
Pauwels, C. and S. De Vinck (2007) 'Can State Aid in the Film Sector Stand the Proof of EU and WTO Liberalisation Efforts?', in K. Sarikakis (ed.), *Media and Cultural Policy in the European Union*, European Studies: An Interdisciplinary Series in European Culture, History and Politics, 24 (Amsterdam and New York: Rodopi), pp. 23–43.
Seeking Alpha (2009) 'Media and Communications Spending to Continue to Outpace US Growth', Seeking Alpha, 25 February 2009, http://seekingalpha.com/article/122641-media-communications-spending-to-continue-to-outpace-u-s-growth, accessed 12 March 2009.
Splichal, S. (1995) 'From State Control to Commodification: Media Democratisation in East and Central Europe', in F. Corcorran and P. Preston (eds), *Democracy and Communication in the New Europe: Change and Continuity in East and West* (Creskill, NJ: Hampton Press).
Stevenson, N. (1999) *The Transformation of the Media: Globalisation, Morality and Ethics* (London and New York: Pearson).
Stevenson, N (2003) *Cultural Citizenship Cosmopolitan Questions* (Maidenhead: Open University Press).
Treaty of the European Community (1997) Treaty of Amsterdam amending the Treaty on European Union, the Treaties establishing the European Communities and related Acts, *Official Journal* C 340, 10 November.
TVinsite (2002) 'Top 25 media Companies', http://www.tvinsite.com/brodcastingcable, accessed 17 October 2002.
UNESCO (2002) Universal Declaration on Cultural Diversity, www.unesco.org/culture, accessed 26 February 2009.
Vartanova E. and Y. Zassoursky (1995) 'Television in Russia: Is the Concept of the PSB Relevant?', in G. F. Lowe and T. Hujanen (eds), *Broadcasting and Convergence: New Articulations of the Public Service Remit* (Göteborg: NORDICOM).
Wallerstein, I. (1974) *The Modern World System* (New York: Academic Press).
Zernetskaya, O. V. (1996) 'Ukraine: Public Broadcasting between State and Market', in M. Raboy (ed.), *Public Broadcasting for the 21st Century* (Luton: John Libbey Media).

8

Public Broadcasters and Transnational Television: Coming to Terms with the New Media Order

Jean K. Chalaby

Introduction: the making of a transnational media order[1]

The close relationship between media and nation has been unravelling over the last two decades. The causes of this disjuncture are complex and include phenomena related to globalisation such as the increasing flow of capital, goods and people crossing borders. Change is also triggered by the unfolding information technology revolution that has further deepened integration between computing, telecommunications and electronic media (Castells, 1996; Forester, 1985). New technology involves a process of convergence between hitherto separate media platforms, the digitisation of broadcasting and satellite systems – making global communication networks more powerful and flexible – and the emergence of new digital media.

The end result is a remapping of media spaces that draws on three related areas: channels, TV formats and media corporations (Chalaby, 2009a; Esser, 2002). The rise of cross-border TV channels lies at the heart of the current regional and global reshaping of media industries and cultures. In the 1980s, transnational TV networks struggled in the grip of a range of problems that included poor satellite transmission, governments reluctant to grant access to their markets and a reception universe that was too small to attract advertisers. They were also searching for a workable model of international broadcasting and a suitable way to address a multinational audience. Facing such difficulties, many of the early cross-border channels were short-lived. But the stars of pan-European television came into alignment in the late 1990s when the transnational shift began to occur in European broadcasting. The commercial, technological and policy context radically changed, and broadcasters progressively understood how to deal with a multinational audience and began to adapt their video feeds to European cultural diversity (Chalaby, 2009b).

Today, cross-border networks count among European television's most prestigious brands and have become dominant in several genres, including international news, business news, factual entertainment and children's television. This chapter analyses the contribution of Public Service Broadcasters (PSBs) to transnational television in Europe and assesses their prospects in the new media environment.

PSBs versus common sense: the Europa débâcle

The development of satellite television was a mixed blessing for PSBs. In the 1980s they were adjusting to the recent emergence of commercial rivals following the break-up of state broadcasting monopolies in several European countries. On the one hand, satellite technology was being recognised as an exciting new medium that would provide an opportunity to disseminate the 'best' of European television to the 'widest possible European audience' (Clarke, 1982: 44). But it was also perceived as a fresh threat to public broadcasters' position, who feared that the technology would allow commercial players to rewrite industry standards. Thus their early interest in satellite technology was as much prompted by their enthusiasm for the new technology as concern over competition from the commercial sector (Barrand, 1986: 11).

Following TV5 and 3sat, PSBs' most ambitious projects were organised under the auspices of the European Broadcasting Union (EBU). Their first project, Eurikon, was an experiment carried out on a broadcaster-to-broadcaster basis for five non-consecutive weeks in 1982 (Clarke, 1983: 29). Each of five EBU members – the IBA (UK), RAI (Italy), ORF (Austria), NOS (Netherlands) and ARD (West Germany) – transmitted to fifteen countries for a week in turn. They tested the pan-European appeal of their programmes, tried to identify a 'pan-national editorial viewpoint' for their news services, and experimented with different methods of communicating simultaneously with a multi-lingual audience (Clarke, 1984: 50).

Following the success of the experiment, six EBU members set up a fund and feasibility study for a new channel. Financial support came from the Dutch government and several European institutions. After lengthy negotiations and deliberations at the EBU five broadcasters from the Netherlands (NOS), Germany (ARD), Italy (RAI), Ireland (RTE) and Portugal (RTP) launched Europa on 5 October 1985.

Europa (based at the NOS studios in Hilversum) started with a handicap since most of the heavyweight EBU members refused to get involved. The French Antenne 2 and FR3 and the German ZDF feared that it would jeopardise their own TV5 and 3sat services. Likewise, the BBC stayed away because it was not convinced by the quality of the project, plus it was involved with ITV's Super Channel and had plans of its own (Papathanassopoulos, 1990: 60). The absence of a British broadcaster was a major drawback in the channel's attempt to reach a European audience. To make matters worse, Europa was moulded in the public broadcasting ethos of its backers: it did not pay enough attention to audience tastes and broadcast too many highbrow programmes. The running costs were high (vast sums were spent on translation) and the channel reached too few homes to attract any significant advertising revenue. The Dutch broadcaster, which was unwilling to carry on assuming much of the financial costs, put its colleagues out of their misery and switched off the signal on 27 November 1986, thirteen months after the launch (Collins, 1998: 143–9; O'Connor, 1986: 54).

TV5Monde and 3sat

The second oldest satellite channel in Europe and the current doyenne of transnational TV channels is the francophone TV5Monde, which started broadcasting as plain TV5 in February 1984. The project was developed by Patrick Imos, a diplomat who realised the role that satellite television could play in asserting French influence in the world. He designed a project of cooperation among five francophone public broadcasters: the three French public television channels (TF1, Antenne 2 and FR3), Wallonia's RTBF and the SSR from French-speaking Switzerland.

TV5 began broadcasting three hours a night with the launch of Europe's first communication satellite, ECS 1, reaching cable networks across the continent. The initial budget was modest and TV5 rebroadcast its partners' programming: Mondays and Thursdays were supplied by Antenne 2, Tuesdays by the SSR, Wednesdays and Sundays by TF1, Fridays by FR3 and Saturdays by the RTBF (*Cable & Satellite Europe*, January 1984: 68–9). In January 1986 a Quebecan consortium of broadcasters joined in, who energised the network by bringing a more commercial approach.

Today, TV5Monde primarily rebroadcasts its partners' programming but it has its own budget, personnel (250 people), newsroom, and identity. Since the 2000 reform, TV5Monde has a strong news strand and produces up to ten daily bulletins. The network has always tried to avoid a French-centric view of the world and most of its bulletins are produced for a specific regional feed. TV5Monde also produces weekly magazines covering Africa, world affairs, the economy, science and culture. The network remains at the heart of the French-speaking world because it brings different francophone communities together and enhances the knowledge and understanding of francophone culture (Cronel, 2008).

3sat followed TV5 on the Eutelsat satellite later in 1984. It was backed by three public broadcasters: ZDF in West Germany, ORF in Austria and SRG in German-speaking Switzerland. The channel was headquartered at ZDF's studios in Mainz, and the German broadcaster, which provided 55 per cent of the programming, was firmly in charge. The Swiss contributed to 10 per cent of the output, the Germans being allegedly wary of the Swiss-German dialect. 3sat began broadcasting to a few thousand cable homes in Germany on 1 December 1984 but reception proved difficult in Switzerland. As soon as the occasion arose, the channel moved to a German satellite, DFS Kopernikus, in order to improve reception (*Cable & Satellite Europe*, December 1984: 45–6; *Cable & Satellite Europe*, March 1987: 26). These three broadcasters were joined by ARD in 1993, and today 75 per cent of 3sat's schedule is provided by these four, of which 66 per cent comes from ZDF and ARD, 25 per cent from ORF and the remaining 9 per cent from SRG. The rest of the programming is produced specifically for 3sat (usually by ZDF) (Fiedler, 2008).

Deutsche Welle Television: from *Volksmusik* to *PopXport*

Deutsche Welle has been an international broadcaster since 1953, transmitting radio programmes in up to twenty-nine languages around the world. When it upgraded to television it turned to an organisation set up for propaganda purposes during the Cold War: Radio in the American Sector (RIAS). RIAS was part of the United States Information Agency (USIA) since 1955 and used to broadcast beyond the Iron Curtain to Eastern Germany and other communist nations. In the late 1980s, under the leadership of USIA director Charles Wick, RIAS added a television channel to its operations (Snyder, 1995). When the Cold War ended RIAS-TV lost its raison d'être and Deutsche Welle took over its staff and headquarters in Berlin. Deutsche Welle Television (DW-TV) began broadcasting in April 1992 and its objectives were later refined by the Deutsche Welle Act of January 2005.

The station's schedule is organised in a similar way to that of BBC World but with a language rotation between German and English every hour. Its news bulletin starts every hour on the hour and the back half-hour is devoted to magazine programmes and documentaries. Together with Deutsche Welle's radio and online services, the channel promotes German language and culture and takes a German perspective on world affairs. DW-TV primarily aims at an international audience although it keeps in touch with expatriates via the German-speaking hour.[2] Magazine programmes like *Made in Germany* (business), *Discover Germany* (tourism) and *Kino* (German movies), promote local industry. With the same objectives in mind, a programme that was popular with German expatriates, *Volksmusik*, was replaced by *PopXport*, a showcase for German pop music (Newel 2008; Ziegele, 2005).

Arte: engineering Franco-German culture

Arte's origins lay in the wide-ranging collaboration programme between France and Germany begun in 1963. Around 1984 both governments reached planning stage for their own cultural channel and the idea of a joint station arose two years later. The project was officially announced by President François Mitterrand and Chancellor Helmut Kohl at the 52nd Franco-German Summit in November 1988. In October 1990 the French Republic and eleven German Länder (responsible for broadcasting in Germany) signed a treatise establishing the legal foundations of Arte. The channel began broadcasting via satellite to cable operators in France and Germany on 30 May 1992.

Three broadcasters were involved in the negotiations: the French cultural channel La Sept, launched in 1989, and the two German public stations, ARD and ZDF. In order to ensure parity between the two countries, ARD and ZDF formed Arte-Deutschland, which became an equal partner in Arte with La Sept. The legal structure chosen for Arte – called a Groupement Européen d'Intérêt Economique in European law – ensured maximum autonomy for the two partners. Negotiations were lengthy and difficult because they brought together two countries with conflicting broadcasting models. The Germans delegate broadcasting matters to regional authorities and place emphasis on cultural diversity, while the French broadcasting policy has statist and nationalist overtones and is centrally controlled from Paris. French news media and journalists remain close – some say subservient – to political personnel while in Germany a Chinese wall separates the two professions. Thus the German partners proceeded with extreme caution and agreed to a deal only when assured that their independence was guaranteed (Utard, 1997: 16–17).

The existence of Arte owes more to diplomacy than public demand. The channel was supposed to strengthen the Franco-German alliance, which used to be considered the engine of European integration. The station was promoted by those in politics who never accepted that resilient national cultures can stand in the way of European cohesion. They recognised that for Europe to prevail as a political entity people would have to feel more European and that a European culture would have to be fabricated *ex nihilo*. Television seemed a reasonable place to start, a view expressed, among others, by Christoph Hauser, Arte's head of programming, who stated in 2006 that the channel's objective was to give Europe 'a soul'.[3]

Based in Strasbourg since October 2003, Arte signed several co-production agreements with European broadcasters between 1995 and the early 2000s, including the BBC and Sweden's SVT. Three public broadcasters: RTBF in Belgium, TVP in Poland and ORF in Austria have become associate members. But for all its European fervour Arte remains essentially a bi-national organisation with the French and Germans solely in charge of the channel's destiny. And although Arte has widespread coverage in six countries (Austria, Belgium, France, Germany, Switzerland and the Netherlands) it draws most of its audience from France and Germany.

Euronews

The European Broadcasting Union drew up plans for an all-news channel in the late 1980s. Like the BBC (see below), its calls for public funding fell on deaf ears until CNN rose to prominence during the Gulf War and the project received financial support from the European Parliament, several governments and participating broadcasters. At first, subsidies lay at the heart of the station's funding, so much so that the channel's commercial income was initially budgeted at only 20 per cent of total revenue.[4]

Euronews went on air on 1 January 1993 and was backed by some twelve EBU members led by a first tier of shareholders comprising France Télévisions (holding 20.9 per cent of shares), RAI (17 per cent) and Spain's RTVE (15.7 per cent). The running costs rapidly proved too high for the European backers and broadcasters. In 1995 the French government somehow managed to convince Alcatel to become the first private investor and acquire 49 per cent of shares through a subsidiary (*Cable & Satellite Europe*, March 1995: 12; Godard and Barker, 1993). The French conglomerate took advantage of a strategic review to pull out soon afterwards and the shares were left on the market with no takers. They were picked up by ITN for £5 million in January 1998, starting an ill-fated joint venture between the British news provider and France Télévisions, the main shareholder. The French public broadcaster's incoming president, Marc Tessier, took against ITN. He resented the fact that a station based in France was run by the 'infidel English' and thought the European news channel stood in his way, occasionally comparing it to a piece of chewing gum stuck to his shoes (Purvis, 2004). He was planning his own domestic news channel and wanted to be involved in the government's project for an international news channel that would eventually become France 24.

The relationship between the two partners deteriorated to the point that Stewart Purvis, ITN's CEO, decided to cut his losses and leave. Not only did Tessier discourage potential partners but he refused to buy back ITN's shares. After six months of legal wrangles France Télévisions agreed to buy ITN's shares for €1, representing a huge shortfall for the British company.

Today Euronews is a much stronger commercial proposition: only 7 per cent of its budget (30 million euros) is covered by subsidies, the rest of its income is derived from advertising and sponsorship (30 per cent), fees received from the participating twenty-two public broadcasters that can air the channel on their frequency (40 per cent), co-production agreements with the European Union (20 per cent) and carriage fees from cable operators. RTR (Russia), with 15.2 per cent of the shares, and SSR (Switzerland), 9.1 per cent, have joined the group of first-tier shareholders.

Despite the handicap of being located in Lyon, Euronews has been a great success with European viewers. Its unique ability to broadcast in eight languages has made it Europe's most-watched international news channel in 2007, ahead of CNN (the previous leader), Sky News, and BBC World (Chalaby, 2009b: 92). The concept of delivering European news to Europeans works well in an age of increasing regional integration. Euronews provides up-to-the-minute news bulletins that cover politics, business, finance, sport, weather and breaking news from a European perspective. Complementary programmes cover the arts, cinema, fashion and travel (*Le Mag* and *Agenda*), business and finance (*Economia* and *Markets*), and European affairs (for example, *Europa*). *No Comment*, launched in the ITN days, is an award-winning signature programme that airs the day's striking images without commentary (Marguin, 2007; Monchenu, 2005; Purvis, 2004).

The chronicles of the BBC in the satellite realm

The BBC began plans for a satellite TV service in the 1980s. Earlier in the decade the BBC World Service suggested that there should be a television version of their radio service but the proposal came to nothing. The idea resurfaced in 1986 when a brave soul pitched the idea for a TV station to the Foreign Office (since they already supported the radio service). This time it got the support of the BBC (and the Foreign Office) but it became apparent that the then Prime Minister, Margaret Thatcher, was not inclined to finance the operation. The plan was kept alive by the Foreign Office but the BBC never received a penny from Whitehall. It later emerged that ITN's Chief Executive, David Nicholas, went to see Mrs Thatcher to complain that the BBC project would damage his own plans for an international news service (which materialised as Super Channel's international news bulletin (Chalaby, 2009b)). Thereafter, although the proposal remained on Geoffrey Howe's Foreign Office desk, he always pushed it to the bottom of his tray because he was not prepared to go to Downing Street and have a row with Mrs Thatcher about it (Clover, 1991; Tusa, 2005).

Since the BBC refused to get involved in Europa and stayed at arm's length from Super Channel (see Chalaby, 2009b: 25–7), the corporation nearly passed the 1980s without getting involved in satellite television. Salvation came from Denmark where two telephone companies, KTAS and JTAS, approached BBC Enterprises (as the corporation's commercial arm was called) with a view to retransmitting BBC One and BBC Two in Scandinavia. BBC Enterprises looked into it and realised that it was impossible to secure the rights to deliberately exploit BBC One and BBC Two in that manner. Thus the corporation created a new channel for the Danes in 1987, BBC TV Europe, which was the live relay of BBC One except for those programmes that the BBC had bought from third parties and for which it was unable to secure international rights. Copyright-sensitive programmes, notably American TV series, were replaced with BBC Two documentaries. The Danish companies, which paid for the transmission costs and the rights for Scandinavia, marketed BBC TV Europe to cable operators in Sweden and Norway but were never able to make money out of the venture. In April 1989 BBC Enterprises assumed the satellite costs and began signing deals with a few other cable operators (Dunsford, 2004; Westcott, 1990).

As BBC TV Europe was building up its distribution and developing a profile on the continent, the corporation recognised that its satellite offering remained fairly crude. Neither the schedule (programmed for UK time) nor the BBC Two fillers were entirely satisfactory. Above all the BBC had not given up its dream of launching an international news channel. The corporation was pressed into action by the Gulf War in 1990–1, when CNN sprung to the world's attention. As elsewhere in Europe the BBC Director General, Michael Checkland, was frustrated to see CNN sweep the floor but his hands were tied financially. It was he who came up with the idea of using the revenue from BBC cable operations to revamp BBC TV Europe. On 15 April 1991 the corporation launched a channel specially created for Europe: BBC World Service Television. Its schedule remained similar to BBC One but with the benefit of two continuity announcers and two news programmes specifically produced for the channel. These were a fifteen-minute World Business Report and a daily half-hour international news bulletin (Clover, 1991). The BBC then looked for a partner to develop its fledgling international news business and Richard Li, Star TV's owner, stepped forward. The two companies created a dedicated 24-hour news channel for Asia, also called BBC World Service Television. The editorial side of the channel was controlled by Bush House and the editor came from the World Service radio newsroom.

As the two channels developed, the situation gradually became clearer for the BBC. The corporation realised that BBC World Service Television was a difficult channel to sell and promote in Europe. In addition to its long name the public did not really understand what the channel was about. The marketplace was also increasingly crowded with stations that were more focused and better defined in terms of what they were offering. By way of contrast the corporation's news channel was doing well in Asia. The BBC decided to drop World Service Television, expand its news channel worldwide and rename it BBC World, and launch an entertainment station called BBC Prime. The corporation financed its project by entering into a joint venture with Pearson and Cox Communications (Attwell, 2005; Dodd, 2007; Tusa, 2005).

BBC World News

The BBC achieved an old ambition when it launched BBC World, a global 24-hour news channel, on 26 January 1995. It invested heavily in the new channel that was based on the low-cost Asian version of BBC World Service Television. Numbers of staff were doubled, new presenters and journalists with television experience were brought in and editorial control was wrestled from Bush House. Because BBC World News – as the channel was renamed in April 2008 – is a commercial property funded by advertising, it contracts the provision of news content to the corporation. From an editorial perspective, the channel is part of the Global News Division that brings together the World Service, BBC Monitoring and the international online news service (the other three divisions are News, Nation & Regions and Sport). BBC World News has its own newsroom, team of producers, presenters and directors, and has access to the unparalleled international newsgathering resources of the BBC.

BBC World's schedule has been imitated countless times. It is based on a news bulletin at the start of every hour followed by the 'back half-hour' devoted to current affairs, business and lifestyle programming. Current affairs programming encompasses news specials, documentaries, talk shows, debates and interviews. Much of it is dedicated to the reporting and analysis of global issues. International affairs and the planet's social and cultural diversity come alive in specially commissioned documentaries and interviews with a wide array of personalities and power brokers.

No one at the BBC denies that the news network reflects European values (Attwell, 2005; Sambrook, 2007). This admission is no reason to accuse the corporation of cultural imperialism and of imposing its own views on the world, a charge that some scholars aim at Western news organisations with a prominent role in the international news flow. While the channel comes with a worldview, it is well equipped to deal with the complexities of the new global order. The World Service heritage has made it attentive to local sensitivities and its long-established network of experienced field correspondents gives it unique insight into local cultures. Evidence shows that BBC World is appreciated by audiences around the globe.[5] Ratings have steadily increased ever since it began broadcasting and the station reaches about 80 million households in Europe and over 300 million homes worldwide (Barnard, 2005; Dodd, 2007).

Even though BBC World News acknowledges a European political heritage and its output reflects the values of free speech, fairness and democracy, it does not make it a 'perspective' channel that pushes a particular national agenda. The editorial staff make a conscious effort to give a platform to diverse arguments and not to project a British point of view. The channel benefits from the World Service tradition of international journalism and prolonged exposure to a variety of cultures. It has enabled BBC World News to gain the trust of viewers in Europe and across the world (see note 5). Overall, the channel's output reveals a news organisation that is remarkably aware of the globalised nature of the world and the cosmopolitan character of the human condition in the twenty-first century.

From BBC Prime to BBC Global Channels

BBC Prime was launched in 1995 and inherited the distribution base of BBC World Service Television, around 2.2 million European subscribers. The news and factual programmes were jettisoned in favour of general entertainment, financed exclusively by subscription revenue. The channel became popular in Scandinavia and Switzerland where there is a taste for British comedy programmes, and in Central and Western Europe, where the channel was positioned as an educational tool to aid English language learning. At its peak in the mid-2000s, BBC Prime reached over 20 million subscribers in Europe and was subtitled in more than ten languages.

In 2006 John Smith, BBC Worldwide's CEO since 2004, devised a new strategy for the commercial arm of the corporation. The company's old priority was to programme and format sales and BBC Prime was only an add-on line of business. The channel controller had to liaise continuously with the sales department in order to check that the rights of a particular programme had not already been sold to a third party, which was usually the case. BBC Prime was a brand with a confused identity and was not a particularly good showcase for the BBC's vast library of programmes. Smith's strategy is based on what other media conglomerates had grasped a while ago: selling programmes is good for the financial spreadsheet but only television channels allow a company to build assets at international level. Accordingly, BBC Worldwide has changed focus from programme sales to development of international channel brands and businesses. A new division was created, Global Channels, and a team headed by Darren Childs was recruited from the best international broadcasters.

BBC Prime is being replaced by five channels: BBC Entertainment is a premium destination offering a mix of contemporary dramas (*Life on Mars*, *Doctor Who*), comedy (*Extras*, *Little Britain*), old classics and light entertainment (for example, *Strictly Come Dancing*). BBC Lifestyle has six key programming strands: food, home, fashion and style, health and body, parenting, and personal development. BBC Knowledge showcases the corporation's factual and documentary programming and its schedule is organised around five broad themes: the world, the past, business, science and technology, and people. CBeebies targets pre-schoolers with programmes like *Tweenies*, *Little Robots* and *Underground Ernie*, and BBC HD offers a mix of programmes in high definition.

Unlike BBC Prime these channels are properly resourced with priority access to BBC programming. The modus operandi has been completely revised: instead of one pan-regional feed covering Europe the new channels are all locally managed and fully adapted to each local market. BBC Worldwide has started deploying its teams across the world. In Europe it launched its first ever non-English channels in Poland in December 2007, managed by a thirty-strong local team. Their impact has been dramatic, multiplying the corporation's Polish audience reach by ten. Similar plans have been drawn up for Scandinavia, Spain, Italy and France. The BBC has converted relatively late to transnational TV networks but because of its formidable programming assets the new strategy should rapidly turn it into a global powerhouse in television entertainment (Dunsford, 2004; Posseniskie, 2008; Young, 2002).

Public broadcasters and transfrontier TV: adapting to the new media order

Transnational public-funded TV channels contribute to the diversity of international television and extend viewers' choice. Public funding can be justified in order to address gaps in channel provision and to reach audiences that may be overlooked by commercial broadcasters. However, these stations face a tough challenge in rapidly changing conditions. They were designed for the twenty-channel environment of the analogue cable platforms that prevailed in Europe in the 1990s. In an era when new TV channels are launched every month and satellite platforms carry hundreds of channels, a new strategy is required.

Many commercial broadcasters, such as MTV or Discovery, have adapted to the extra competition by multiplying the number of channels (Chalaby, 2009b). Where once one station represented 5 per cent of a twenty-channel cable platform, now ten are needed on a digital satellite platform to achieve the same presence and combat audience fragmentation. Public broadcasters' stand-alone international channels struggle to keep an audience because they lack visibility and are slipping out of viewers' sight. In the case of general interest channels, such as Arte and TV5Monde, their offerings might be too broad in an era when cable and satellite viewers are getting accustomed to niche stations. The BBC alone among public broadcasters has come to terms with this problem by launching its own portfolio of channels. For the other public broadcasters a solution might be to discuss distribution agreements and bundle their channels together in order to break their isolation on satellite platforms and increase brand awareness.

Too many public broadcasters are still locked in a national mindset. They need to engage more in transnational production and distribution arrangements. Cross-border thematic channels offer a solid ground for constructive collaborations. In the past, public broadcasters' collaborative efforts under the aegis of the EBU have not always been crowned by success. Although Arte itself should have thought about enlarging its ownership structure, as a transnational organisation it can provide a model for future joint ventures set up by two or more public corporations. Public broadcasters should pay particular attention to genres in which they could make an original contribution such as factual entertainment and children's television. By all accounts, unless they adapt their strategies to the new media order they risk disappearing into the ether of the satellite realm.

Acknowledgements

The author would like to express his deep gratitude to all the interviewees for their time and cooperation. Job titles and company names at time of interview.

Notes

1. To a very large extent, this chapter presents material already published in Chalaby (2009b).
2. In partnership with ARD and ZDF, DW launched German TV in 2002, a channel specifically aimed at expatriates, but it was discontinued in 2006.
3. http://arte-tv.com/fr/Impression/4982,CmC=890532,CmStyle=900344.html, accessed March 2006.
4. For many years, a third of the station's income was provided by various European institutions such as the European Parliament.
5. Based on independently conducted research commissioned by BBC World, 2007. The rapid growth of BBC World and the World Service in Muslim countries in Asia, Africa and the Middle East in 2007 attests to the trust that Muslim audiences have placed in the corporation's international services (Hillier, 2007: 1).

References

Attwell, R. (2005) (Deputy Head of Television News, BBC), interview with author, 15 June.

Barnard, A. (2005) (Chief Operating Officer, BBC World), interview with author, 5 April.

Barrand, C. (1986) 'Europa Television: the Transponder's Eye View', *EBU Review: Programmes, Administration, Law*, 37(2): 11.

Cable & Satellite Europe (1984) 'TV-5: the Free French Footprint', January: 68–9.

Cable & Satellite Europe (1984) '3SAT Makes its Bid', December: 45–6.

Cable & Satellite Europe (1987) 'High German', March: 24–9.

Cable & Satellite Europe (1995) 'EuroNews Receives First Private Backing', March: 12.

Castells, M. (1996) *The Information Age: Economy, Society and Culture*, vol. I: *The Rise of the Network Society* (Oxford: Wiley-Blackwell).

Chalaby, J. (2009a) 'Broadcasting in a Post-National Environment: the Rise of Transnational TV Groups', *Critical Studies in Television*, 4(1): 39–64.

Chalaby, J. (2009b) *Transnational Television in Europe: Reconfiguring Global Communications Networks* (London: I. B. Tauris).

Clarke, N. (1982) 'Pan-European TV?', *Irish Broadcasting Review*, 14: 42–7.

Clarke, N. (1983) 'Eurikon Promises a New Programme Concept', *Intermedia*, 11(3): 28–30.

Clarke, N. (1984) 'The Birth of the Infant Eurikon', *Intermedia*, 12(2): 50–1.

Clover, J. (1991) 'London Calling the World', *Cable & Satellite Euope*, May: 20.

Collins, R. (1998) *From Satellite to Single Market: New Communication Technology and European Public Service Television* (London: Routledge).

Cronel, J.-L. (2008) (Directeur du réseau commercial, TV5), interview with author, 25 January.

Dodd, D. (2007) (Head of Strategy for Journalism, BBC), interview with author, 1 May.

Dunsford, W. (2004) (Director of Channels, BBC Worldwide), interview with author, 13 December.

Esser, A. (2002) 'The Transnationalization of European Television', *Journal of European Area Studies*, 10(1): 13–29.

Fiedler, D. (2008) (Coordinator, 3sat), interview with author, 11 February.

Forester, T. (1985) *The Information Technology Revolution* (Oxford: Blackwell).

Godard, F. and P. Barker (1993) 'Nouvelles cuisine', *Cable & Satellite Europe*, February: 32.

Hillier, S. (2007) 'Good Figures at Global', *Ariel*, 22 May: 1.

Marguin, S. (2007) (Head of Research, Euronews), interview with author, 29 July.

Monchenu, O. de (2005) (Sales and Distribution Director, Euronews), interview with author, 9 March.
Newel, A. (2008) (Head of Distribution, Deutsche Welle-TV), interview with author, 3 April.
O'Connor, V. (1986) 'Climbing a Euro-mountain', *Cable & Satellite Europe*, July: 54–5.
Papathanassopoulos, S. (1990) 'Towards European Television: the Case of Europa-TV', *Media Information Australia*, 56: 57–63.
Possenniskie, D. (2008) (Senior Vice President, Global Channels – EMEA, BBC Worldwide), interview with author, 11 February.
Purvis, S. (2004) (CEO, 1995–2003, ITN, and President, SOCEMIE), interview with author, 4 August.
Sambrook, R. (2007) (Director, BBC World Service and Global News), interview with author, 12 June.
Snyder, A. A. (1995) *Warriors of Disinformation: American Propaganda, Soviet Lies, and the Winning of the Cold War – An Insider's Account* (New York: Arcade).
Tusa, Sir John (2005) (Managing Director, BBC World Service, 1986–92), interview with author, 25 March.
Utard, J.-M. (1997) 'Arte: Information télévisée et construction d'un point de vue transnational – étude d'un corpus franco-allemand', PhD thesis, Strasbourg III.
Westcott, T. (1990) 'London Calling', *Cable & Satellite Europe*, September: 28–32.
Young, M. (2002) (Managing Director, BBC Prime), interview with author, 11 July.
Ziegele, M. (2005) (Editor assigned to the Editor-in-Chief, Deutsche Welle-TV), interview with author, 22 July.

9

Public Service Media and Children: Serving the Digital Citizens of the Future

Alessandro D'Arma and Jeanette Steemers

Introduction[1]

This chapter examines the policy framework shaping the provision of children's content by three Public Service Broadcasting (PSB) organisations – the BBC in Britain, PBS in the US and RAI in Italy. It shows how political decisions regarding the remit and funding of PSBs have traceable consequences for the ability of these organisations to make a real public service contribution in this area, by providing a service that is at the same time distinctive and popular, two essential goals for any Public Service Media (PSM) organisation.

A focus on children's media, we suggest, offers an interesting entry point to debates about PSBs as they reconfigure as PSM organisations. This is not only because children's provision has always been, and remains, a key element of the PSM remit. More importantly, we note that in recent years several PSB organisations have singled out children's media as a strategic field of activity. The decision is likely to have been informed by two considerations. The first is that 'children constitute the public service users and "digital citizens" of the future' (Steemers, forthcoming (a)). From the standpoint of PSBs trying to legitimise their future existence, it is essential that younger generations grow up using their content, developing a sense of 'loyalty' towards their brands. Second, children's media is arguably an area where PSBs can be seen as making a public service contribution that is distinct from commercial offerings. Activities in the field of children's media, therefore, can be showcased to demonstrate public service credentials.

The three PSB organisations included in our study, of course, differ greatly in a number of key respects, in terms of their governance, funding and the market and regulatory environment in which they operate. They were chosen because they clearly illustrate how different policy contexts influence PSM's ability to serve children and develop an adequate response to market challenges.

The chapter proceeds as follows. Before examining the remit and obligations of the BBC, RAI and PBS in relation to children's provision and funding, we start by making a number of observations about traditional regulatory concerns relating to children's content, the rationale for public service intervention and differences in policy approaches. We conclude with an assessment of the performance of our PSM in children's media and relate this to the policy context.

Children and broadcasting regulation

Nowhere is the dual nature of broadcasting content regulation – 'negative' regulation (regulating against negative effects) and 'positive' regulation (regulating for positive outcomes) – more apparent than in rules formulated with the well-being of children in mind. Both sets of rules rest on an assumption of television's powerful social influence, either for good or for bad. As economists would have it, there is a widely shared belief that the positive and negative externalities associated with television affect children more than any other group in society.

Negative rules – typically, but not necessarily, bans and prohibitions – arise out of societal pressures to address the negative effects of television. They are designed to protect children from violence, pornography, over-commercialisation and other 'harmful' material, and are justified on the grounds that children are a vulnerable audience. Special advertising restrictions applied during children's airtime are a prime example of 'negative' rules. Their goal is to ensure that children are not exposed to an over-commercialised environment. Concerns about the negative effects of television on children tend to dominate public debates and media coverage, especially in America, where concerns about children's television have been 'very much focussed on violence, health issues and excessive commercialism, and expressing anxiety about a purely consumerist model of the child' (Messenger Davies, 2007: 8).

But television is also widely recognised as a potentially positive social influence on children (ibid.). Testament to this is the range of public policy tools designed to create the conditions for television to contribute positively to children's development. Chief among these tools in Europe is PSB, namely a service provided by publicly owned and, at least partly, publicly funded organisations which are institutionally mandated to cater for children through content designed to entertain and educate them and contribute to their moral development.

Policy approaches to children's television

PSB, however, is not the only form of 'positive' intervention. In countries with some form of public support for children's content, two broad policy approaches are identifiable (Ofcom, 2007b: 22). In several countries, publicly funded broadcasters are the only (or the main) intervention mechanism for children's content. A second, less numerous, group of countries use a wider range of policy tools alongside publicly funded broadcasting. These include quotas, qualitative public service obligations for commercial broadcasters, production subsidies and tax breaks. With a large and generously funded public broadcasting sector and virtually no other forms of public intervention, Germany is the clearest example of the first group of countries (ibid.: 35–6; see also Humphreys, 2008). By contrast, France is the paradigmatic example of the second approach. Scheduling and/or investment quotas apply, to varying degrees, to French private terrestrial broadcasters TF1 and M 6, as well as to cable and satellite services – the latter are required to invest a proportion of their revenues in animation. French animation studios are generously subsidised and are the beneficiaries of various tax incentives (Ofcom, 2007b: 15).

Arguably the three countries included in this study – America, Britain and Italy – all belong to the first group. They rely primarily on publicly funded broadcasters to achieve public service goals in children's television. There are, however, very important differences. Italy represents an extreme case. Italian commercial channels have never been subject to quotas or public service obligations on children programming (Mazzoleni and Vigevani, 2008).

The collocation of the US and UK in the first group of countries is far less clear. British commercially funded terrestrial broadcasters have always been seen as part of the PSB system, alongside the BBC. In fact, the term 'public service broadcasting' has traditionally referred to all terrestrial broadcasting organisations, with commercially funded channels subject to extensive public service obligations. This is, however, far less true today. Children's television provides a dramatic illustration. Before 2003, British commercial terrestrial broadcasters had to meet detailed scheduling quotas for children's programming, a measure designed to ensure provision of children's services beyond the BBC (ITC, 2003). The children's quotas, however, were abolished by the deregulatory 2003 Communication Act to allow commercially funded terrestrial broadcasters more flexibility in an increasingly competitive multi-channel environment. Under a new co-regulatory regime, commercial terrestrial broadcasters can now determine levels of provision themselves in consultation with the regulatory body, Ofcom. Facing increasing commercial pressures, over the last three years ITV, Britain's main commercial network, drastically reduced its investment in children's television, arguing that the new competitive scenario no longer makes children's content a viable commercial endeavour (Ofcom, 2007a). Plurality of PSB provision remains a core normative value of British television, as policy debates about how to ensure that the BBC will not remain the only PSB provider have shown (ibid.). However, in children's programming, it could be argued that Britain has moved closer to those countries that rely primarily on publicly funded organisations to achieve public service goals (Steemers, forthcoming (b)).

Finally, like Britain, but for different reasons, the US is somewhat of a borderline case. Historically, PBS has been a key provider of educational children's programmes, especially for young children (Morrow, 2006). However, its impact in the market is limited as the vast majority of children's viewing goes to commercial television (Ofcom, 2007b: 59). Cable networks (Nickelodeon, Disney Channel, Cartoon Network) are the most popular channels and these are not required to show educational programming. By contrast, in order to qualify for licence renewal, commercial terrestrial stations are required under the 1990 Children's Television Act and a 1996 amendment to air at least three hours of educational programming a week (Kunkel, 1999; Lisosky, 2001). In 2007, this requirement was extended to digital terrestrial stations (Kunkel, 2007). However, there are serious question marks about the effectiveness and impact of the educational requirement (ibid.: 210–17). It is fair to say, therefore, that provision of educational programming still rests primarily with PBS – with the exception of educational preschool programming on commercial cable networks that satisfy parental expectations.

Children's television as a 'market failure' genre?

A final important question needs to be settled, before we move on to analysing the policy framework shaping PSM provision of children's content. Children's television is often thought of as falling within those public service genres that are defined as 'market failures' – such as arts or cultural programming – that a free market would under-provide. The very definition of children's television as a genre, however, is problematic. Children's television refers to 'programs targeted primarily to children and designed to attract a majority of viewers who are children' (Alexander and Owers, 2007: 57). What distinguishes children's television from other programming is its target audience, not a particular language, topic or format as the term 'genre' suggests. In fact, within what we commonly refer to as children's television, we find a variety of programming types (drama, comedy, news, factual, variety shows) and forms (animation, live action).

The unqualified statement that a free market in television services would not be capable of delivering children's content is also patently inaccurate. In advanced multi-channel television markets, children's television is one of the most competitive and crowded market segments (Ofcom, 2007b; Papathanassopoulos, 2002: 227–43). By 2007 in Britain, for instance, there were over twenty commercial children's channels (Ofcom, 2007a). Even in the old age of 'spectrum scarcity' when only a limited number of generalist channels were available, it was not immediately evident that a free market system would necessarily under-provide children's programmes. One clear, if somewhat extreme, illustration of this is Italy. Operating in a virtually free-market environment, the new commercial channels which launched in the 1970s relied heavily on cheap children's programming to fill their daytime schedules (Fenati and Rizza, 1992; Richeri, 1986). For while the advertising revenues that can be generated from children's programming are comparatively small, the costs of acquired children's programmes can be far lower than those associated with programming for adults. Under certain circumstances, it may still be profitable for a generalist broadcaster to air children's programmes at times of the day (in the morning and afternoon) when the overall audience is relatively small and a good proportion of it is made up of children.

The statement that a free market in television services is not capable of delivering children's programming, therefore, needs to be qualified. What the historical and contemporary evidence suggests is that certain types of children's programmes – particularly domestically produced programmes and non-animated content – are under-provided by commercial broadcasters. In 1980s Italy, virtually all children's programmes on the new commercial channels were imported animation from Japan (Richeri, 1986: 30). Currently, the children's fare on the channels owned by US conglomerates Disney, Viacom (Nickelodeon) and Time Warner/Turner Broadcasting (Cartoon Network) is unsurprisingly made up of a high volume of US-originated animation (Ofcom, 2007a: 37, 43).

The definition of children's television as a 'market failure' obscures, rather than illuminates, what the rationale of a PSM approach to children's provision should be. It is not just about providing niche content. It is rather a question of providing a range of children's genres as well as high levels of domestically produced programmes. High levels of domestic originations, genre range, as well as high quality – even though quality is a notoriously elusive concept – are three key aspects of any public service provision for children.

PSM and children's provision

It could be argued, as Humphreys does (2008: 3), that the multiplication of media technologies and outlets makes regulation increasingly difficult, reinforcing the case for strong multi-platform PSM organisations 'to compensate for the declining scope for traditional content regulation' and 'ensure that a universal service continues to respond to all of society's communication needs and to perform a socially integrative function in an age when audiences are fragmenting' (ibid.: 3–4). Peter Humphreys' argument appears to be borne out by developments in British children's television, where the virtual withdrawal of ITV from commissioning new programming 'has reinforced the BBC's position as the key supporter of and outlet for UK-originated children's content' (Steemers, forthcoming (b)).

The extent to which PSB organisations will continue to be able to serve children with a range of high-quality, diverse and indigenous content depends on 'political choices regarding the public service remit and, most importantly, funding' (Humphreys, 2008: 6). The following sections consider these policy aspects shaping PSB provision of children's content by comparing and contrasting the situation in Italy, Britain and the US. Despite the very different regulatory, cultural and institutional histories, what these countries have in common is that they all currently seem to rely mostly on PSB organisations to promote quality, range and diversity in children's media.

Remit

Historically, catering for children has been regarded as a core element of the PSB remit (Blumler, 1992: 37–8; Ofcom, 2007b: 11–14). The BBC Trust, for instance, has recently emphatically restated the importance of children's provision: 'we believe that strong children's content is a fundamental aspect of public service broadcasting and essential if the BBC is to successfully create public value' (BBC Trust, 2009: 15). Serving children can be seen both as part of PSB's broader commitment to provide a comprehensive service catering for the needs, tastes and interests of different audience groups, and as a recognition of children's special status as vulnerable viewers. There are, however, important differences in the remit and obligations of PSB organisations, which in turn reflect different governance arrangements as well as different public expectations about the role of PSB in children's provision. The three countries in our study exemplify this broad spectrum of possibilities.

Over the last fifteen years, a defining feature of European policy approaches to PSB has been greater emphasis on 'auditing public broadcasting performance' (Brants, 2003; Coppens and Sayes, 2006). Compared to the general mandates of the past, European PSBs are now subject to closer political scrutiny and their performance is assessed through more narrowly defined performance criteria. Views differ as to whether this should be seen as a positive development improving PSBs' responsiveness or something to be opposed on the grounds that it constrains PSBs' independence, creativity and ability to innovate (for contrasting views see Jakubowicz, 2003; McQuail, 2003; Picard, 2003). The core of this new system is that the government assigns a number of quite specific tasks and targets to the public broadcaster through contracts or service licences, which are periodically assessed.

Such a new approach to governing PSM has been developed the furthest in Britain. Under new governance arrangements introduced in 2006, all the BBC's public service activities operate under an individual service licence, which defines the key characteristics of the service (remit, scope of delivery, annual budget and aims) and its contribution to the BBC's public purposes (BBC Trust, 2008a). Each licence also explicates the criteria and indicators that the BBC Trust uses to measure the performance of a service. The Trust is then in charge of monitoring BBC compliance with its licence conditions and statutory commitments. Under the new service licences approved in April 2008, the BBC's two analogue terrestrial channels, BBC One and BBC Two, are expected to 'bring children's programmes to analogue viewers, by showing a range of output ... at times convenient for children's viewing' (BBC Trust, 2008b: 5). The two channels share a commitment to broadcast at least 1500 hours of children's programmes annually combined. This commitment is higher than previous service licences, when it stood at 500 hours. The increase came about in response to concerns raised by Ofcom (2007a: 1) that the BBC could in theory reduce significantly its output on its two analogue channels, since BBC One and BBC Two's service licences did not reflect their actual delivery of children's programming (1941 hours combined in 2006–7). BBC's two digital children's services, CBeebies (for preschoolers) and CBBC (older children) also have service licences. The expectation is that the vast majority of the content on these services should be domestically produced, and a commitment to range, diversity and citizenship is reflected in stipulations that require CBBC to broadcast 85 hours of news, 665 hours of drama and 550 hours of factual programming annually (BBC Trust, 2008c).

In keeping with this emerging 'auditing philosophy', a new governance system was introduced for RAI in the early 1990s amid criticisms that RAI was failing to live up to its public service duties (D'Arma, 2007). Since then RAI has been held accountable through a series of three-year service contracts agreed with successive governments. Under the current contract (2007–9), RAI's three generalist terrestrial channels (RAI1, RAI2 and RAI3) are required to devote 10 per cent of their combined output between 7 a.m. and 10.30 p.m. to 'entertainment programmes for minors and formative and informational programmes for children and young people' (RAI, 2007a: 9). These programmes must be shown between 4 p.m. and 8 p.m. RAI is also required to invest 0.75 per cent of its turnover in children's animation, as part of a 15 per cent commitment to Italian and European audio-visual content (ibid.). Unlike the BBC, there are no individual service licences. Also, RAI's service contract appears to place less emphasis on diversity and range than the BBC's service licences – the 10 per cent programming quotas concerning 'entertainment programmes for minors and formative and informational programmes for children and young people' appear rather vague and imprecise in Italian too! Finally, there appear to be serious problems with monitoring compliance. Mazzoleni and Vigevani (2008: 35) comment that 'it is not known whether RAI fulfils these quotas, as no reports are available from AGCOM [the communications regulatory body] or RAI on this topic'.

Children's provision by PBS, unlike BBC and RAI, is not based on any explicit obligations set out in licence conditions or any other type of external or internal regulation, reflecting the very different institutional model which characterises American public television (GAO, 2007). Children's activities by PBS are expected to contribute to the goals set out in PBS Kids' own mission statement, namely, to make 'a positive impact on the lives of children through curriculum-based entertainment', 'to build children's knowledge, critical thinking, imagination and curiosity', and to encourage them to interact as 'respectful citizens in a diverse society' (cited in Ofcom, 2007b: 14).

PSM's expansion into new digital media

A widely debated issue, with very important implications for PSBs' ability to serve their child audience in future, is the degree of freedom that they are given by governments to expand into digital, online and interactive media (see Aslama and Syversten, 2007; Nissen, 2006: 26–8). Despite sustained lobbying by private competitors to persuade governments to keep PSM out of these new fields of activity, and despite a generally hostile political and ideological climate, most European governments appear to have agreed on the necessity to allow PSBs to develop a multi-platform strategy, a key step towards their transformation into PSM organisations (Leurdjik, 2007: 82). At the European level, on paper at least, both the European Commission and the Council of Europe have endorsed the idea that PSBs should be allowed to use the latest media technologies to discharge their public service remit (Jakubowicz, 2007: 29).

The first important step in the digital strategy of several PSBs has been the development of a family of digital television channels. Children's programming has been identified as a key area for expansion. The three PSBs in our study have all launched digital television channels targeted at children within the last ten years, even though there are important differences in the nature of their services.

More recently, of course, online media have become a new focus of attention for PSBs, as the take-up of broadband internet grows and new applications are developed. Research shows that (older) children are at the forefront in the use of new media services, including social networking, file sharing, online gaming and mobile applications (Ofcom, 2007a). PSB organisations have been eager to develop their online offerings, aiming to establish themselves as 'full portfolio' content providers in order to connect especially with younger audiences who are increasingly drawn to new media. Online activities have become an integral part of their remit and of their public service obligations. Children's provision in turn is an important part of these activities. The service licences of CBBC and CBeebies stipulate that their broadcast content 'may be simulcast on fixed and mobile internet protocol networks', it may also be offered on the same platforms 'for seven days after it has been broadcast', and should be complemented with 'programme-related content on bbc.co.uk' (BBC Trust, 2008c, 2008d). In 2007–8 CBBC and CBeebies had access to an online and interactive content budget of £6.1 million (about 6.9 million euros) (BBC Trust, 2009: 49). The service licence for the BBC's online activities states that 'as a key priority', the BBC should continue to develop 'a comprehensive service for children, to ensure availability of UK online content for children, directed towards learning outcomes and promoting safer use of the internet' (BBC Trust, 2007). In Italy, while there is no specific reference to RAI's online provision for children, the general framework regulating RAI's activities in digital media is one which encourages and requires RAI to expand into online services. Under its current service contract, RAI is expected to make its content available across digital distribution platforms, including the internet (RAI, 2007a). In January 2007, RAI launched a new video portal, rai.tv. 'Junior' is the gateway to children's on-demand services, streamed video, podcasts, online games and activities, and communities.

Funding

Financial, rather than regulatory, constraints are more likely to limit the expansion of PSBs into digital media, negatively affecting the nature, purpose and ultimately effectiveness of their overall provision for children. Having access to different levels of public funding and dependent to varying degrees on commercial revenues to supplement their public income, the three PSBs in our study exemplify the different financial challenges and constraints facing PSM organisations.

RAI exemplarily illustrates the problems facing any PSB heavily dependent on advertising and limited public funding. The TV licence fee paid by Italian households is one of the lowest in Western Europe (107.5 euros in 2009). It is also one of the most widely evaded (RAI, 2007b: 28). Nearly half of RAI's income is generated from advertising. At one level, therefore, RAI is facing the same problems as commercial broadcasters, arising from the recent decline of television advertising revenues. At the same time, over the past few years Italian governments have either refused to increase the level of the licence fee or have accorded increases below, or just in line with, inflation. As a consequence RAI finds itself having to spread a budget which is decreasing in real value over an expanding array of services. Children's provision is likely to be affected less than other content areas, because the limited budget for children has little impact on RAI's overall budget. However, commercial pressures were apparently behind RAI's recent decision to reduce children's airtime on its three analogue terrestrial channels and to move one of the two children's blocks from the first and more popular channel, RAI1, to the third and more public service-oriented RAI3 (Starcom, 2007). Pressure to maximise commercial revenues also explains RAI's rather questionable choice, from a public service perspective, to distribute two of its three digital children's channels on a subscription basis only.

In the US, PBS's children's provision has always been associated with educational programming and non-commercialism. The American public television system is funded through a combination of federal funding, donations and sponsorships from businesses and foundations (GAO, 2007). Overall, when one considers the size of the US market and the scale of competition that PBS faces from mainstream commercial media, it is clear that the system is vastly under-financed. Despite this, in recent years, under the Republican administration, there were repeated attempts to cut federal funding, which accounts for 15 per cent of public television's revenues (Freedman, 2008: 167–9). As argued by Freedman (2008: 168), the prospect of reduced federal funding is forcing PBS 'increasingly to turn to other sources of income, in particular merchandising tie-ins, sponsorship and branding', as well as business ventures. This is well illustrated by PBS Kids Sprout, a preschool digital television channel and VOD service launched in 2005 as a joint venture with commercial partners. Unlike children's programming shown on PBS-affiliated stations, PBS Kids Sprout shows advertising between programmes and is a profit-making enterprise.

Not even the BBC – a PSB that by international standards is very generously funded and carefully insulated from commercial pressures – escapes budgetary pressures. These have the potential to undermine the corporation's ability to perform its public service functions. In 2007, the BBC was awarded a lower than expected licence fee increase, which requires a 5 per cent budget cut every year over five years (Steemers, forthcoming (b)). Across all areas, including children's services, the BBC is expected to deliver efficiency savings, totalling £1.9 billion (2.4 billion euros) over five years by 2012–13 (BBC, 2009). There are serious concerns that efficiency savings may have a negative impact on the quality of children's output. This was acknowledged by the BBC Trust itself (2009: 58). Investment in children's content is expected to fall in real terms over the next five years (BBC Trust, 2009: 57). One response by the BBC to budgetary constraints and increasing competition has been the 'Fewer, Bigger, Better' strategy, involving the concentration of resources on fewer programmes 'to compete on quality rather than quantity' (ibid.: 7). In its recent review of BBC's children's services, the BBC Trust acknowledged that 'taken too far the Fewer, Bigger, Better strategy could begin to have a negative impact on performance', citing increasing repeat levels and less diversity as main concerns. In fact, repeats made up 95 per cent of transmissions on CBeebies in 2007, up from 63 per cent in 2004 (Steemers, forthcoming (b)). The Trust also expressed some reservations about the BBC's intention to exploit commercial sources of income from co-productions and secondary revenues twelve-fold by 2010 to help finance children's content. The Trust remarked that increasing reliance on commercial revenues may inhibit the corporation's ability to satisfy its core purposes, by focusing attention on a more limited range of commercially and internationally appealing programming, which may be less relevant to British child audiences (BBC Trust, 2009: 58–9).

Conclusion

We began by noting that distinctiveness and popularity are two essential objectives for any PSM organisation, and that the policy context has a determining influence on PSM's ability to serve children.

In the area of children's provision, the three PSMs in our study – the BBC, RAI and PBS – appear to be making, to different degrees and in different ways, a distinctive contribution, compared to what is available on commercial channels. PBS in the US continues to be the key provider of educational children's programming, especially preschool content, with children's programmes accounting for a significant proportion of the total broadcast output of PBS-affiliated stations (GAO, 2007: 19) in an advertising-free environment. Now that ITV has virtually withdrawn from the origination of children's content, the BBC has been left, by a long distance, as the main funder and broadcast outlet for UK-originated children's content (Ofcom, 2007a), sustaining range and diversity in the process. For example, as noted by the BBC Trust (2009: 23), the schedule of preschool channel CBeebies comprises 76 per cent non-animated content, compared with levels of animation reaching 70 per cent on rival channels. Frequently criticised for not being sufficiently distinctive, even RAI appears to be playing a valuable role as the main outlet, commissioner and funder of original children's programming across genres, nurturing talent and supporting the domestic animation industry (Sarra, 2006; Screen Digest, 2006), in a market which is dominated by American and Japanese animation.

More problematic, however is the achievement of the second condition which we have identified as essential for successful PSM organisations – popularity. The BBC is the only one of our three PSBs that appears to be 'delivering popular children's content which successfully appeals to a wide audience' both offline and online (BBC Trust, 2009: 3). The BBC's two digital channels, especially CBeebies, are performing strongly in terms of audience reach and share (BBC Trust, 2009; Ofcom, 2007a). BBC children's websites, again CBeebies more than CBBC, are also widely used and compare favourably with usage figures for other children's broadcasters' websites (BBC Trust, 2009: 3–4). The other two PSMs in our study are clearly failing to achieve the same level of popularity. Publicly available data (Starcom, 2007, 2009) suggest that RAI is far less popular with children than its main commercial rivals (Disney and Mediaset). PBS has historically had 'had little influence on the wider broadcasting landscape and its effectiveness' has been 'largely limited to a small number of preschool flagship shows' (Steemers, forthcoming (c)).

As we noted, funding, and not regulation is the critical issue. RAI's regulatory framework, like the BBC's, does not limit RAI's ability to expand into new digital media. Under its service contract, RAI is actually required to distribute its content, including its children's content, across digital platforms.

Constraints are rather of a financial nature. This is most evidently the case for RAI and PBS, two PSM organisations which have always suffered from inadequate public funding and have therefore been exposed to the risk of marginalisation and/or commercialisation. But even a well-resourced PSB like the BBC is far from being unaffected by financial constraints. As we have seen, the budgetary restrictions that the BBC is currently facing, and the strategic responses that it is adopting, have the power to potentially undermine its ability to continue to deliver the same range and diversity of children's output.

Note

1. The research for this piece was supported by a grant from the Arts and Humanities Research Council (AHRC) (Grant number: 119149).

References

Alexander, A. and J. Owers (2007) 'The Economics of Children's Television', in A. Bryant (ed.), *The Children's Television Community* (Mahwah, NJ: Lawrence Erlbaum), pp. 57–74.

Aslama, M. and T. Syvertsen (2007) 'Public Service Broadcasting and New Technologies: Marginalisation or Re-monopolisation', in E. de Bens (ed.), *Media Between Culture and Commerce* (London: Intellect Books), pp. 167–78.

BBC (2009) 'BBC Trust Approves BBC Budget for 2009/10', 20 March, http://www.bbc.co.uk/bbctrust/news/press_releases/2009/bbc_budget.html, accessed 17 May 2009.

BBC Trust (2007) 'bbc.co.uk Service Licence', 30 April, http://www.bbc.co.uk/bbctrust/framework/bbc_service_licences/bbc_co_uk_service_licence.html, accessed 17 May 2009.

BBC Trust (2008a) 'BBC Service Licences: Operating Framework', http://www.bbc.co.uk/bbctrust/assets/files/pdf/regulatory_framework/service_licences/operating_framework_10_08.pdf, accessed 17 May 2009.

BBC Trust (2008b) 'BBC One Service Licence', 7 April, http://www.bbc.co.uk/bbctrust/assets/files/pdf/regulatory_framework/service_licences/tv/2008/bbc_one_Apr08.pdf, accessed 17 May 2009.

BBC Trust (2008c) 'CBBC Service Licence', 7 April, http://www.bbc.co.uk/bbctrust/assets/files/pdf/regulatory_framework/service_licences/tv/2008/cbbc_Apr08.pdf, accessed 17 May 2009.

BBC Trust (2008d) 'CBeebies Service Licence', 7 April, http://www.bbc.co.uk/bbctrust/assets/files/pdf/regulatory_framework/service_licences/tv/2008/CBeebies_Apr08.pdf, accessed 17 May 2009.

BBC Trust (2009) 'Service Review: Review of Children's Services and Content', February, http://www.bbc.co.uk/bbctrust/assets/files/pdf/regulatory_framework/service_licences/service_reviews/childrens/childrens_review.pdf, accessed 17 May 2009.

Blumler, J. G. (1992) 'Vulnerable Values at Stake', in J. G. Blumler (ed.), *Television and the Public Interest* (London: Sage), pp. 22–42.

Brants, K. (2003) 'Auditing Public Service Broadcasting Performance', *Javnost/The Public*, 10(3): 5–10.

Coppens, T. and F. Sayes (2006) 'Enforcing Performance: New Approaches to Govern Public Service Broadcasting', *Media, Culture & Society*, 28(2): 261–84.

D'Arma, A. (2007) 'Broadcasting Policy in Italy's Second Republic 1994–2006', unpublished doctoral thesis (University of Westminster).

Fenati, B. and N. Rizza (1992) 'Schedules and Programmes on Television in Italy', in A. Silj (ed.), *The New Television in Europe* (London: John Libbey), pp. 151–216.

Freedman, D. (2008) *The Politics of Media Policy* (Cambridge: Polity).

GAO (Government Accountability Office) (2007) *Issues Related to the Structure and Funding of Public Television*, http://www.gao.gov/products/GAO-07-150, accessed 17 May 2009.

Humphreys, P. (1996) *Mass Media and Media Policy in Western Europe* (Manchester: Manchester University Press).

Humphreys, P. (2008) 'Redefining Public Service Media: a Comparative Study of France, Germany and the UK', paper presented at the RIPE Conference 8–11 October, Mainz, Germany, http://www.uta.fi/jour/ripe/2008/papers/Humphreys_P.pdf, accessed 17 May 2009.

ITC (Independent Television Commission) (2003) 'Children's Television', *ITC Notes*, Note 17, www.ofcom.org.uk/static/archive/itc/itc_publications/itc_notes/view_note73.html, accessed 17 May 2009.

Jakubowicz, K. (2003) 'Endgame? Contracts, Audits, and the Future of Public Service Broadcasting', *Javnost/The Public*, 10(3): 45–62.

Jakubowicz, K. (2007) 'Public Service Broadcasting in the 21st Century: What Chance for a New Beginning?', in G. Lowe and J. Bardoel (eds), *From Public Service Broadcasting to Public Service Media* (Göteborg: NORDICOM), pp. 29–49.

Kunkel, D. (1999) 'Children's Television Policy in the United States: an Ongoing Legacy of Change', *Media International Australia*, 93: 51–63.

Kunkel, D (2007) 'Kids' Media Policy Goes Digital: Current Developments in Children's Television Regulation', in A. Bryant (ed.), *The Children's Television Community* (Mahwah, NJ: Lawrence Erlbaum), pp. 203–28.

Leurdjik, A. (2007) 'Public Service Media Dilemmas and Regulation in a Converging Media Landscape', in G. Lowe and J. Bardoel (eds), *From Public Service Broadcasting to Public Service Media* (Göteborg: NORDICOM), pp. 71–85.

Lisosky, J. (2001) 'For all Kids' Sakes: Comparing Children's Television Policy-Making in Australia, Canada and the United States', *Media, Culture & Society*, 23(6): 821–42.

Mazzoleni, G. and G. E. Vigevani (2008) 'Television across Europe: Follow-up Reports 2008, Italy', OSI/EU Monitoring and Advocacy Program, www.mediapolicy.org/tv-across-europe/follow-up-reports-2008-country/italy-web.pdf.pdf/at_download/file, accessed 17 May 2009.

McQuail, D. (2003) 'Public Service Broadcasting: Both Free and Accountable', *Javnost/The Public*, 10(3): 13–23.

Messenger Davies, M. with H. Thornham (2007) 'Academic Literature Review: the Future of Children's Television Programming', http://www.ofcom.org.uk/consult/condocs/kidstv/litreview.pdf, accessed 17 May 2009.

Morrow, R. (2006) *Sesame Street and the Reform of Children's Television* (Baltimore: Johns Hopkins University Press).

Nissen, C. (2006) 'Public Service Media in the Information Society', report prepared for the Council of Europe's Group of Specialists on Public Service Broadcasting in the Information Society (MC-S-PSB), http://www.obercom.pt/client/?newsId=105&fileName=conselho_da_europa_media_de_servi_o_p_bl_na_si_fev06.pdf, accessed 17 May 2009.

Ofcom (2007a) *The Future of Children's Television Programming* (London: Office of Communications).

Ofcom (2007b) 'The International Perspective: the Future of Children's Programming Research Report', http://www.ofcom.org.uk/consult/condocs/kidstv/international.pdf, accessed 17 May 2009.

Papathanassopoulos, S. (2002) *European Television in the Digital Age* (Oxford: Polity).

Picard, R. (2003) 'Assessment of Public Service Broadcasting: Economic and Managerial Performance Criteria', *Javnost/The Public*, 10(3): 29–44.

RAI (2007a) 'Contratto di Servizio 2007-2009', 5 April 2007, http://www.comunicazioni.it/binary/min_comunicazioni/televisione_rai/contratto_servizio_5_aprile_2007.pdf, accessed 17 May 2009.

RAI (2007b) 'Piano Industriale 2008–2010', 24 October, http://www.mcreporter.info/documenti/rai_pianoind08-10.pdf, accessed 17 May 2009.

Richeri, G. (1986) 'Television from Service to Business: European Trends and the Italian Case', in P. Drummond and R. Paterson (eds), *Television in Transition* (London: British Film Institute), pp. 21–35.

Sarra, A. (2006) 'Animazione e Cartoni: Una Piccola Industria "Globale"', *L'industria: Rivista di Economia e Politica Industriale*, 27(4): 623–54.

Screen Digest (2006) 'Europe's Animation Industry Grows: Television Output Stable while Features Struggle for Distribution', 1 December.

Starcom (2007) 'Scenario TV Bambini', http://www.primaonline.it/wp-content/plugins/Flutter/files_flutter/1229616326file153059102881989.pdf, accessed 17 May 2009.

Starcom (2009) 'TV Bambini', http://www.primaonline.it/wp-content/plugins/Flutter/files_flutter/1237377550LaTvdeibambiniFeb09.pdf, accessed 17 May 2009.

Steemers, J. (forthcoming (a)) 'Little Kids TV: Downloading, Sampling and Multiplatforming the Preschool TV Experiences of the Digital Era', in J. Bennett and N. Strange (eds), *Television as Digital Media* (Durham, NC: Duke University Press).

Steemers, J. (forthcoming (b)) 'The BBC's Role in the Changing Production Ecology of Preschool Television in Britain', in *Television and New Media*.

Steemers, J. (forthcoming (c)) *Creating Preschool Television* (Basingstoke: Palgrave Macmillan).

10
Heritage Brand Management in Public Service Broadcasting

Gregory Ferrell Lowe and Teemu Palokangas

Introduction

Growing complexity in branding literature reflects increasing maturity in scholarship and disagreement about how branding works and what matters most. The difficulty is partly due to complexity and partly to what Brown (2006: 50) characterised as 'logorrhoea where the target word is becoming increasingly festooned with add-ons and modifiers'. Hulberg (2006) conducted a comprehensive literature review and found two schools of thought.

The 'American instrumental' school is historically oldest and still most common. It is functionalist, rationalist, formulaic and features a managerial bias. Aaker (1996), Aaker and Joachimsthaler (2000) and Keller (2003) are especially influential. The 'European interpretive' school is characterised as an alternative but can be understood as complementary. It is the result of research that understands branding primarily as a communication process emphasising the social construction of meaning. Because branding is a co-production, much is outside management control. This approach emphasises customer interaction and accommodates resistance. Kapferer (2004) and de Chernatony (2001) are leading lights.

This chapter favours a postmodern view that bridges the two schools. Strategic brand management is about a lot more than marketing; it is about continually developing meaningful differentiation which is always some combination of emotional attachment and functional performance (Bennett and Rundle-Thiele, 2005). Branding is essential for any company in an industry that is dependent on consumers in a market characterised by abundant choice. That certainly characterises broadcast media (Ellis, 2000) and suggests that branding Public Service Broadcasting (PSB) is necessary. This view is contrary to Hoynes (2003) who, for normative reasons, was critical of branding in American public broadcasting. While respecting normative theory as our approach rests on the social responsibility premise, our view is more pragmatic.

PSB must master branding to compete and to develop because the social environment is an increasingly comprehensive brand culture. Branding facilitates differentiation of companies and products from competitors, provides an assurance of standardised quality, and is instrumental for developing loyalty (Picard, 2008). Strategic brand management can accomplish much of proven benefit. The challenge is how to ensure alignment of this commoditised framework with PSB's core values rooted in a unique non-profit heritage. Squaring this circle is where emerging research on 'heritage brands' with managerial emphasis on 'brand stewardship' is particularly useful (Urde et al., 2007). The heritage brand perspective is quite new. This chapter discusses its instrumental and symbolic value for strategic brand management in PSB today.

Branding media

The origins of branding coincide with the penny press which gave rise to mass media. Branding has always been a tool for differentiation in competitive markets where products have high functional equivalency. Branding provides an assurance of standardised quality while imbuing products with some intentional character making them memorable and personable. This is the familiar idea of 'positioning' (Ries and Trout, 1986).

Although branding is nothing new *in* media, it is quite new *for* media. As late as the mid-1990s, branding was not common among media firms (McDowell, 2006). This changed as media grew faster than audience size, producing fragmentation that drives the need for differentiation (Drinkwater and Uncles, 2007). A second driver is low switching costs in broadcasting, evident in browsing and zapping. Arguably media firms need branding for an assurance of quality more than many industries because these products are 'credence goods' (Siegert, 2008); audiences cannot know their worth until after consuming them. Trust is a key success factor.

Branding media started earlier in the US than in Europe because it was highly competitive (Blumenthal and Goodenough, 2006). Until recently TV branding was largely confined to logos, for example the NBC 'peacock'. Radio was faster off the mark than television in the US as well as Europe, again due to high competition earlier.

In the 1980s Music Television (MTV) arrived like a branding tsunami. Even the logo had personality. The entire enterprise was about attitude and sub-cultural affiliation. By the late 1990s dozens of profiled cable and satellite channels had followed suit, especially in light of MTV's success with youth audiences. As branding swept pay-TV the domestic majors began developing their brands to compete. In Europe that meant PSB companies as well as private commercial operators. Building fresh, vibrant brands has been more difficult for these older mainstream companies in both sectors. Long histories have deeply conditioned audience perspectives in two directions: (a) perspectives on audiences inside media organisations, and (b) perceptions among audiences of what these companies are and mean.

In the first decade of the twenty-first century PSB brand portfolios became increasingly complex. Hierarchy rooted in one master brand reflects the dominant structural model where public service is provided by a 'single, coherent broadcasting organisation dedicated to delivering programmes of quality and distinction across all media', as formulated in the BBC's *Extending Choice* (1992: 48). Coherence means each sub-brand is associated with the master (or parent) brand for good and ill. This was a crucial issue in the YLE radio reform in Finland in 2003 when the youth channel, Radio Mafia, was rebranded as YLEX. Although never any secret that Radio Mafia was owned by public broadcaster YLE, its managers and makers considered the association a liability for brand image. In contrast, YLEX is explicit about this association.

The portfolio challenge is about identifying each respective brand and clarifying roles (Aaker, 1996). Given the proliferation of platforms and channels in the digital environment, this is increasingly complex, as the BBC's breakdown illustrates. According to the *Brand Manual* (BBC, 2009), the BBC has *public service brands* that 'define all the services, media platforms and content delivered by the BBC' (for example, BBC One and bbc.co.uk), *commercial brands* (BBC Worldwide, *Good Food* magazine), *channel brands* which are a 'mix of editorial content' in separate media (BBC One, BBC Radio Two), *content brands* defined by 'editorial content' instead of medium (*EastEnders, Fat Nation* initiative), *event brands* (BBC Proms, One Big Weekend), *genre brands* (BBC News, BBC Sport), *initiative brands* (*Fat Nation, The Big Read*), *platform brands* (bbc.co.uk, BBCi), *programme brands* (*The Office, The Apprentice*), *publishing brands* (*Good Food, Good Homes* magazines), and *service brands* (BBC Shop, BBC Costumes & Wigs).

> Every BBC sub-brand contributes to some part of the BBC's brand mission; to enrich people's lives through great content. But a BBC sub-brand obviously has a life of its own and may even have values not shared by the BBC. That's fine. So long as a BBC sub-brand always helps to shape the BBC brand, that's what's important. (BBC *Brand Manual*, 2009)

Often particular sub-brands are most important for the future of a company because they embody values the company wants to be known for. In YLE, political entertainment programmes have long enjoyed prominence because they contrast unique attributes of public service provision of information and entertainment programmes. YLE recently acquired rights to broadcast HBO television series in part because this endorses contemporary quality drama without commercial interruption. Such decisions are fundamental to strategic brand management.

Brand extensions have similarly become important in media, as the BBC case illustrates. America's TV network NBC is one of the most successful as evident in the development of MSNBC online and CNBC in pay-television (on cable and satellite channels). PSB operators have been quite active as well, launching 'brand new' profiled channels, profiling existing channels and structuring 'channel bouquets'. In the current decade the BBC and YLE are among many that have branded all channels and services with a corporate identity (for example, YLE Teema is a new digital TV channel and BBC One in radio).

Branding has been everywhere fuelled by the imperative to differentiate where products have functional equivalency. An increasingly common criticism of PSB has two relevant dimensions. Critics suggest that too often PSB programmes are not distinctive, that they mimic popular success in the commercial sector. The second dimension is that commercial operators could do most of what PSB does at least as well if they enjoyed the same protected revenue, that is, the argument about 'contestable funding'. Both dimensions are evident in recent reports from the UK's Office of Communication (Ofcom, 2006) as well as an international comparative study of PSB funding (Thompson et al., 2005). Both have also been issues in the Swedish market (Humphreys and Norbäck, 2008) while also in Finland at the moment YLE's full service role is being criticised as market interference. All of this illustrates the continuing importance of Nissen's (2006) argument that PSB must be careful to balance its competitive need for popularity with distinctiveness.

Strategic brand management: aligning identity and image

Branding in media companies has mainly focused on marketing (Ots, 2008). Its potential for grounding strategic management is newer and complicated by perceptions of branding as expense rather than investment (Chan-Olmsted, 2006). In strategic brand management the brand becomes the lens for focusing development of corporate strategy and the ground for decision-making about content and services. Success in strategic brand management depends on (1) ensuring the 'proper' alignment of brand identity with brand image, and (2) gearing all business processes and resource management to guarantee consistency in fulfilling identity claims (de Chernatony, 2001; Ind, 2001). The 'brand ecology' is pictured in Figure 10.1.

Beginning at the outermost circle, the social environment is marketised and characterised as a 'brand culture'. Schroeder and Salzer-Mörling (2006: 10) stated that 'the cultural landscape has been profoundly transformed into a commercial brandscape in which the production and consumption of signs rivals the production and consumption of physical products'. This has applicability for PSB. All media are engaged in a continual competition for attention, construed as a 'time market' (Albarran and Arrese, 2003). As credence goods, the reputation of a media company is crucial to competitive success (Wolinsky, 1995). And media content is influential in producing

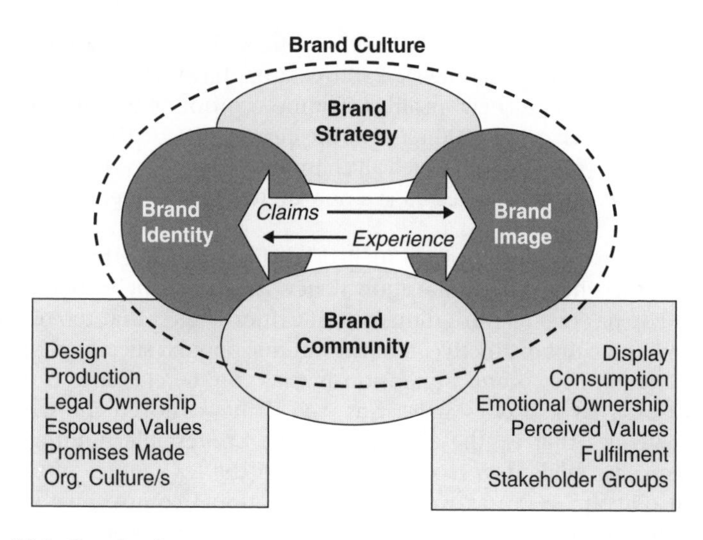

Figure 10.1 Brand culture
Source: Authors' analysis.

cultural images that become popular for identity modelling (Elliot and Davis, 2006; Farholt and Bengtsson, 2006).

Thus, as Balmer (2006) observed, 'brands are not made in a factory – they are constructed in people's minds'. As such they are only partly about *legal ownership*; they are as much about *emotional ownership*. The emotional experience of consumers is a crucial determinant of competitive success in today's 'experience economy' (Pine and Gilmore, 1999), which certainly includes broadcasting. Brand strength depends on emotional associations and attachments (Aaker, 1996). This is precisely the problem for PSB companies where older audiences are attached but younger audiences are not.

Brands are not only about what the company 'owns'; they are about what consumers embrace, construe and enact. That is why no brand is ever completely under control (Bergvall, 2006), a sore spot for a management paradigm so focused on control (Hamel, 2007). The brand lives and dies in its relations with consumers (Pickton, 2008). That's why brand management is an ongoing communication project (Hatch and Rubin, 2006) focused on ensuring the alignment of brand identity and brand image (Kapferer, 2004). PSB companies have oddly not been very good at this project; 'oddly' because communication is an obvious core competence. But the legacy in mass media which emphasised transmission rather than dialogue, combined with historic monopoly which conditioned internal perspectives, ill-prepared PSB for today's interdependent environment (Lowe and Daschmann, forthcoming).

Paraphrasing Keller and Richey (2006), a company's brand image is a composite of (a) what it makes, (b) the actions it takes and (c) the ways it relates. PSB has had problems in each aspect. What it makes has often been considered old-fashioned and pedagogic. The actions it takes have been understood as self-serving with questionable commitment to public service ideals in recent years. The ways it relates have been criticised for arrogance linked to excessive insularity (Lowe and Bardoel, 2007).

Credibility in the promise of quality is not based on corporate claims but on consumer experiences. Credibility depends on fulfilling every identity claim. Among the most common causes of brand management failure is making decisions about what a company will do and be in strategic design but failing to create the organisational structures necessary to achieve this in practice (Shultz and Hatch, 2006). Every espoused value must have strategic purchase in a specific identity claim, and every claim must be embedded in concrete business processes. Although dated, Balmer (1994) demonstrated how claims made by the BBC in the early 1990s about being accessible and accountable were more about marketing myth than operational reality.

A brand can satisfy three kinds of needs (Nandan, 2005). Functional needs are utilitarian, symbolic needs are emotional associations important for self-identity and group affiliation, and experiential needs are about how brand use turns out in practice. Branding strategy will often specialise in one of these needs, but a company must fulfil all three as these are components of an 'expectation set' (Banerjee, 2008). Some expectations are generic to the industry. These are core competencies because consumers expect every firm to perform at what eventually becomes a general standard. This is why companies in an industry converge to be more alike over time – consumers expect every firm in the industry to be equally competent at 'the basics'. That explains why PSBs must become public service media (about more than traditional radio and television broadcasting). Other expectations in the set will be specific and depend on fulfilling unique claims. Thus, PSB has typically done well in credibility perceptions of news as trustworthy and reliable. This has been consistently found in many places where PSB has a long history, for example over decades of experience in Finland (Jääsaari, 2004; Snell et al., 2003). The European Broadcasting Union's Digital Strategy Group found the PSB brand is almost everywhere associated with reliability and quality and understood as embodying 'irreplaceable values' (EBU, 2002: 19).

But brand image is never altogether under control. It is affected by how the industry is perceived (Burmann et al., 2008) and by perceptions of sectors within industries. Brand meaning is contingent on a 'brand constellation' (Jevons et al., 2005), a concept highlighting comparative experiences with other brands in the same category. Microsoft is judged in relation to Apple, whether it wants to be or not (SMH, 2008). The problem here is that PSB as a sector tends to be viewed as old-fashioned while the private sector is seen as exciting (Cunningham and Turner, 2006). These associations are based on practical experience with companies and their products over decades, a point sometimes ignored by managers who think audiences are merely misinformed or labouring under misconception. In fairness, however, such associations are also based on legacy perceptions that may be inaccurate today but which the company has not yet succeeded in changing. In Finland, for example, YLE has succeeded in its new media strategy because two of its most popular services are YLE Areena for late or repeat viewing of broadcast programmes and YLE Elävä Arkisto which offers digitised content of historic broadcast programming. Young Finns like both services yet still tend to think of YLE as a bit old-fashioned.

Moving up another level of analysis, Dowling (2000) demonstrated that a corporate brand is the product of a 'network' of associations which include the country image, the industry image, the corporate brand image and its product brand images. How a programme is perceived may not only be a function of content but also of the channel on which it airs, the image of the industry and sector that produced it, the company that programmes it, and even the country and language of original broadcast. American programming is attractive in part because of perceptions about America's wealth and vibrancy, and earlier pleasant experiences consuming American culture industry products. Of all influential associations, none is more important than the historic *experience* of a PSB company and its products. That is the history that frames a heritage.

The PSB heritage brand and its managerial stewardship

The importance of a company's social profile has grown apace with (a) competition between firms and (b) functional equivalence between products. It has become increasingly difficult to differentiate on the basis of functional features. This is why corporate policy is today often situated as social policy; brand differentiation is keyed to supporting some social issue attuned with the firm's core values, relevant to the firm's core business and of demonstrated importance to core consumers (Hulberg, 2006). That is why oil companies are 'going green' and clothing manufacturers are vocal in opposing child labour. This highlights corporate commitment to social responsibility, a situation likely to deepen given popular perceptions of causes for the current economic crisis in capitalism. That should resonate with strategic managers in PSB which is rooted in the 'social responsibility' model (Christians et al., 2009; McQuail, 2005). Social policy needs to be at the heart of PSB brand policy to ensure its contemporary legitimacy and secure developmental support.

It is not enough to make claims – policies must be operationalised to secure credibility. Research from Borgerson et al. (2006: 172) concluded that 'a corporate identity aspiring to ethical characteristics or a socially responsible image cannot simply be an isolated slogan, a collection of phrases; rather, such underlying identities ... require tangibility, visibility, and consistency'.

The core value of PSB brand identity as being about social responsibility is only meaningful when it is evident in practical ways. Too often claims from PSB firms mainly support the *institution as an organisation* rather than the *institution as an orientation*. Wilmott (2003: 362–3) reviewed the literature and concluded:

> Study after study has shown that being a good corporate citizen really does translate into longer term business success ... via the mechanism of branding. This is where the consumer research is so illuminating because it shows that people are increasingly judging companies on their values and wider roles and behaviour in society. In this way, brands – and brand equity – increasingly incorporate a feeling, a sense, of how in touch with the world a company or product is.

This is an issue of contemporary relevance for PSB companies, indeed for the enterprise as a mission. It is vital that PSB discovers, develops and deploys the core values that define social responsibility in media today. PSB is in danger of sacrificing its social responsibility values when strategic managers embrace market-based and competition-oriented perspectives to such a degree that they 'forget' the legacy they are entrusted to nurture. Each identity claim must be rooted in core values, must be manifestly true, and every promise must be fulfilled in practical ways to an evident extent. This is where strategic brand management can be of essential value for renewal. Renewing PSB is not even mainly about mastering new media technologies; it is about revitalising core values for redeveloping social legitimacy.

This is not to say that every value that grounded PSB historically ought to be 'salvaged'. Each company must distinguish between core values that are peripheral and those that are fundamental (de Chernatony and Cottam, 2008). There are ethical values that root a company in the present by binding it with its past and which prescribe a path for its future. Other values are no longer socially relevant. The trick is in knowing which is what. It has long been clear that the pedagogical approach to the enlightenment mission is not a value worth keeping, but that is not to say the enlightenment mission is irrelevant. The challenge here lies in developing a 'newly enlightened mission' (Lowe and Bardoel, 2007). Getting the balance right is crucial. A PSB company that cannot synchronise with its evolving environment will stagnate; a company that cannot consolidate its core will collapse.

A company with heritage value is a recognised leader and its heritage is jealously guarded. Examples include Rolex in watches, Rolls-Royce in automobiles and Coca-Cola in soft drinks. It seems odd that in broadcasting PSB companies have been fleeing their heritage as though such is a liability. In their efforts to become something else they threaten to be nothing special. This despite the fact that PSB companies have been among the most successful online Web services in part because the brand heritage has facilitated successful extensions. PSB strategic managers should reconsider. Their heritage brand can be quite valuable. That is why 'brand stewardship' is an aspect of excellence in 'management innovation' (Hamel, 2007) for PSB.

One needs to distinguish between history and heritage. History is about explaining the past; heritage is about clarifying contemporary relevance. Many companies have a history but only a comparative handful have heritage. This is not to be confused with a 'retro brand' where the company is associated with an era infused with nostalgia. A heritage brand can be timeless because it is based on 'traditions [that] have a salience for the present; value is still being invested in the brand as well as extracted from it. Heritage brands are distinct in that they are about both history and history in the making' (Urde et al., 2007: 7).

PSB companies trying to distance themselves from their heritage obviously do not value it or, at least, have managers who fail to recognise heritage as an asset. That is a mistake because heritage adds depth to value propositions and authenticity to identity claims, is fundamental to brand equity, and is the irreplaceable platform for strengthening relationships with varied stakeholders. Heritage cannot be imitated.

Urde et al. (2007) propose a model for evaluating the firm's 'Heritage Quotient'. Three dimensions are especially pertinent for PSB. First, the company has a track record which means it has lived up to espoused values over time. That accounts for credibility and trust. Second, the company has not simply existed but has demonstrated consistency in living up to its core values. That accounts for assurance of quality standards. Third, the company bases everything in strategy and product development on its core values. That accounts for synergy and consistency. Notice that this is a customer-centric exercise; analysis is about discovering how to best guarantee continuation and improvement in delivering value.

Although many PSB companies have the potential to be heritage brands, most have a history of insularity. Far from being customer-centric, they are product-centric at best and self-centric at worst (Lowe, 2008). The focusing lens for strategic brand management is clarifying what customers value and understanding why. Some aspects of PSB history may even be dysfunctional, for example the pedagogic mission as traditionally practised. The objective of analysis is not simply to identify heritage but to establish the basis of that as differentiation that can be translated meaningfully in the marketplace.

Associations are many and deeply established for heritage brands. Investigation of PSB heritage needs to focus on distinctiveness as the key issue for corporate policy and PSB governance. What is meaningful in the differentiation, and how has PSB created that? What is both meaningful and unique in the functional, experiential and symbolic benefits one gets from consuming PSB? What associations matter most, to whom and why? Where are the associations problematic and how did they become part of the heritage? What needs to be not only continued but also developed as the core values that are essential assets? What needs to be ended or evolved? The work of PSB strategic management is about brand *stewardship*.

The principles that define brand stewardship (Urde et al., 2007) are highly pertinent to senior management teams in the PSB firm:

- You know the brand is 'bigger' than you.
- You know you are a link in a long chain.
- You would like to leave an 'even stronger brand after you'.
- You take a long-term perspective with retrospective knowledge.
- You treat what has been done before with respect.
- You understand and focus on the core values and their link to heritage.
- You recognise the value and importance of symbols and symbolic actions.
- You know when to accommodate change that involves the heritage.
- You are prepared to say 'no' to preserve heritage and reputation.

All of this is distilled into four practical duties that are core competencies in brand stewardship for a heritage company. The management team

- Maintains a strong, consistent sense of personal responsibility for the brand heritage and its future.
- Is actively engaged in guaranteeing long-term continuity while at the same time adding something that is relevant for improvement of the brand.
- Takes seriously their obligation to safeguard the brand in light of heritage and for heritage.
- Is committed to adaptability without sacrificing the underlying values that give the brand its deep meaning.

At issue, then, is how media firms can guarantee the preservation of integrity. This is the greatest asset any media firm possesses because its future success depends, above all, on the trust of audiences (Tungate, 2004). That is why a company's heritage transcends its history. Especially in the context of PSB this is about continuing to build the enterprise as a 'citizen brand' which is essential to every corporate brand premised on public trust. Wilmott (2003: 369) found that being a citizen brand 'means understanding society and the problems and issues that are engaging people ... It is about being outward looking, not inward looking; it is about actively participating in society rather than passively ignoring it. *It is about putting society at the heart of the company*' (emphasis added).

Discussion

Brand loyalty is a constant challenge when media competition is strong, functional equivalence is high, and consumer preferences are not static. As new platforms transform the ways people use media, strategic brand management is essentially about maintaining core values while evolving operation and orientation. The key is to continually strengthen meaningful differentiation. That is the standard and rule of measure for every strategic decision, especially when the company is a heritage brand.

Strategic brand management is a tough balancing act which includes the entire organisation – not only the marketing department. Two fundamental adjustments are required. First, brands must be understood as an asset and branding as investment. Broadcasters have been slow to realise this (Chan-Olmsted, 2006) and that is crucial for nurturing heritage brands. Strategic decision-making needs to be aligned with the legacy of the brand. Second, although the brand cannot be fully controlled it can be developed systematically. The citizen branding paradigm (Gobé, 2002) illustrates why most successful brands emphasise people. To make a brand identity claim real in daily operations and in the lives of consumers, three elements must be balanced: personnel, products and public image. Meaningful differentiation comes in their interaction. Critical consumers will call the bluff whenever espoused promises are not manifest in daily operations.

Although positioning PSB in the marketplace is vital today, managers must be careful not to take market logic for granted or seek to position themselves exclusively in competitive terms. That risks losing core values that differentiate PSB as a vital constituent of the public sphere: non-commercialism, market independence and social responsibility for socio-cultural needs. The public service heritage provides a unique opportunity for harnessing strategic brand management to align corporate development with growing trends in preferences for companies committed to social responsibility. PSB's position as a civil society organisation independent of both the state and the marketplace is the most significant of PSB brand assets.

Understanding the heritage value of PSB is critical to identifying the role these companies should have in the emerging media marketplace. It is unlikely to be the same role in all respects that was developed historically. Beyond the separate programmes, varied genres and respective platforms, only the public service ethos can provide the crucial insights needed to define a special purpose and logic that is fully contemporary but of heritage mission value. This is the core task for PSB brand differentiation. Successful branding is not about 'making it up'; it is about making it work.

PSB faces unique challenges when trying to achieve the goals in this chapter. Such companies are mandated by charters (BBC) or law (YLE). Their purpose is externally defined to a great extent. Moreover, PSB has indirect benefits that are difficult to realise in branding because they are intangibles. Finally, PSB in most countries is building brand hierarchies on a scale that will require skill and talent in strategic brand management at the highest level. Some are engaging in commercial endeavours which add another layer of complexity. Handling the PSB portfolio as a heritage brand with consistency, clarity and credibility is a tremendous challenge but a crucial frontier for PSB management development and innovation.

The dilemmas for PSB in today's environment are substantial. Developing the strategic brand management capability, with its emphasis on understanding and nurturing PSB as a heritage brand from a customer-centric perspective, offers some solutions. There are opportunities for nurturing PSB distinctiveness as a core asset in the continual project of brand differentiation that guarantees a standard of quality and builds loyalty relations. In consideration of the fact that consumers are increasingly concerned about corporate social responsibility, stewardship of PSB heritage brands is an essential aspect.

References

Aaker, D. A. (1996) *Building Strong Brands* (New York: The Free Press).

Aaker, D. A. and E. Joachimsthaler (2000) *Brand Leadership* (New York: The Free Press).

Albarran, A. and A. Arrese (2003) *Time and Media Markets* (Mahwah, NJ: Lawrence Erlbaum Associates).

Balmer, J. T. (1994) 'The BBC's Corporate Identity: Myth, Paradox and Reality', *Journal of General Management*, 19(3): 33–49.

Balmer, J. T. (2006) 'Corporate Brand Cultures and Communities', in J. E. Schroeder and M. Salzer-Mörling (eds), *Brand Culture* (New York: Routledge), pp. 34–49.

Banerjee, S. (2008) 'Strategic Brand-Culture Fit: a Conceptual Framework for Brand Management', *Journal of Brand Management*, 15(5): 312–21.

BBC (1992) *Extending Choice: the BBC's Role in the New Broadcasting Age* (London: BBC).

BBC Brand Manual (Not dated) Available at http://www.bbc.co.uk/branding/manual/index.shtml, accessed 9 May 2009.

Bennett, R. and S. Rundle-Thiele (2005) 'The Brand Loyalty Life Cycle: Implications for Marketers', *Journal of Brand Management*, 12(4): 250–63.

Bergvall, S. (2006) 'Brand Ecosystems: Multilevel Brand Interaction', in J. E. Schroeder and M. Salzer-Mörling (eds), *Brand Culture* (New York: Routledge), pp. 186–97.

Blumenthal, H. J. and O. R. Goodenough (2006) *This Business of Television* (3rd edn) (New York: Billboard Books).

Borgerson, J. L., M. Escudero Magnusson and F. Magnusson (2006) 'Branding Ethics', in J. E. Schroeder and M. Salzer-Mörling (eds), *Brand Culture* (New York: Routledge), pp. 171–85.

Brown, S. (2006) 'Ambi-brand Culture: On a Wing and a Swear with Ryanair', in J. E. Schroeder and M. Salzer-Mörling (eds), *Brand Culture* (New York: Routledge), pp. 50–66.

Burmann, C., K. Schaefer and P. Maloney (2008) 'Industry Image: Its Impact on the Brand Image of Potential Employees', *Journal of Brand Management*, 15(3): 157–76.

Chan-Olmsted, S. (2006) *Competitive Strategy for Media Firms: Strategic and Brand Management in Changing Media Markets* (Mahwah, NJ: Lawrence Erlbaum Associates).

Christians, C. G., T. L. Glaser, D. McQuail, K. Nordenstreng and R. A. White (2009) *Normative Theories of the Media: Journalism in Democratic Societies* (Urbana and Chicago: University of Illinois Press).

Cunningham, S. and G. Turner (eds) (2006) *The Media and Communications in Australia* (2nd edn) (Crow's Nest, Australia: Allen & Unwin).

de Chernatony, L. (2001) *From Brand Vision to Brand Evaluation* (Oxford: Butterworth Heinemann).

de Chernatony, L. and S. Cottam (2008) 'Interactions Between Organisational Cultures and Corporate Brands', *Journal of Product & Brand Management*, 17(1): 13–24.
Dowling, G. (2000) *Creating Corporate Reputations: Identity, Image and Performance* (Oxford: Oxford University Press).
Drinkwater, P. and M. Uncles (2007) 'The Impact of Program Brands on Consumer Evaluations of Television and Radio Broadcaster Brands', *Journal of Product & Brand Management*, 16(3): 178–87.
EBU (2002) *Media with a Purpose: Public Service Broadcasting in the Digital Era*, The Report of the Digital Strategy Group of the European Broadcasting Union, available at http://www.ebu.ch/CMSimages/en/DSG_final_report_E_tcm 6-5090.pdf, accessed 8 May 2009.
Elliot, R. and A. Davis (2006) 'Symbolic Brands and Authenticity of Identity Performance', in J. E. Schroeder and M. Salzer-Mörling (eds), *Brand Culture* (New York: Routledge), pp. 155–70.
Ellis, J. (2000) *Seeing Things: Television in the Age of Uncertainty* (London: I. B. Tauris).
Farholt Csaba, F. and A. Bengtsson (2006) 'Rethinking Identity in Brand Management', in J. E. Schroeder and M. Salzer-Mörling (eds), *Brand Culture* (New York: Routledge), pp. 118–35.
Gobé, M. (2002) *Citizen Brand: 10 Commandments for Transforming Brand Culture in a Consumer Democracy* (New York: Allworth Press).
Hamel, G. (2007) *The Future of Management* (Boston: Harvard Business School Press).
Hatch, M. J. and J. Rubin (2006) 'The Hermeneutics of Branding', *Journal of Brand Management*, 14(1/2): 40–59.
Hoynes, W. (2003) 'Branding Public Service: the "New PBS" and the Privatization of Public Television', *Television & New Media*, 4(2): 117–30.
Hulberg, J. (2006) 'Integrating Corporate Branding and Sociological Paradigms: a Literature Study', *Journal of Brand Management*, 14(1/2): 60–73.
Humphreys, E. and M. Norbäck (2008) 'Brand Ownership of Public Service', paper for the RIPE@2008 conference, Public Service Media in the 21st Century: Participation, Partnership and Media Development, Mainz, Germany. Available at: http://www.uta.fi/jour/ripe/2008/papers/Humphreys_& _Norback.pdf, accessed 8 May 2009.
Ind, N. (2001) *Living the Brand: How to Transform Every Member of Your Organization into a Brand Ambassador* (London: Kogan Page).
Jääsaari, J. (2004) *Yle yleisön ehdoilla? Tutkimus suomalaisten television nykytilaa ja tulevaisuutta koskevista arvostuksista* (Helsinki: Yleisradio).
Jevons, C., M. Gabbott and L. de Chernatony (2005) 'Customer and Brand Manager Perspectives on Brand Relationships: a Conceptual Framework', *Journal of Product & Brand Management*, 14(5): 300–9.
Kapferer, J. N. (2004) *Strategic Brand Management: Creating and Sustaining Brand Equity Long Term* (London: Kogan Page).
Keller, K. L. (2003) *Strategic Brand Management: Building, Measuring and Managing Brand Equity* (Upper Saddle River, NJ: Prentice Hall).
Keller, K. L. and K. Richey (2006) 'The Importance of Corporate Brand Personality Traits to a Successful 21st Century Business', *Journal of Brand Management*, 14(1/2): 74–81.
Lowe, G. F. (2008) 'Customer Differentiation and Interaction for Innovation: Two CRM Challenges for PSB', *Journal of Media Business Studies*, 5(2): 1–22.
Lowe, G. F. and J. Bardoel (2007) *From Public Service Broadcasting to Public Service Media*, RIPE@2007 (Göteborg, Sweden: NORDICOM).
Lowe, G. F. and G. Daschmann (forthcoming) *The Public in Public Service Media* (Göteborg, Sweden: NORDICOM).

McDowell, W. S. (2006) 'Issues in Marketing and Branding', in A. B. Albarran, S. M. Chan-Olmsted and M. O. Wirth (eds), *Handbook of Media Management and Economics* (Mahwah, NJ: Lawrence Erlbaum Associates), pp. 229–50.

McQuail, D. (2005) *McQuail's Mass Communication Theory* (5th edn) (London: Sage).

Nandan, S. (2005) 'An Exploration of the Brand Identity-Brand Image Linkage: a Communications Perspective', *Journal of Brand Management*, 12(4): 264–70.

Nissen, C. S. (ed.) (2006) *Making a Difference: Public Service Broadcasting in the European Media Landscape* (Eastleigh, UK: John Libbey Publishing).

Norbäck, M. (2008) 'The Making of Public Service in Collaborative Content Production', paper for the RIPE@2008 conference, Public Service Media in the 21st Century: Participation, Partnership and Media Development, Mainz, Germany. Available at: http://www.uta.fi/jour/ripe/2008/papers/Norback.pdf, accessed 8 May 2009.

Ofcom (2006) *Communications – The Next Decade: a Collection of Essays Prepared for the UK Office of Communications*, ed. E. Richards, R. Foster and T. Kiedrowski. Available at: http://www.ofcom.org.uk/research/commsdecade, accessed 28 November 2008.

Ots, M. (ed.) (2008) *Media Brands and Branding* (Jönköping, Sweden: JIBS Research Reports).

Picard, R. G. (2008) 'Foreword', in M. Ots (ed.), *Media Brands and Branding* (Jönköping, Sweden: JIBS Research Reports).

Pickton, D. (2008) 'Editorial: What is a Brand Worth?', *Journal of Brand Management*, 15(3): 155–6.

Pine, B. J. II and J. H. Gilmore (1999) *The Experience Economy: Work is a Theatre and All Want to Buy* (London: Harper & Row).

Ries, A. and J. Trout (1986) *Positioning: the Battle for Your Mind* (New York: McGraw-Hill).

Schroeder, J. E. and M. Salzer-Mörling (eds) (2006) *Brand Culture* (New York: Routledge).

Schultz, M. and M. J. Hatch (2006) 'A Cultural Perspective on Corporate Branding: the Case of LEGO Group', in J. E. Schroeder and M. Salzer-Mörling (eds), *Brand Culture* (New York: Routledge), pp. 15–33.

Siegert, G. (2008) 'Self Promotion: Pole Position in Media Brand Management', in M. Ots (ed.), *Media Brands and Branding* (Jönköping, Sweden: JIBS Research Reports), pp. 11–26.

SMH (2008) 'Apple's Cool Factor Spreads', *The Sidney Morning Herald*, 18 January. Available at http://www.smh.com.au/news/technology/apples-cool-factor-spreads/2008/01/17/1200419953601.html, accessed 29 April 2009.

Snell, S., A. Lahelma and P. Toppinen (2003) *Parempia ohjelmia: TV-ohjelmatestien satoa 2001–2003* (Helsinki: Yleisradio)

Thompson, P. A., J. Williams, E. Mason, L. Beamer, C. Yeardley-Davern, J. Harrison, D. Henson, D. Mead, S. Monteiro and D. Rolland (2005) *Mechanisms for Setting Broadcasting Funding Levels in OECD Countries*. Report for the New Zealand Ministry for Culture and Heritage, December 2005. Available at http://www.mch.govt.nz/publications/broadcast-funding/index.html, accessed 22 December 2008.

Tungate, M. (2004) *Media Monoliths: How Great Media Brands Thrive and Survive* (London: Kogan Page).

Urde, M., S. A. Greyser and J. M. T. Balmer (2007) 'Corporate Brands with a Heritage', *Journal of Brand Management*, 15(1): 4–19.

Willmott, M. (2003) 'Citizen Brands: Corporate Citizenship, Trust and Branding', *Journal of Brand Management*, 10(4/5): 362–9.

Wolinsky, A. (1995) 'Competition in Markets for Credence Goods', *Journal of Institutional and Theoretical Economics*, 151: 117–31.

Part II

11
The BBC and UK Public Service Broadcasting

Jeremy Tunstall

Introduction

With typical national modesty, UK citizens often refer to the BBC as the world's best, or most admired, broadcasting organisation. The BBC played an important role in twentieth-century British history; it became part of national mythology. But BBC 'Public Service Broadcasting' has lacked a single neat definition. UK PSB resembles the British constitution in being unwritten or, more accurately, written in several slightly different laws, documents, reports and decisions.

In the UK, broadcasting policy and PSB have tended to develop incrementally. The only two big policy leaps were the 1955 introduction of a commercial channel to compete with BBC TV; and secondly the 1989 arrival of Rupert Murdoch's Sky television (and its satellite monopoly of 1990).

Broadcasting policy has usually been made, and PSB has been largely redefined, on a loosely bi-partisan (Conservative and Labour) basis. Westminster politics and London-based national newspapers have been major influences across nine decades.

The BBC has attracted the hostility of three long service prime ministers – Winston Churchill (1940–5 and 1951–5), Margaret Thatcher (1979–90) and Tony Blair (1997–2007). Consequently, top BBC executives tend to 'run scared'. But UK prime ministers in general have not devoted much of their time to attacking the BBC or to redefining PSB. Most British broadcasting policy has been amateurish, and many policy initiatives have had unanticipated consequences.

In recent decades there have been three main categories of PSB, each of which has depended on a different form of finance:

- Domestic BBC TV, radio and other services have been funded by a licence fee, paid by households.
- 'Commercial' TV (Channels 3, 4 and 5) and commercial radio have some public service obligations and are financed by advertising.

- The BBC's foreign TV and radio services are funded by an annual government grant. In recent years this has involved not only BBC World but other dedicated television channels such as BBC America, BBC Canada, BBC Arabic and BBC Persian.

All three of these types of PSB require a significant output of news and other serious, artistic, and educational programming. For example, ITV 1 (one of the world's leading commercial TV channels) puts out 1.5 hours of news each weekday evening in the 4.5 hours between 6.00 p.m. and 10.30 p.m. Channel 4, especially, has had major PSB obligations since its launch in 1982. This includes an hour of news each (Monday–Thursday) evening at 7.00–8.00 p.m. Only cable and satellite, funded primarily by subscription, have no significant PSB obligations.

Policies overt, covert and absent

Advocacy of 'Public Service' involves a rhetoric of non-profit public benefit and public enlightenment. But overt policies and goals are not the whole story. Elements of covert policy often accompany the PSB rhetoric. Some advocacy of PSB is really intended to damage the BBC – by, for example, demanding that the BBC avoid popular entertainment (and thus fail in terms of ratings competition). Both Conservative and Labour governments have supported BSkyB and its satellite dominance, because these governments feared electoral opposition and character assassination by the Murdoch national newspapers (notably *The Sun, The Times, News of the World* and *Sunday Times*). Margaret Thatcher in 1990 (her last year as prime minister) deployed a remarkable mix of overt policy and contradictory covert policy. Overt policy consisted of the Broadcasting Act 1990, which, after five years of debate, broadly supported most of the existing arrangements, including a big element of PSB.

However, during 1990 two UK satellite TV operations were engaged in expensive competition. The officially UK licensed company was British Satellite Broadcasting (BSB) which used a high technology system, transmitted expensive programming and was making big losses. Its competitor was Sky Television, launched into the UK market by Rupert Murdoch; Sky was using cheaper technology, was licensed in Luxembourg and thus subject to almost no regulation. On 29 October 1990 Murdoch visited Margaret Thatcher, telling her of a planned merger between Sky and BSB. Mrs Thatcher indicated to Murdoch that she did not object to the merger, which was duly announced four days later, on 2 November. The surviving company (controlled by Murdoch whose News Corporation company was close to bankruptcy at the time) was named British Sky Broadcasting (BSkyB).

This Thatcher–Murdoch agreement was remarkable in several ways. Firstly, Thatcher was ignoring public service and endorsing a 'British' system whose sole initial attraction was Hollywood movies. Secondly, Thatcher was using powers which no British prime minister actually possesses. Thirdly, there were no complaints from the official regulator of UK commercial television. The Broadcasting Act of 1990 in fact turned the old Independent Broadcasting Authority (IBA) into the new Independent Television Commission (ITC). The IBA and ITC had the same chairman (Sir George Russell). Both Sir George Russell and the ITC's then Chief Executive (David Glencross) subsequently told this writer that, on the days in question, they had lacked the necessary legal powers to challenge the Thatcher decision. This bizarre Thatcher mixture of overt and covert policy was seen, and is seen, by PSB advocates as a policy disaster.

However, some other covert (or seldom articulated) policies favour the BBC. This is especially true of BBC national radio, which retains the cream of the *national* FM frequencies, while commercial radio's FM frequencies are predominantly *local* ones. This helps the BBC to use its radio output as a leading example of Public Service. Radio Three is a serious music station. BBC Radio Four is the dominant UK news and talk service, and its early morning news sequence is a leading agenda setter in UK national politics. BBC Radios One and Two are national pop music channels which have some PSB element. BBC Radio Five Live offers sports, news and talk. Meanwhile the BBC is also the UK leader in digital radio, and BBC radio (national and local) had 57 per cent of the total UK radio audience in 2008.

Supposedly 'PSB' policies can be contradictory. The BBC is repeatedly urged to appeal to the mass audience, but when it does compete fiercely (especially in new digital services) it is accused of being unfairly competitive, and is also accused of neglecting public (that is non-profit) service.

In other key policy areas there is no policy at all. Public Service, when supported by the national government, is inevitably also National Service. As elsewhere in Europe, UK PSB supports and finances national programming output and national production. However, British politicians also expect British broadcasting to export successfully; but there is no realistic export policy. The main exporter, the PSB BBC, is successful in prestige exporting, but relatively unsuccessful in commercial earnings. 'BBC Worldwide' and BBC services do reach many foreign markets (with radio, TV, online, DVDs and books) but most of the extra earnings are used up in extra costs. 'Profits' are small.

Like other European national media, British national media (and successive governments) have never developed a Hollywood policy. There has never been a coordinated policy to compete with Hollywood (film and TV) in the UK, in the US, or in the world. This lack of a UK policy to compete with Hollywood goes alongside the fact that American electronic media are exceptionally strong in the UK in several ways:

- BSkyB is an American-owned company which dominates both satellite and cable in the UK. BSkyB, whose main premium (expensive) offerings

continue to be live UK sport and recent Hollywood movies, now has gross revenues higher than those of the BBC.
- Google and other American internet companies are exceptionally strong in the UK, not least in attracting advertising.
- While Hollywood programming has little or no presence in the top hundred UK TV ratings each week, Hollywood products have a big place in daytime TV on the main channels and on at least 200 small-audience digital channels.

In terms of exports to the rest of world, UK producers provide only weak competition for Hollywood. Most British TV series are 'too British' (not least because of PSB and other pressures to appeal to the UK market). This is especially true of super-popular UK soap and semi-soap drama. Most other British series have too low an output (such as six or eight episodes a year) to compete with big Hollywood series which typically produce about twenty-three episodes per year. Britain has been very successful in recent years in selling formats to the US market, but formats earn neither prestige nor much money. In 2007 UK television exports, against imports, generated the negligible surplus of £62 million.

One unintended consequence is that the BBC's and British broadcasting's most prestigious and public service efforts (such as BBC Natural History, Classic Drama, and Documentary) rely on co-production deals with American organisations such as, in recent years, Discovery and WGBH. Even the BBC's large book publishing imprint has been sold to (German/US) Random House.

Lack of national media policy was especially noticeable in the 2008 acquisition of Reuters by Thomson of Canada. Since the late nineteenth century, Reuters was Britain's leading news provider on the world scene. Since the 1920s the BBC has been a major customer for Reuters news. Both Reuters TV news and the international TV news operations of Associated Press (of the US) are based in London. Most of the foreign video news on all of the world's TV screens comes through these two London-based operations. Yet the UK government made no attempt to stop – or even delay – the sale of Reuters to Thomson. It's hard to imagine the French, German or Spanish governments allowing AFP or DPA or EFE to be sold to a North American company.

Continuing tradition and continuous redefinition of UK PSB

British Public Service Broadcasting obviously has many characteristics found in other PSB systems across Europe. However, British PSB has some distinctive features, found neither in most broadcasting systems in East or West Europe nor in North or South America.

- The UK shares the main language of the United States, making the UK a 'natural' market for US programming. The BBC has since the 1930s

welcomed some imports (such as radio and TV comedies) from the US, but has always tried to limit direct imports and to dress imported formats in distinctively British clothing.
- The BBC is unusual in its unbroken public service tradition, stretching back to 1922, and strengthened (not weakened or destroyed) during 1939–45.
- The BBC has long tried to incorporate itself, in effect, into Britain's unwritten constitution. For example, the BBC enthusiastically celebrates all major British national anniversaries; a classic example was the 200th anniversary in 2009 of Charles Darwin's birth in 1809; the BBC produced several short TV and radio series to celebrate the 'Greatest Scientist' in British history. The BBC also seeks to weave its own history and tradition into UK national history and tradition.
- Both the BBC and the commercial channels have remained remarkably unpoliticised by European standards. Although many British politicians have tried to steer the BBC in particular partisan directions, most such attempts have been somewhat hesitant and of limited success. Both the relevant politicians, and the relevant senior civil servants, recognise that the BBC usually scores more highly in opinion polls than does the incumbent government.
- The BBC has been somewhat unusual (in Europe) in being dependent solely on the licence fee for its domestic services. This means that BBC finances are a high-wire act with no advertising safety net. However, the BBC is distanced from the commercial pressures of advertising finance; the BBC also does not compete for advertising revenue against either the press or commercial broadcasting. This separation between different forms of revenue has often meant that BBC licence fee finance, and commercial broadcasting advertising finance, have moved in different directions. This tends to result in frequent 'funding crises' as BBC finance either falls behind, or leaps ahead of, commercial broadcasting finance.

UK PSB is gradually and continuously redefined. John Reith, who managed the BBC for its first sixteen years (1922–38), welcomed 'entertainment', including popular music, as well as 'education' and 'information'. This entertainment ingredient has, of course, changed radically over the decades. In its early years the BBC did not collect its own news and did not employ reporters; today the BBC operates a large news and newsgathering operation, both in the UK and across the world.

Definitions of UK PSB have usually been expressed in rather general terms, and even these general terms require regular updating. Des Freedman (2008: 147–8) suggests a five-item definition. UK Public Service broadcasting, he states:

- Rejects market definitions.
- Assumes the audience to consist of rational citizens with a broad range of interests.

- Attempts to foster a shared public life.
- Supports social amelioration.
- Reflects and cements public opinion.

The Office of Communications, the newly established super-regulator of British media and telecommunications, produced its own list of 'Public Service Broadcasting purposes' (Ofcom, 2008b: 13):

- 'Informing our understanding of the world'.
- 'Stimulating knowledge and learning'.
- 'Reflecting UK cultural identity'.
- 'Representing diversity and alternative viewpoints'.

Not satisfied merely by four 'Purposes', Ofcom also listed six 'Characteristics' of public service broadcasting:

- 'High quality' – well funded and well produced.
- 'Original' new UK content.
- 'Innovative' – breaking new ideas or reinventing exciting approaches.
- 'Challenging' – making viewers think.
- 'Engaging' – remaining accessible and attractive to viewers.
- 'Widely available' – a large majority of citizens need to be able to watch publicly funded content.

Ofcom, in September 2008, was most worried not about BBC Public Service but about the lesser PSB elements within the main commercial Channels, 3, 4 and Five. Most at risk were ITV Regional News (because it requires so many separate editions), UK Children's programming, Education (non-school) programming, Religion and UK Comedy.

However autumn 2008 marked the escalation of the international financial crisis and big falls in advertising expenditure. By early 2009, it was already clear that almost all 2008 predictions were too optimistic. As with other 'funding crises' in the past, the two systems (BBC and commercial) were diverging rapidly in terms of financial prospects. ITV, and Channels 4 and Five, all faced massive drops in funding and new difficulties in meeting their (modest) public service obligations. By contrast, the BBC had been awarded a relatively favourable (and slowly rising) licence fee from 2007 to April 2013.

The BBC quickly offered yet another redefinition of Public Service. The BBC in early 2009 announced that it might be willing to help ITV (its traditional competitor) by some arrangement for providing ITV with BBC regional news.

470 television channels and comeback auntie BBC

The BBC was in some respects in a stronger position in 2009–10 than it had been a decade earlier. But how was this possible, given that already by 2007 the UK had 470 separate TV channels, which together transmitted 2.1 million hours in the year, or 5750 hours of TV per day? (Ofcom, 2008a). About 1 million hours were shopping channels, 'adult' sex channels, and other offerings outside conventional definitions of television. Of the remaining 1.1 million hours, only about 10 per cent was first run, new, programming. Digital satellite and cable households are offered big numbers of niche channels – for example, in large and small Asian languages. Many are niche American channels which attract very small UK audiences.

BSkyB is the main supplier of both sports and movie channels; the same or similar offerings are supplied by Virgin Cable. Both BSkyB and Virgin have focused on raising revenue per household and not on providing fresh programming across the genre range. By 2009 Sky and cable digital services were in about half of UK households. Sky offers three general entertainment channels, four Sky sports channels and eleven Sky movie channels. But only two of Sky's own channels (Sky One and Sky Sports One) have more than a 1 per cent share of the total UK TV audience. Sky News, however, offers effective competition to the BBC News channel.

The BBC's entire family of TV channels (including ten BBC half-owned 'UKTV' channels) had about one third of the total UK TV audience in 2009. BBC Radio (both national and local) had over half of the UK radio audience. The BBC had over 40 per cent of combined TV and radio audience hours. BBC Online has been another strength, and it competes successfully at home and abroad against several strong online operations run by UK national newspapers.

In the multi-platform era the BBC is the UK's multi-platform leader. The BBC vigorously cross-promotes between its TV, radio, online, DVDs and books. The other PSB/commercial operators (ITV 1, Channel 4 and Five) have had a much harder time – not least because they lack the BBC's cross-promotional and other advantages.

From amateur to professional regulation of UK broadcasting

Until the 1990s, British broadcasting policy and regulation were conducted in a distinctly amateur style. But during the 1990s this changed; and after the year 2000 communications policy-making and regulation followed a much more professional pattern.

Until the 1990s five separate categories of people contributed to amateurish UK broadcasting policy-making and regulation. Firstly, over some four decades, four major broadcasting committees produced reports, each of which led to one additional television channel:

- The Beveridge Committee (1945–51) led to ITV (1955).
- The Pilkington Committee (1960–2) led to BBC Two (1964).

- The Annan Committee (1974–77) led to Channel 4 (1982).
- The Peacock Committee (1985–6) led to Channel Five (1996).

None of these four chairmen (Beveridge, Pilkington, Annan or Peacock) had any significant previous professional connection with broadcasting. The committee memberships consisted of 'The Great and the Good' – typically public people in late middle age who could devote one day a week to a committee, serviced by a very small secretariat. All four committees favoured both more competition and arrangements to maintain a substantial element of serious programming.

Secondly, the BBC was presided over by a Board of Governors who legally controlled the entire BBC. These Governors were expected to devote one day per week to BBC business. The biggest categories of BBC Governors over a 51-year period (1927–78) were former politicians, finance and business people, ex-diplomats, social services people, trade union leaders, authors, journalists and retired military men (Briggs, 1979: 32–4). While the Governors presided, the BBC was operated on a day-to-day basis by a Board of Management – which included executives in charge of TV, radio, engineering, finance, and so on. The BBC was effectively run by the chief executive (the Director General) and the chairman of the Governors. But all major decisions had to be approved by the full 'amateur' Board of Governors.

Thirdly, commercial television and radio were regulated by separate regulators. Over a 50-year period (1954–2004) there were in fact five regulatory bodies which covered one or more of commercial TV, commercial radio and cable. The final phase of these regulators (1991–2004) had the Independent Television Commission (ITC) covering TV and cable, while the Radio Authority (RA) regulated commercial radio. All five of these regulatory bodies had had Members (loosely modelled on the BBC Governors) and a Chief Executive or Director General.

Fourthly, the government ministers, responsible for broadcasting policy, were equally amateur and largely lacking in professional knowledge or expertise. For example, Margaret Thatcher's three administrations (1979–90) adopted a deregulatory approach to broadcasting. But in a period of fifteen years (1979–94) a succession of eight senior ministers held the top communications policy post. Only one of these senior ministers – William Whitelaw (1979–83) – left a significant legacy, namely Channel 4 (launched in 1982).

Fifthly, the relevant civil servants also tended to move quickly in and out of broadcasting policy. It was possible in the 1980s to find relevant senior civil servants who were supposedly introducing Washington-style or European-style broadcasting deregulation, but who knew very little indeed about US or European media policy and regulation.

The year 1990 marked a turning point in UK broadcasting policy. This was the year of Margaret Thatcher's departure, and of her awarding BSkyB its monopoly of UK satellite television. 1990 also saw the passage of the

Broadcasting Act which led to a radical rearrangement of the fifteen companies of the ITV system. The events of 1990 were seen inside the BBC as a warning that the BBC should become more expert and more professional in legislative and regulatory lobbying. The then BBC Director General, Michael Checkland, established a sizeable BBC 'Policy and Planning Unit' in 1987 and this was expanded by John Birt, who served as BBC Director General from December 1992 to January 2000. The Policy and Planning Unit was run by Patricia Hodgson and described by John Birt as 'the most powerful capability of its kind anywhere in European broadcasting' (Birt, 2002: 333–4). Patricia Hodgson subsequently described to this writer how she and John Birt developed an elite policy group within BBC central management.

The professionalising of broadcasting policy in the 1990s was also advanced by some 150 London law firms specialising in UK, US and European media law. Management consultancy firms (especially McKinsey) had been used by the BBC since the 1960s. But in the 1990s John Birt's BBC was reported by various observers to be spending about £25 million per year on consultants – with McKinsey, PA Consulting, Ernst and Young and KPMG Management the most heavily used. When Greg Dyke became BBC Director General in early 2000, one of his first decisions (he told me in an interview) was to end the use of consultants. But by 2000 the BBC itself employed a substantial team of young policy professionals and ex-consultants. John Birt himself had an (Oxford) engineering degree and the 'Birt Revolution' involved emphasis on several types of quantification. Hidden costs were made transparent; audience research (which had existed since the 1930s) became much more central to BBC commissioning and scheduling of programmes; and PSB itself began to be quantified and measured in new ways.

BBC policy lobbying is now in constant and active operation. Licence fee negotiations have always been salient, especially in times of high inflation. The BBC operates under a special legal instrument, a Royal Charter, which guarantees the BBC's continued existence for another ten or fifteen years; the most recent Charter renewals were in 1981, 1996 and 2006. In recent years the BBC has had to negotiate in considerable detail and depth over new digital channels, and new digital technologies and services. The legal and regulatory position of independent production companies is another major policy focus. Meanwhile the House of Commons, the House of Lords and the European Commission are increasingly active in broadcasting policy. The BBC now faces in Ofcom a new broad spectrum regulator whose regulatory remit includes not only commercial broadcasting, but also the BBC.

The BBC Governors have been replaced by BBC 'Trustees', all of whom supposedly devote two days (no longer one day) a week to the BBC. The Chairman of the Trustees is now full-time; and the Trustees (unlike the Governors) now have their own offices, a small secretariat and a small research budget. Paradoxically the increased policy professionalisation within the

BBC may have increased the power of the Trustees. The more quantified approach to policy and planning means that the Trustees can more clearly specify how much money they want spent, especially on particular new digital services.

However, Ofcom (unlike the old commercial TV and radio regulators) has several new powers over the BBC. Ofcom in practice defines and measures PSB; it has a big role in the growth of digital services, super-fast broadband, the allocation of frequencies and so on. Ofcom can fine the BBC for breaking agreed programming guidelines; other 'compliance matters' in which the BBC must follow the Ofcom lead include an array of significant issues from equal opportunities to subtitling of programmes.

Advertising-financed public broadcasters and their digital families

UK 'PSB' officially encompasses not only all BBC TV and radio output but also the main advertising-financed TV channels – ITV 1, and Channels 4 and Five. Ofcom now officially monitors their output in general and their more serious output in particular. Of these channels, only Channel Five depends very heavily on imported entertainment – such as Hollywood crime series and Australian soaps. Since its 1987 launch the channel has never obtained more than 6 per cent of the UK TV audience. Channel 4 is much more obviously a public broadcaster. In its first twenty-seven years (1982–2009) it averaged a national audience share of about 9 per cent. It carried substantial quantities of news, current affairs, documentary, comedy and original drama. It also relied on a lot of factual entertainment formats and imported Hollywood comedies. Channel 4 has always had a younger audience than its competitor, BBC Two.

The old ITV was more effective than the old BBC at public relations. Several recent books present a somewhat nostalgic account of the old ITV in general and of Granada (the Manchester company) in particular (Finch, 2003; Fitzwalter, 2008; Goddard et al., 2007). In the 1970s ITV continued to have a 50 per cent audience share (against just two BBC channels) despite some high-profile ITV public service programming. ITV was ahead of the BBC in launching its half hour *News at Ten* in 1967. ITV also had two long-running half-hour current affairs shows – *World in Action* (Granada, 1963–98) and *This Week* (London, 1956–92). Granada was especially proud of two (book-based) drama series – *Brideshead Revisited* (1981, eleven hours) and *The Jewel in the Crown* (1984, fifteen hours). These high-profile short prestige series certainly pleased the (amateur) regulators of the day. But ITV was mainly an entertainment producer. ITV's (and Granada's) most successful programme was *Coronation Street* which transmitted its 7,000th half-hour episode in 2008.

The salience of new British programming (as opposed to Hollywood imports) on ITV owed much to the strength of the broadcasting trade unions, which – until 1990 – used their industrial strength to insist on British programming and wasteful over-manning.

ITV seemed after 1990 to have considerable difficulty in sustaining its significant public service requirements. ITV was required to carry a high proportion of original programming from its in-house production and from independent producers. ITV must be a major broadcaster of news and current affairs; it must produce significant quantities of programming in studios outside London; and it must transmit special programming specifically for the English regions and for Scotland, Wales and Northern Ireland.

ITV's UK audience share fell from about 50 per cent in the 1970s to 19.1 per cent in 2008. But this was only the ITV 1 channel. After 2000, ITV successfully built up its ITV 2, ITV 3 and ITV 4 digital channels; by 2009 these three additional ITV digital channels were together getting about 5 per cent of the total UK audience. Channel 4 was also doing well with its own digital family members – E 4 and More 4. Nevertheless, during 2007 ITV, and Channels 4 and Five, were all complaining that they would be unable both to fulfil existing PSB requirements and to avoid bankruptcy. There were large-scale sackings at all three channels even before the escalation of the international financial crisis in autumn 2008. In March 2009 the German Bertelsmann company (the majority owner of the UK's Channel Five), announced that it did not believe a lone Channel Five to be viable and that a merger with ITV would be most suitable.

The BBC's energetic march into the digital world has both strengthened and weakened its public service viability. The BBC outplayed ITV in the emergence of a terrestrial digital service, namely Freeview (followed by FreeSat). Freeview plus BSkyB and Virgin Cable ensured that, by early 2008, 87 per cent of UK households had some digital channels and, in particular, most or all of the BBC's new digital channels. While this seemed like a triumph for BBC Public Service, the new BBC digital channels carried more PSB than entertainment. The BBC had two children's channels (CBBC and CBeebies), a Parliament channel, and a BBC 24-hour TV news channel. Even BBC Three and BBC Four were defined as public service channels which attracted low audience shares. However the BBC also owned 50 per cent of ten UKTV channels, which were relaunched and rebranded; UKTV Gold 2 relaunched as Dave, UK History became Yesterday, and People became Blighty. In 2008 BBC television had an audience share of 36 per cent, if one combined the two main BBC channels, the small digital channels and the ten UKTV offerings.

Eager debate seems certain to continue as to what PSB means and also as to what PSB quotas should be required of particular channels. The BBC is certain to argue that a PSB element should be retained within the advertising funded channels. Spokespersons at ITV, and Channels 4 and Five, will argue that they have too many PSB quota requirements, while the BBC should do more Public Service Broadcasting and less popular entertainment.

Ofcom: policy and regulation professionalised, PSB quantified

The Office of Communications, which was the merger of five regulatory agencies, came into existence in December 2003 and during 2004 employed over 900 people. At any point in time Ofcom seemed to be working on a bewildering range of technically and financially complex telecommunications issues, and on a wide range of politically and financially complex cable, satellite, television, radio, advertising, internet and mobile communications issues.

Traditional media lobbies have become more vigorous. PACT (the independent producers' organisation) lobbied with great success and in the Communications Act of 2003 obtained the legal right for independent producers to hold on to the copyright of their own commissioned programming. This decision helped the UK independent sector to grow and to achieve annual revenues of over £2 billion in each of 2007 and 2008.

The emergence of the professional media regulator was exemplified by the careers of four men – Stephen Carter, James Purnell, Ed Richards and Andy Burnham – all born between 1964 and 1970. All four had meteoric careers. All four had connections with Labour prime ministers (Tony Blair and Gordon Brown). Two became Cabinet ministers in charge of Culture, Media and Sport, while still aged under 40. Three had very youthful experience in media policy. Two worked at the BBC. Two of the four were the first two chief executives of Ofcom. All four of these youngish men had economics and/or financial expertise.

All four of these professional media regulators seemed to believe in PSB. They all also believed that PSB could be split into defined components and then measured. Ofcom's numerous reports are packed with data and statistical tables. In the Ofcom era, PSB has been quantified.

References

Annan, Lord (1977) *Report of the Committee on the Future of Broadcasting* (London: HMSO), Cmnd 6753.

Barnett, S. and A. Curry (1994) *The Battle for the BBC* (London: Aurum Press).

BBC Trust (2008) *On-Screen and On-Air Talent* (London: BBC).

Beveridge, W. (1951) *The Report of the Broadcasting Committee, 1949* (London: HMSO), Cmnd 8116.

Birt, J. (2002) *The Harder Path: the Autobiography* (London: Time Warner).

Bonner, P. and L. Aston (1998) *Independent Television in Britain*, volume V: *ITV and the IBA: the Old Relationship Changes* (Basingstoke: Macmillan).

Born, G. (2004) *Uncertain Vision: Birt, Dyke and the Reinvention of the BBC* (London: Secker & Warburg).

Briggs, A. (1961, 1965, 1970, 1979, 1995) *The History of Broadcasting in the United Kingdom*, 5 vols. (Oxford: Oxford University Press).

Briggs, A. (1979) *Governing the BBC* (London: BBC).

Briggs, A. (1985) *The BBC: the First Fifty Years* (Oxford: Oxford University Press).

Brown, M. (2007) *A Licence to be Different: the Story of Channel 4* (London: British Film Institute).

Congdon, R. et al. (1992) *Paying for Broadcasting: the Handbook* (London: Routledge).
Curran, C. (1979) *A Seamless Robe, Broadcasting: Philosophy and Practice* (London: Collins).
DCMS (2006) *A Public Service For All* (London: Department for Culture, Media and Sport).
Dyke, G. (2004) *Inside Story* (London: HarperCollins).
Finch, J. (2003) (ed.) *Granada Television: the First Generation* (Manchester: Manchester University Press).
Fitzwalter, R. (2008) *The Dream that Died: the Rise and Fall of ITV* (Leicester, UK: Matador).
Freedman, D. (2008) *The Politics of Media Policy* (Cambridge: Polity Press).
Goddard, P., J. Corner and K. Richardson (2007) *Public Issue Television: World in Action, 1963–98* (Manchester: Manchester University Press).
Hendy, D. (2007) *Life on Air: a History of Radio Four* (Oxford: Oxford University Press).
Hill, C. (1964) *Both Sides of the Hill* (London: Heinemann).
Hill, C. (1974) *Behind the Screen* (London: Sidgwick & Jackson).
Horrie, C. and S. Clare (1994) *Fuzzy Monsters: Fear and Loathing at the BBC* (London: Heinemann).
Hussey, M. (2001) *Chance Governs All* (Basingstoke: Macmillan).
ITC (2002) *UK Programme Supply Review* (London: Independent Television Commission).
Leapman, M. (1987) *The Last Days of the Beeb* (London: Coronet).
Milne, A. (1969) *DG: the Memoirs of a British Broadcaster* (London: Coronet).
Ofcom (2008a) *Communications Market Report* (London: Office of Communications).
Ofcom (2008b) *Ofcom's Second Public Service Broadcasting Review* (London: Office of Communications).
Ofcom (2009) *Ofcom's Second Public Service Broadcasting Review: Putting Viewers First* (London: Office of Communications).
Palmer, M. and J. Tunstall (1990) *Liberating Communications: Policymaking in France and Britain* (Oxford: Blackwell).
Peacock, A. (1986) *Report of the Committee on Financing the BBC* (London: HMSO).
Pilkington, H. (1962) *Report of the Committee on Broadcasting 1960* (London: HMSO), Cmnd 1753.
Trethowan, I. (1984) *Split Screen* (London: Hamish Hamilton).
Tracey, M. (1998) *The Decline and Fall of Public Service Broadcasting* (Oxford: Oxford University Press).
Tunstall, J. (1983) *The Media in Britain* (London: Constable).
Tunstall, J. (1986) *Communications Deregulation: the Unleashing of America's Communications Industry* (Oxford: Blackwell).
Tunstall, J. (1993) *Television Producers* (London: Routledge).
Tunstall, J. (2007) *The Media Were American: US Mass Media in Decline* (New York: Oxford University Press).
Tunstall, J. and D. Machin (1999) *The Anglo-American Media Connection* (Oxford: Oxford University Press).
Wyatt, W. (2003) *The Fun Factory: a Life in the BBC* (London: Aurum Press).

12
France: Presidential Assault on the Public Service

Raymond Kuhn

Introduction

In the French media landscape the values associated with the mission of public service are most strongly embedded in the public broadcasting companies. This does not mean that elements of public service are wholly absent in other media sectors and outlets. In the press, for instance, it could be argued that in the breadth and quality of its news coverage a high quality newspaper title such as *Le Monde* aspires to perform a public service information function for its readers. More recently, in the online sector one can find websites, such as those of central and local government, that provide essential non-partisan public information to citizens on a range of issues from social welfare to cultural activities.

The picture is further complicated in the French context because the claims of public broadcasters to public service status have to be qualified by their long historical association with the political executive that has often negatively impacted on their capacity to provide independent and balanced news. Moreover, since the opening-up of the broadcasting system to commercial providers in the 1980s, public broadcasters have had to compete in a market for audiences and funding. In short, elements of public service can be found across different media sectors and outlets, while in broadcasting the notion of public service applies solely to the public providers.

With these caveats in mind, in examining the state of public service media in contemporary France this chapter focuses on those outlets that in terms of their formal mission have the strongest prima facie claim to public service status – the public broadcasters Radio France and France Télévisions. The chapter is divided into four parts. The first provides a broad introduction to the evolution of French public broadcasting since the end of World War Two in terms of three consecutive phases of development. Part two examines issues concerning the contemporary organisation, funding and regulation of public broadcasting, including the recent proposals of President Sarkozy to reform public television. Part three focuses on the interdependence between political elites and public broadcasters with regard to political content. The concluding part addresses some of the key issues currently facing public broadcasting in France.

History of public broadcasting

The history of public broadcasting from its origins after the end of World War Two up until the present day can be usefully analysed in terms of three successive phases of development, with each era characterised by the prevalence of particular structural and operational features.

The first phase, which lasted from 1945 to 1982, was marked by the total dominance of a French variant of public service values and institutions. Broadcasting remained a state monopoly throughout the whole of this period. In the television sector no private commercial competition was allowed to enter the market for the supply of programming to viewers. In contrast, public radio faced competition from commercial stations such as Europe 1 that transmitted to French audiences from outside the national territory and whose existence was tolerated by the French state. Until 1975 both public service radio and television were organised in a single, unitary corporation – its last institutional embodiment being the Office de la radiodiffusion-télévision française (ORTF) which was set up in 1964.

The ORTF was a large organisation with a stake in all aspects of broadcasting – production, programming and transmission – in both radio and television. In the eyes of its critics it had grown to dysfunctional proportions, become difficult to manage and was prone to service disruptions by powerful trade unions. Politically it was a strong symbol of Gaullist control of the state. Under these circumstances it is not surprising that reform of the broadcasting sector was a priority for the first non-Gaullist president of the Fifth Republic, Valéry Giscard d'Estaing. The 1974 Giscardian reform dismantled the ORTF and created separate organisational entities, including three public television channels (TF1, Antenne 2 and FR3) and a public radio company (Radio France). Although encouraging a degree of competition between the three channels for audiences and advertising revenue, the reform maintained the monopoly status of public broadcasting.

The second phase, which lasted from 1982 up until the mid-1990s, was marked by a more competitive era in French broadcasting. The Socialist government's 1982 broadcasting reform abolished the state monopoly. Competition with the established public television companies came first from a pay-TV channel, Canal+, and then from new free-to-air terrestrial channels, La Cinq and M6. Niche channels were also made available through cable and satellite, though audience take-up of these alternative means of programme distribution was comparatively low by the standards of some other Western European countries.

Despite these initiatives, it was the privatisation by a conservative government of the main public channel, TF1, in 1987 that radically altered the market conditions for the supply of television programming during the second age. For several years afterwards, the transfer of TF1 into the private sector skewed the whole television system in favour of commercial providers, with private channels led by TF1 in the ascendancy in terms of audience ratings, and the two main public channels, Antenne 2 and FR3, placed on the defensive. Public television provision grew with the creation of the Franco-German cultural channel, Arte, and the educational channel, La Cinquième, both of which came to public prominence when they took over the terrestrial transmission network vacated by the bankrupt La Cinq in 1992.

The radio sector also became more competitive in the 1980s with the legalisation of private local radio stations. While the original intention of the Socialist government was to promote small-scale community stations funded from donations and public subsidies, in practice it was not long before advertising-funded national private networks came to dominate. Yet because Radio France had been long accustomed to competition with commercial networks, the culture shock of the new broadcasting landscape was much less for public radio than for public television.

The third phase, which began in the mid-1990s and is still ongoing, has been characterised by the transition from analogue to digital technology. The move to digital began with satellite and cable delivery platforms in the 1990s, with a terrestrial platform starting to come on stream only as late as 2005. The most obvious consequence of digitisation from the supply side has been yet another increase in the amount of broadcast content made available, including eighteen free and eleven pay television channels on the terrestrial platform and around 200 channels on the top cable, satellite and ADSL packages. On the demand side, there has been a significant growth in the number of households accessing multi-channel television. Public service providers in France, as elsewhere in Europe, now operate in a television market in which there is increased variety and competition on the supply side with the result that on the demand side viewer loyalty can no longer be taken for granted. To a new generation of French audiences, socialised to believe in the virtues of consumer sovereignty and market choice, the highly restricted state monopoly system of less than thirty years ago must now seem like a historic relic from the Jurassic age of broadcasting.

Organisation, funding and regulation

Key macro-level features of the media environment within which French Public Service Broadcasters (PSBs) now function – more competition in the distribution of content, audience fragmentation, the spread of digital media, the impact of the internet – are not specific to the French case, but can be found in other media systems across Europe and beyond. In these circumstances it may be tempting to paint a picture of French public broadcasters (and other media policy stakeholders) as subject to the push and pull of extraneous factors such as technological convergence and globalisation that are wholly or largely outside of their control. This would be a mistake. The capacity of public service providers in France to respond to these macro-level factors is to a large extent still determined by the *national* policy context and in particular by the vital role played by state actors, most notably the president, in media policy-making.

In this context, the terms of the current debate on PSB in France were spectacularly set by President Sarkozy at a press conference in January 2008, when he stunned the assembled audience of journalists with the revelation that he intended to remove commercial advertising as a funding stream for France Télévisions. This initiative became part of a broader reform – the biggest policy shake-up in broadcasting since the 1980s – that was finally passed by the legislature at the beginning of 2009. The three main provisions of the new legislation on public television are as follows:

- The establishment of a single corporation to manage digital channels and online services.
- The withdrawal of commercial advertising from France Télévisions, initially in the period between 8 p.m. and 6 a.m., and later to be extended to the rest of the schedules.
- The appointment of the Director General of France Télévisions directly by the president.

Although defended by its proponents as a forward-looking piece of legislation, in these three key aspects the reform returns public television to an earlier age of French broadcasting. First, the establishment of a single corporation re-establishes the institutional arrangements that existed prior to the 1974 Giscardian reform and completes a process of moving towards the formation of a single public television organisation that began in the late 1980s. The public channels incorporated in this reorganisation include France 2 (generalist), France 3 (regional), France 4 (youth), France 5 (educational) and France Ô (overseas *départements* and territories). The argument in favour of this reorganisation is that in a more competitive and fragmented media landscape the services of public television require better coordination to ensure a strong public sector presence against an expanding array of generalist and niche-oriented commercial channels and new content providers. The main difference between the new organisational set-up and that of the Gaullist ORTF is that the new framework does not include public radio, whose local and national services continue to be organised in a separate company.

Second, the withdrawal of commercial advertising from France Télévisions takes public television back to the financial arrangements of the 1960s when the ORTF was funded overwhelmingly from licence fee revenue. In stark contrast to the BBC's domestic television services, public television in France has been part funded from commercial advertising since 1968. The contribution of advertising as a share of public television's finances grew substantially in the 1970s and 1980s and then fell back in recent years. In 2004, for example, advertising accounted for just over 29 per cent of public television's total revenue, down from nearly 39 per cent in 1998.

Nonetheless, in 2007 advertising still represented about 30 per cent of total revenue for France Télévisions and its withdrawal represented a major financial challenge for the company's management. This is particularly the case as the level of the licence fee in France (118 euros in 2009) is low compared to that of most other European broadcasting systems and President Sarkozy has shown no inclination to increase it significantly to help make up any financial shortfall. Instead the 2009 reform has introduced a new tax on internet service providers and mobile phone operators, the proceeds of which will go to France Télévisions. In addition, commercial television channels are to be taxed on their advertising profits and the resultant revenues invested in public television. It remains to be seen how this system will function in practice, although the management of France Télévisions have already expressed their concerns.

This central aspect of the reform was defended by President Sarkozy on the grounds that because of their dependence on advertising revenue for an important part of their income stream the programme output of the public channels was insufficiently differentiated from that of their commercial rivals, with the tyranny of the ratings influencing both the substantive content of programming and the allocation of programmes to particular time slots (Risser, 2004). To remedy this situation, Sarkozy proposed that France Télévisions should be liberated from its dependence on advertising to become a French-style BBC.

Was the president's allegation accurate? Certainly it was fiercely contested by the management of France Télévisions who argued that their output was different from that of their competitors. The answer is complex. In some programme genres there has been considerable similarity in content between public and private providers, while in other areas public television has explicitly chosen not to compete with its commercial rivals. For instance, despite the popularity of some reality television shows with audiences in recent years, including *Loft Story*, a French-style *Big Brother* first shown on M6 in 2001, the public channels have steered clear of investing in this prime time genre as they do not regard it as falling within their public service mission.

The withdrawal of advertising predictably caused uproar when it was first announced. Yet in political terms the initiative was shrewdly calculated since the opposition parties of the left, including the Socialist party, had long campaigned against the perverse impact of advertising on programming without ever going so far as to legislate for its abandonment. The broadcasting trade unions were similarly put on the back foot – opposed to advertising on public television in principle, but fearful of the consequences of its withdrawal in practice, notably on levels of employment. Some television professionals welcomed the change, arguing that it would remove the pressure of the ratings system and open up new possibilities for cultural creativity. The problem in the eyes of those opposed to this aspect of Sarkozy's reform is not the principle of the initiative – there is nothing unacceptable in having an advertising-free public service provider of television content – but rather the government's reluctance to raise the level of the licence fee and the problematic nature of the proposed additional revenue streams.

Finally, the president has made it clear that he regards it as part of his remit to have direct responsibility for the appointment of the head of France Télévisions (and of Radio France), with the regulatory authority reduced to exercising a purely consultative role. This direct mode of appointment returns French broadcasting to the era prior to the 1982 reform that established a regulatory authority for broadcasting, one of whose powers was to appoint the heads of the public broadcasting companies. The avowed intention of the Socialist government was that this would help cut the umbilical cord that had tied broadcasting to the state during the Gaullist and Giscardian eras. Successive reforms of broadcasting since 1982, introduced by governments of both right and left, maintained the role of the regulatory authority as the source of appointment to the top managerial posts in public broadcasting.

Why did Sarkozy propose a return to the status quo ante 1982? First, it was argued that since the president enjoys the power of appointment of the chief executive in other spheres of the public sector such as the railways, it was logically coherent that he should enjoy a similar power in the case of public broadcasting. Second, it was alleged that the system of appointment by the regulatory authority was essentially hypocritical in that while it appeared to depoliticise the process, in reality the regulatory authority bowed to the wishes of the political executive. Direct appointment of the Director General of France Télévisions and Radio France by the president would, it was claimed, simply bring the legislation in line with prevailing practice. Such a view rather depressingly presupposes that in France the regulatory authority can never secure a satisfactory degree of operational independence from the political executive.

The 2009 reorganisation of public television is motivated above all by political concerns. It forms part of a series of reforms that have been undertaken by Sarkozy since he won the presidency in 2007. Sarkozy campaigned on a platform of 'quiet change' and has been at pains since his election to give expression to his desire for reform across a broad range of domestic and foreign policy areas. Television is an area where the impact of change can be quickly seen by voters. Both Giscard d'Estaing and Mitterrand introduced broadcasting reforms near the start of their presidential terms and so in paying attention to reform in this area Sarkozy is maintaining a Fifth Republic tradition of presidential interventionism and executive power.

There was also an element of revanchism in the president's concern to reform France Télévisions. On various occasions during the 2007 campaign and before, Sarkozy had complained about the supposed inefficiencies of the public broadcasting organisation and the allegedly discourteous attitude shown to him by some of its journalists, notably on the regional channel France 3. As president he has not sought to develop a fruitful working relationship with the Director General, Patrick de Carolis, who is not generally regarded as a Sarkozy supporter. In contrast, Sarkozy enjoys close relations with several media bosses in the private sector, including the owner of TF1, Martin Bouygues. In the run-up to the 2008 presidential press conference TF1 actively lobbied the president and his immediate entourage in favour of the type of reform that Sarkozy subsequently announced, leading to charges in the press that the reform of public television was motivated primarily by a desire to help the big private channels prosper in the more competitive digital environment and to create French media groups capable of competing in European and global markets.

Certainly the withdrawal of advertising from public television seemed expressly designed to bolster the balance sheets of the generalist channels TF1 and M6. The audience share of TF1, for long France's most popular channel, has been in decline since the introduction of digital terrestrial television in 2005: from a 31 per cent market share in 2007 to 27 per cent in 2008. However, this is not the only problem it faces. Its programming is increasingly regarded as old-fashioned by certain sections of the audience, notably the young, while the downturn in the advertising market as a result of the economic recession has hit the channel badly. Moreover, although a major player in France, the company has little presence in international markets. In this context a political decision to stop advertising on France Télévisions must have looked like a quick fix solution to TF1 management. At the same time all commercial channels in France stand to benefit from the introduction of additional advertising breaks in programme schedules as a result of regulatory changes. The resultant virtuous circle for commercial television is the possibility on the supply side to increase the number and length of advertising breaks, including during prime-time schedules, coupled with greater demand from advertisers, who can no longer place their advertisements on the public channels.

Within this increasingly competitive market, public broadcasting remains subject to a high degree of content regulation. Even after the abolition of the state monopoly, France never embraced the savage deregulation which affected Italy after Silvio Berlusconi's entry into the market in the 1970s. Currently public television is subject to a higher degree of regulation than its private competitors, for instance with regard to the provision of specific programming for religious and political expression. In addition, quotas to sustain domestic programme production and to help the French film industry have been a traditional means of cultural and industrial protectionism. Indeed, one of the main strengths of public service broadcasting in the past has been as a showcase for domestic product in a society where elites across the political spectrum have traditionally valued the importance of national cultural dissemination and have regulated to protect and promote national cultural product in broadcasting.

Since 1989 regulation of the broadcasting sector has been undertaken by the Higher Audiovisual Council, which consists of nine members – three appointed by the president, three by the chairman of the National Assembly and three by the chairman of the Senate. The Council is the latest in a line of regulatory authorities whose origins go back to the 1982 reform (Chauvau, 1997; Franceschini, 1995). While the establishment of a regulatory authority represented an important symbolic break with previous practices of direct political control, it took some time – and the establishment of no fewer than three consecutive regulatory bodies in the 1980s – for the notion of an independent regulatory authority to be accepted across the political class.

It is true that politicians from all sides have frequently been reluctant to accept the substantive implications of a regulatory authority separate from the political executive. Appointments to the Council are frequently informed more by party political considerations than professional competence and the Council currently has a very strong bias towards the ruling political right. Nonetheless, prior to the start of the Sarkozy presidency the Council had survived several alterations in government between left and right – no mean feat in a domain marked by high political controversy – and, while one should not overstate this, its existence had to some degree constrained the freedom of manoeuvre of president and government to interfere directly in the management of public broadcasting.

Editorial independence and political coverage

Television has long been the single most important medium of political information in France and this continues to be the case, with surveys putting television well ahead of the press, radio or the internet as the public's primary source of national and international news. For many years political control of television news was a notable feature of the public broadcasting system. During the de Gaulle presidency (1958–69), for instance, close direct control of television news content was maintained by the Ministry of Information, while key managerial and editorial appointments were made with reference to partisan political loyalties. As a result, for several years de Gaulle's political opponents were largely absent from the television screen outside of formal election campaigns (Bourdon, 1990; Chalaby, 2002). The political control exercised over public television during the period of state monopoly prevented the embedding of a UK-style professional model of broadcasting in France (Hallin and Mancini, 2004: 30–3). Instead, this period represented one of the classic examples of the government model of broadcast organisation (ibid., p. 106).

In part because of the huge expansion in the number of broadcast outlets, this top-down, command-and-control approach by politicians is now inappropriate as a means of news management. As a result, the power relationship between political elites and broadcasters is now characterised by different modes of interdependence. For example, the closeness of the interpersonal linkages between top politicians and journalists in France has often been remarked upon. While leading political journalists and top politicians may not go through the same tertiary educational institutions, they do inhabit the same milieu and get to know each other well – some would argue too well. Those who see close collusion between mainstream politics and broadcasting point to the relationship between politicians and journalists in the Fifth Republic as having been marked by a high level of deference on the part of the latter towards the former, most notably evident in broadcast interviews, and by close cooperation, even connivance, between the two sets of actors (Carton, 2003), resulting in what one commentator has called a 'journalisme de révérence' (Halimi, 2005: 17–48).

Like their counterparts in other advanced democracies, leading French politicians now employ a wide range of news management and public relations techniques to secure access and get their message across in the news media. They engage in negotiation and bargaining with journalists, exploiting their legitimacy and mobilising resources in a competitive struggle with other political sources for access to help structure the news agenda and frame issue coverage. Sarkozy has long had a reputation for his news management skills, acquiring a considerable reputation for the way in which he incorporated a campaigning communication perspective into his political activities as government minister, party leader and president (Artufel and Duroux, 2006).

In this process of interdependence between politicians and broadcast journalists, any difference between public and private broadcasting is more one of degree than of kind. It would be wholly unsustainable, for example, to argue that the commercial channels are less the object of political seduction and pressure than their public counterparts. TF1, for instance, is far too important a news provider not to be courted by politicians. Equally, it would be completely misleading to contend that public television is subservient to political elites or – an even less valid hypothesis – that it can simply be equated with 'government television'. Within mainstream political coverage on public television news there is no evidence of overt and intended partisan bias. Moreover, even in the more qualitative elements of an evaluation of public television's political coverage, it would be difficult to demonstrate with any conviction a significant difference in, say, journalistic tone between news items or interviews featuring politicians of the governing majority on the one hand and those representing the mainstream opposition on the other.

One of the responsibilities of the Higher Audiovisual Council is to monitor the allocation of time by broadcasters to different political actors in news coverage. The Council's findings are made public on a regular basis. It is fair to say that there has been an institutionalised degree of pluralism and equity in the amount of coverage accorded the parties of the mainstream opposition alongside that given the government and the parties of the governing majority. Smaller parties, however, still complain about the amount of coverage they receive – testimony to the difficulties of defining and then operationalising the concept of equity in what is from one perspective a highly fragmented multi-party system. Opposition parties have also condemned the fact that the many television appearances of the president – Sarkozy has been called the *téléprésident* – are not taken into account in the CSA's calculations regarding political balance. The concern of opposition politicians in this respect is quite understandable. While the principle of omitting the television appearances of the head of state in the CSA's calculations is not new, opposition parties justifiably claim that under Sarkozy these have reached an unprecedented level.

One charge frequently made against political coverage on television concerns the representation of views outside of – and in contradiction to – those of the political mainstream. For some commentators, political output on French television – both private and public – largely reflects a loose ideological consensus on the part of mainstream political elites. This has led to the accusation that as a result television may fail to give appropriate weight to alternative and oppositional perspectives. This is not just a question of how television should cover the views of minority parties from both extremes of the ideological spectrum, although this is clearly one aspect. It also incorporates what some regard as a dislocation between sections of French society and the mainstream media. For instance, the decisive victory of the 'No' vote in the 2005 referendum on the EU constitution has been put forward as evidence of the distance between a majority of public opinion and the broadcast media, the vast bulk of whose formal referendum coverage supported a 'Yes' vote.

Public broadcasting: problems and prospects

In the short term the management of public television will be concerned primarily with the impact of the new funding regime on programming. Yet important as it may be, this is not the only issue on the agenda of public broadcasters in France. Three others are briefly examined in this concluding section: the distinctive role of PSBs in a multi-channel competitive market; the transition from PSB to public service communications media; and the importance for PSBs of reflecting the cultural diversity of contemporary French society.

First, there is the question of the mission of PSBs in a media landscape increasingly dominated by private providers and market concerns. At one level this is simply about public broadcasters' securing audiences in a more competitive market. The expansion in the supply of programming as a result of digital switchover, especially via the terrestrial platform, will inevitably lead to a reduction in the audience share of public television. Yet even without the financial imperative of securing commercial advertising revenue it will be impossible for public television simply to isolate itself from audience-driven market pressures. In part this is because the political case for the continued existence of a licence fee system to fund public television is dependent to some extent on the success of the programme output of its mainstream channels, notably during prime time, in attracting mass audiences. While it is true that not all public television output is obliged constantly to aim simply for high ratings or to attract a particular socio-demographic section of the audience, it still remains the case that consistently low ratings make public broadcasters vulnerable to criticism from viewers, sections of the media and political elites.

In short, France Télévisions will have to face up to the dilemma that all public broadcasters have to negotiate in a system characterised by market competition and audience fragmentation: what is the distinctive contribution of the public service provider? If public broadcasting increasingly emulates the management practices and resembles the output of commercial rivals, then the defence of a specific public service component in broadcasting and communications on the grounds of its particular contribution to the achievement of socially desirable objectives is severely weakened, along with the claim to a secure source of public funding (Iosifidis et al., 2005: 11). Conversely, if France Télévisions retires to a small public service ghetto of 'worthy' broadcast ouput, it runs the risk of losing more viewers and so the case for public funding via the licence fee is again undermined. With its emphasis on educational and cultural programming, the 2009 reform seems to push France Télévisions in this latter direction. The possibility in this scenario is that public television in France becomes a less central part of the broadcasting system, playing a supporting and complementary role in a landscape where the tone is increasingly set by commercial operators.

Second, there is the question of the transition from public service broadcasters to public service communications media. The public broadcasting organisations in France do not enjoy a status or legitimacy in their national media landscape equivalent to that of the BBC in the UK. In part this is because the experience of the government model in the formative years of its development prevented public broadcasting from developing a tradition of political independence which would allow it to foster a positive relationship with civil society and embed itself in popular consciousness as a national icon. In addition, the break-up of the ORTF in 1975 drove a wedge between the provision of public radio and television services. Even after the implementation of the 2009 reform there will be no single public broadcasting organisation in France. Moreover, while the BBC has for long regarded news and information provision as a central, defining component of its mission, this has been less true of French public broadcasters. Thus, while the contemporary BBC regards itself not just as a *broadcaster* but as a major public service *communication* actor, embracing radio, television and online services, Radio France and France Télévisions are still in many respects traditional suppliers of radio and television programming respectively. The online provision of news and related information and educational services by French public broadcasters is underdeveloped compared to that of the BBC. Yet – and here we return to the funding issue – a successful transition from broadcaster to multi-platform communications media requires a stable financial foundation currently lacking.

Finally, there is the capacity of public broadcasters to reflect the cultural diversity of contemporary French society by adapting output to the reality of a multi-ethnic and multicultural Republic. Public broadcasting has long lacked role models from ethnic minority communities, while much programming fails to take account of the variegated ethnic composition of French society. This is a difficult issue for French media policy-stakeholders since it forms part of a much broader and highly contentious socio-political debate on how the tradition of universal Republicanism can be reconciled with greater cultural pluralism. Anglo-American notions of multiculturalism are widely rejected among French political elites, as they are seen as promoting group loyalties that break the bond between the individual and society, between the citizen and the state. This has led in the past to an under-representation of ethnic minorities both in appointments and in programme content. In recent years there has been growing awareness of the need for broadcasters to reflect this national cultural diversity, with the Higher Audiovisual Council inviting broadcasters to ensure that in their news coverage and fictional programming the diversity of French society is better represented.

French public broadcasters are certainly not exceptional among their European counterparts in having to face up to these socio-economic challenges. However, in the French case what makes the situation particularly difficult for public service broadcasting institutions is the combination of the hostility of an interventionist president and the lobbying influence of commercial broadcasters on government policy. The current political climate in France is scarcely conducive to the well-being of public broadcasting institutions or the values they aspire to represent.

References

Artufel, C. and M. Duroux (2006) *Nicolas Sarkozy et la communication* (Paris: Éditions Pepper).
Bourdon, J. (1990) *Histoire de la télévision sous de Gaulle* (Paris: Anthropos/INA).
Carton, D. (2003) *'Bien entendu . . . c'est off'* (Paris: Albin Michel).
Chalaby, J. K. (2002) *The de Gaulle Presidency and the Media* (Basingstoke: Palgrave Macmillan).
Chauvau, A. (1997) *L'Audiovisuel en liberté?* (Paris: Presses de Sciences Po).
Franceschini, L. (1995) *La régulation audiovisuelle en France* (Paris: Presses Universitaires de France).
Halimi, S. (2005) *Les nouveaux chiens de garde* (2nd edn) (Paris: Raisons d'Agir).
Hallin, D. C. and P. Mancini (2004) *Comparing Media Systems* (Cambridge: Cambridge University Press).
Iosifidis, P., J. Steemers and M. Wheeler (2005) *European Television Industries* (London: British Film Institute).
Risser, H. (2004) *L'Audimat à mort* (Paris: Éditions du Seuil).

13

Public Service Broadcasting in Germany: Stumbling Blocks on the Digital Highway

Runar Woldt

Introduction

After World War Two, Public Service Broadcasting (PSB) in West Germany was re-established after the model of the British Broadcasting Corporation (BBC). However, the German system developed its own peculiarities from the beginning: PSB was set up in a federal structure, following the political structure of West Germany. The federal states ('Länder') have sole responsibility for culture and the media. The German federal constitution guarantees freedom of speech, freedom of broadcasting, and non-interference of the state in broadcasting (Grundgesetz, 1949, Art. 5). The Federal Constitutional Court was instrumental in strengthening and developing this system. The principal legal foundation of broadcasting lies in the broadcasting laws of each of the sixteen German federal states. In addition, the Länder join together in inter-state treaties, which establish a quasi-national structure in broadcasting regulation (Rundfunkstaatsvertrag, 2007).

Regional PSBs serve individual Länder, or groups of Länder (based on inter-state treaties). A network of these regional broadcasters was established in 1950 under the title ARD (Arbeitsgemeinschaft der öffentlich-rechtlichen Rundfunkanstalten). A second public service television channel, ZDF (Zweites Deutsches Fernsehen), was launched in 1961 as a nationwide broadcaster. The governing bodies of ARD and ZDF consist of representatives of political parties, unions, trade and industry, churches, universities, cultural institutions and non-governmental organisations (NGOs) in order to ensure that broadcasters are accountable to society and enjoy political independence.

Since the establishment of the so-called 'dual system' in the late 1980s, the Constitutional Court has ruled that reduced obligations for private broadcasting would be acceptable as long as PSB provided a sufficient range of public service programming (Bundesverfassungsgericht, 1986, 1991). PSBs have been successful in stabilising their positions in recent years, but the private sector has gained considerable support from politicians in a call to limit

PSBs' initiatives in digital television and the evolving online media ecology. The European Commission (EC) has also been exerting pressure on Germany to lay down the mandate of PSB more explicitly and ensure that PSBs do not get in the way of private enterprise in the online sector.

This chapter discusses the problems facing PSBs in Germany at this crucial phase of technological and market development. It provides an overview of the current market situation, with particular emphasis on digital and online developments. Further, it analyses the related political issues that were at the centre of the negotiations between Germany and the EC between 2004 and 2007. It is argued that the most important questions have yet to be addressed: What can (and should) society expect from PSB in a digital, multimedia landscape? How will PSB be able to fulfil its remit in the future, given the trend towards technological convergence and demographic change?

Market developments

With 35.3 million television households the German broadcasting market is the largest in Europe. Only 4.6 per cent of these households receive their television signal exclusively via terrestrial means, while 44.1 per cent rely on satellite signals and 51.3 per cent on cable networks (AGF/GfK, 2009a). The high percentage of cable and satellite multi-channel households explains why Germany is also the most competitive of the European television markets. The most recent list of television channels licensed in Germany contains 131 private channels (KEK, 2008: 70). Among these are a large number of domestic free-to-air channels broadcasting in the German language.

There are twelve PSB organisations. Of these, eleven are members of ARD: nine regional broadcasters, Deutsche Welle (Germany's international broadcaster funded by the federal government) and DeutschlandRadio, the national public service radio broadcaster. Alongside this there is ZDF, the second public service television system. ARD and ZDF are funded through a mixed system, including licence fees, advertising, sponsoring and other means, such as programme sales and merchandising. In 2007, ARD's annual income from the licence fee was 5.2 billion euros, while net advertising income was 413 million euros. ZDF received 1.7 billion euros from the licence fee and 117 million euros from advertising. While income from advertising seems marginal compared to the total budget, it is still regarded as vital by PSBs as it provides a certain room to manoeuvre outside the more politically sensitive licence fee.

The national channels, ARD/Das Erste (the 'first channel') and ZDF, plus ARD's seven regional channels form the core of the public service television sector. Two special interest channels are joint ventures of ARD and ZDF: KI.KA, a children's channel, and Phoenix, an information channel. Both ARD and ZDF run their own digital bouquets, and are partners in the international cultural channels 3sat and Arte. In all, these PSB channels

Table 13.1 Programme genres on main German television channels (2007) in %

	ARD/Das Erste	*ZDF*	*RTL*	*Sat.1*	*ProSieben*
News	9.3	9.5	3.9	2.9	0.8
Magazine, consumer info	24.1	30.7	13.7	13.8	13.0
Reportage, documentary	11.5	11.5	11.0	2.0	14.1
Docu-fiction, docu-soap	0.5	0.3	7.9	9.9	0.3
Live event	4.1	3.8	1.1	0.6	0.6
Talk, debate	2.7	5.2	3.4	7.2	5.8
Quiz, game show	1.7	0.8	1.0	9.3	3.5
Non-fictional entertainment	3.2	2.5	6.8	5.5	8.7
Drama, film	21.2	15.8	6.9	8.1	19.4
Series	17.2	15.7	18.7	18.1	13.0
Other editorial content	0.6	0.3	0.3	0.1	0.2
Trailers, fillers, etc.	2.4	2.5	4.7	5.3	5.4
Advertising	1.4	1.4	20.8	17.2	15.4

Source: Krüger and Zapf-Schramm (2008: 167).

accounted for approximately 115,000 hours of television broadcasts in 2007 (Basisdaten, 2008). By comparison, the BBC's television output in 2007–8 was approximately 50,000 hours (BBC, 2008; Ofcom, 2008a).

ARD and ZDF are by far the leaders in the provision of informational programmes, while private channels have a stronger output of entertainment programmes and various talk formats (see Table 13.1).

Referring to television news in particular, there seems to be a clear focus in the main PSBs' TV news programmes on political news, whereas private news programmes show a much higher percentage of news in the categories accident/disaster, crime and human interest/everyday life (see Table 13.2). Research conducted on behalf of the regulatory authorities for private broadcasting confirmed an increase of non-political news on private television in recent years (Maier et al., 2009). It can be assumed therefore that the contribution of private television to the public discourse is decreasing, at least as far as the political content is concerned.

Despite the large number of television channels available in the country, in 2008 only ten channels accounted for 80 per cent of the viewing. As of year end 2008, public service television channels held a combined audience share of almost 44 per cent. ARD's national channel, Das Erste, was the market leader with 13.4 per cent, with ARD's regional channels (the so-called 'third channels' or Dritte) coming second with a combined share of 13.2 per cent, followed by ZDF in third place with 13.1 per cent and RTL, the leading private channel, with 11.7 per cent (see Table 13.3).

As the comparison between audience shares in 2000 and 2008 shows, ARD and ZDF have been remarkably resilient in establishing their positions in the audio-visual landscape. However, changes in viewing behaviour seem to

Table 13.2 Thematic categories in leading German TV news programmes (2008) in %

'Tagesschau'	*Public*		*Private*	
	'heute' (ARD)	*'RTL aktuell' (ZDF)*	*'Sat.1 Nachrichten' (RTL)*	*(Sat.1)*
Politics	48	38	18	27
Economy	11	10	8	11
Society/legal	9	8	7	8
Science/culture	4	5	4	5
Accident/disaster	4	5	8	9
Crime	3	4	9	7
Human interest/ everyday life	1	5	14	15
Sport	9	11	18	9
Weather	7	8	7	4
Other	5	7	7	5

Source: InfoMonitor (2008).

Table 13.3 Audience shares of German television channels (2000 and 2008), viewers aged 3 years and older, in %

	Public service/Private	*2000*	*2008*
ARD/Das Erste	ps	14.3	13.4
ARD Regional/Dritte	ps	12.7	13.2
ZDF	ps	13.3	13.1
RTL	pr	14.3	11.7
Sat.1	pr	10.2	10.3
ProSieben	pr	8.2	6.6
VOX	pr	2.8	5.4
RTL II	pr	4.8	3.8
kabel eins	pr	5.5	3.6
Super RTL	pr	2.8	2.4
KI.KA	ps	2.8	2.0
3sat	ps	0.9	1.1
N24	pr	–	1.0
Phoenix	ps	0.4	0.9
DSF	pr	1.2	0.9
Eurosport	pr	1.0	0.9
n-tv	pr	0.7	0.8
Arte	pr	0.6	0.6
Other		3.5	8.8

Source: Darschin and Gerhard (2003: 159–60); Zubayr and Gerhard (2009: 103–4).

work against them in several ways. Firstly, their most loyal audiences are getting older. In 2008, ARD had an audience share of only 7.5 per cent among 14- to 49-year-olds, but 18.8 per cent among those of 50 years and older (Zubayr and Gerhard, 2009: 105). In direct contrast, RTL, the leading private channel, had an audience share of 15.8 per cent among 14- to 49-year-olds, but only 8.8 per cent among over 50-year-olds. The average age of viewers of both ARD and ZDF has increased; reaching younger audiences has therefore been one of their main concerns in recent years. Furthermore, although television viewing is still very popular among teenagers and young adults, the internet is becoming a 'medium' which young people use for entertainment, information, news and communication with their peers. ARD and ZDF still enjoy a strong image as providers of news and information also among younger age groups, but given the trends described above these users may in future increasingly turn to online sources for news and orientation. In order to fulfil its remit, PSB seems to have no choice other than to expand into the digital and online sector. This, however, has been a contentious issue in German media politics for a number of years.

PSB in the digital era

In 2000, the Federal Ministry of Economics and Labour outlined steps towards a nationwide switch-off of analogue television by 2010 as well as switch-off of analogue radio between 2010 and 2015 (BMWA, 2000). Initially, the responsibility for putting the plan into action was left to the market and no real progress was made for a couple of years. With coordinated initiatives involving regulatory authorities and PSB in particular, digital television over terrestrial networks (DVB-T) has nevertheless become a reality. In August 2003, the region of Berlin/Brandenburg was the first in the world to switch off terrestrial analogue television transmission. Subsequently, DVB-T was introduced in other regions, and by the end of 2008 the switch-off of analogue terrestrial television was completed – ahead of schedule (Überall-TV, 2009). In January 2009, there were approximately 12 million television households in Germany connected to a digital receiver, of which two-thirds were satellite households (AGF/GfK, 2009b).

ARD and ZDF have been the forerunners of digital television in Germany. Both launched their digital channels as early as 1997. The PSBs were instrumental in the successful switch-off of terrestrial analogue transmission; they are also supporting the development of digital radio (DAB) as well as multimedia standards (MHP). Despite the early start, PSBs' digital strategy has been fairly cautious, because of the costs involved with simulcasting, the investments in digital equipment (production and distribution), and also because of the slow growth of digitalisation in Germany compared to other countries. In early 2009, the share of television households in Germany equipped with digital receivers was 34 per cent (AGF/GfK, 2009b), a percentage Britain had already reached in 2001 (Ofcom, 2008b). Currently, ARD's digital bouquet consists of ARD's analogue television channels, three special digital television channels, and over sixty radio channels. ARD also offers an interactive 'online channel', which provides an EPG, additional interactive information on broadcast programmes and various multimedia services. Of the three special digital channels, 'EinsFestival' offers a programme mix with a focus on culture, knowledge and entertainment; 'EinsPlus' is a service and educational channel; 'EinsExtra' is devoted to news and current affairs.

ZDF's digital package contains its main analogue channel, television channels produced mainly in cooperation with ARD, three special digital channels and several radio channels. ZDF's special digital channels are devoted to news and current affairs ('ZDFinfokanal'), documentaries and background information ('ZDFdokukanal'), and theatre ('ZDFtheaterkanal'). Viewing shares of ARD's and ZDF's digital channels are marginal, because of the slow growth in digital households, the dominance of mainstream channels, and also because the digital channels offer only a small percentage of original material. Large parts of their schedules are repeats of programmes already shown on analogue channels.

Both broadcasters are working on plans to increase the appeal of their digital channels. ARD is planning to put more original programming on its informational channel 'EinsExtra'; the service channel 'EinsPlus' and drama and entertainment channel 'EinsFestival' will gradually be upgraded with formats aiming especially at younger audiences. ARD has recently started its own channel on the Web platform YouTube, offering clips from its analogue information and entertainment television programmes. ZDF is planning a major revamp of its digital bouquet. 'ZDFdokukanal' will be relaunched as 'Familienkanal', a general interest channel geared towards 'young families'; 'ZDFtheaterkanal' will be renamed 'Kulturkanal', offering a larger range of cultural programmes, again for younger audiences.

PSB and the internet

Research about the adaptation and usage of the internet in Germany shows that after a period characterised by a rapid increase in user numbers, especially among younger and better educated groups of society, the internet is now becoming part of everyday life for a large number of people of all ages and backgrounds. In 2008, 66 per cent of all Germans aged 14 years and above were using the internet at least 'occasionally' (van Eimeren and Frees, 2008: 331).

The growing importance of the internet for an increasing number of people from all socio-demographic groups also means that, nowadays, no television channel can do without its own presence on the World Wide Web. Websites of television broadcasters are among the most popular with internet users in Germany. In a ranking of major German language websites, research organisation AGOF lists RTL.de and ProSieben.de at seventh and eleventh place respectively. Top of the list, however, are big internet portals such as T-Online.de, Web.de, MSN.de and Yahoo! Deutschland (AGOF, 2008: 6).

Websites of PSBs do not appear on the AGOF list as they are not carrying advertising – ARD and ZDF's online activities are funded through the licence fee. However, KEF, the commission for the assessment of the funding of PSB, reports that in 2006 ARD and ZDF's websites had 4 billion and 1.63 billion page impressions respectively (KEF, 2007: 42). The websites of ARD and ZDF have a strong emphasis on news and information for all age groups. In contrast, the websites of the private television broadcasters typically focus on entertainment and commerce, clearly targeting younger user groups. Until new regulation comes into force in 2009, PSBs are only allowed to put content on the internet which has a connection with broadcasting programmes. ARD for instance runs tagesschau.de and boerse.ARD.de, two websites closely linked with the most important TV news and stock market programmes respectively. On the other hand, ARD.de, DasErste.de and radio.ARD.de are portal websites which offer comprehensive information on the whole of ARD's output. In addition, each regional member of ARD as well as national broadcasters Deutsche Welle and DeutschlandRadio provide their own internet content. By contrast, ZDF's output on the internet is more limited in size, but not necessarily in scope – it offers a general portal, ZDF.de, a news website, heute.de, as well as a children's website, tivi.de.

In 2007–8 both PSBs commenced their respective catch-up service for broadcast material on the internet. 'ZDFmediathek' (mediathek.zdf.de) offers hundreds of ZDF programmes on-demand. ARD's similar service (ardmediathek.de) contains television programmes as well as a large number of radio broadcasts. In addition, some of the regional ARD members offer on-demand services on their own websites. ARD and ZDF increasingly apply a multimedia approach to their operations, integrating online with their traditional production of radio and television, at the same time developing formats specific to the new medium of the internet with the aim of reaching younger audience groups.

The scope, size and purpose of PSBs' presence on the internet have been subject to heated debate in the country for quite some time. Complaints by commercial broadcasters and (even more vehemently) by print publishers addressed the effects that online activities of ARD and ZDF allegedly have on commercial investments in online markets. Due to intensive commercial lobbying and political pressures, ARD and ZDF in 2003 committed themselves to a maximum online budget of 0.75 per cent of their total budgets. It is worth mentioning that for the year 2006 they reported expenditures of 47.5 million euros (ARD) and 14.6 million euros (ZDF) for their various internet sites (KEF, 2007). The online engagement of ARD and ZDF has been the subject of a political struggle between Germany and the EC, too, which will be discussed in the following section.

Political struggles

PSB has frequently been subject to political struggles in Germany. While in the past these concerned for instance allegations of biased reporting or disputes between political parties over the appointment of senior broadcasting staff, in more recent years these struggles reached new dimensions and touched upon the foundations of PSB, that is, the question of its funding and the scope of its remit in a changing multimedia environment.

In 2004, a controversy erupted between PSB and the governments of the German states over the issue of the increase of the level of the licence fee. Germany has developed a unique system for settling the level of the licence fee. An independent commission, KEF, is responsible for assessing the financial needs of PSB. Every two years, KEF considers the budgetary plans presented by ARD, ZDF and DeutschlandRadio and submits a recommendation to the Länder concerning the future level of the licence fee. In 2004, a number of Länder governments refused to accept the proposal made by KEF and decided that any increase in the licence fee be subject to certain conditions. Basically, they wanted ARD and ZDF to improve efficiency, cut costs and streamline their offers, especially in the fields of the internet and new media. This intervention was unprecedented, and it ignored the reason why KEF had been set up in the first place: to ensure that the licence fee is not decided upon on political terms.

Following heavy political wrangling, the Länder governments in the end agreed on an increase of the monthly licence fee substantially below KEF's initial proposal (0.88 euros instead of 1.09 euros). After some hesitation ARD and ZDF finally took the case to the Federal Constitutional Court which, in 2007, ruled that the politically motivated handling of the licence fee issue by the Länder had been a violation of the principle of broadcasting freedom (Bundesverfassungsgericht, 2007). This ruling was regarded a major victory for ARD and ZDF. Although the Court upheld the right of the Länder to define the scope of the PSB remit and to settle the level of the licence fee, it disapproved of the procedure and political argumentation by the Länder. The Court also underlined the obligation of the Länder to fund PSB in a way that it can function properly in a changing media environment.

Politically, this conflict was very risky for ARD and ZDF because it culminated at the same time that negotiations between Germany and the EC over PSB had reached a critical stage. Taking the case to the Constitutional Court meant risking alienating the Länder, important allies that ARD and ZDF needed in order to defend their positions against the European Commission. Following on from a complaint that VPRT, the German lobby organisation of private broadcasters, had filed with the EC in 2003 over competition from ARD and ZDF in the online sector, the Commission was pressing Germany hard for a tighter regime for PSB. The Commission argued that the funding of PSB through the licence fee constitutes state aid, thereby giving the EC the right to interfere should the licence fee be used by PSBs in a way that distorts competition in the market. ARD, ZDF and the German Länder, referring to the Amsterdam Protocol to the Treaty on the European Union (European Communities, 1997), argued that the broadcasting licence fee does not constitute state aid, that the remit of PSB was sufficiently defined by existing regulation, and that the Commission should leave the regulation of PSB to the member states.

After extensive negotiations between the German Länder and the EC, a compromise was reached in 2007. Germany agreed to formulate new rules for PSB in the forthcoming inter-state Treaty on broadcasting, defining more clearly the remit of PSB, and the limits of its online activities in particular. The European Commission on the other hand promised to close the pending case against PSB in Germany. Apparently, it regards elements of the new regulations for PSB in Germany as a 'blueprint' for regulations in other EU member states (Knappmann and Hönighaus, 2008).

The new inter-state Treaty was formally agreed by the heads of the Länder governments in December 2008 and was expected to come into force on 1 June 2009 (Rundfunkstaatsvertrag, 2008). The new regulation for the first time makes reference to 'telemedia' (internet and mobile communications) as part of the PSB remit. In its Article 11(d) the Treaty specifies that telemedia offered by ARD and ZDF should promote participation of citizens in the information society, provide orientation and support media literacy. This short paragraph, however, is the only reference in the regulation to a changing remit of PSB in the digital media landscape. Otherwise restrictions are tough and largely reflect the concerns of the print media.

More specifically, ARD and ZDF will not be allowed to offer online content which is 'similar to that provided by print media' (meaning text-based content). Radio or television programmes that are made available as streaming media over the internet can only be offered for seven days after their original broadcast; for sports programmes, the limit is only twenty-four hours. The same 'seven-day' rule also applies to additional material directly related to a programme. Archives of broadcast material which is of 'historical value' can be offered online without time limitation, but on prior approval only. Finally, a 'three-steps test' has to be applied to all online content of ARD and ZDF which is not directly related to a programme or which broadcasters want to offer online for longer than the seven-day period. This test will be carried out by the governing bodies of the PSBs to ensure that the telemedia fall under the PSB remit (step 1), contribute to pluralism and diversity (step 2), and are provided with reasonable budgets (step 3). Private competitors get a chance to present their views during this process.

This three-steps test not only applies to new online ventures, for all existing online content also had to be cleared by August 2010, according to the new rules. ARD and ZDF are required to formulate comprehensive 'telemedia concepts' outlining in detail the aims and scope of their online activities.

An enormous bureaucratic procedure has been put in train in order to meet the deadlines. How the governing bodies of the PSBs will manage this task, and how the new regulations will affect existing online content, remains to be seen, though representatives of PSBs as well as independent observers expect a series of court cases over the interpretation and application of the new rules (Peters, 2009). In the meantime, ARD and ZDF have revoked their voluntary commitment to limit their online expenses to 0.75 per cent of their total budgets.

Conclusion

The German PSB system is the largest in the world, both in terms of income and output. Viewers in Germany probably enjoy a larger choice of free-to-air channels than those in any other country in Europe, thanks to the availability of a good number of private channels and to the strong position enjoyed by PSBs. Since the contribution of private channels to the public discourse is declining as far as information on social and political issues is concerned, it has become even more important that PSBs continue to fulfil this function.

However, for a number of years there has been a hostile climate towards PSB both by political actors and media (press and broadcasting) interests. From the beginning of the so-called 'dual system', private broadcasters have been complaining about unfair competition, because PSB is funded by both public and commercial means. These complaints have gained momentum in the current economic downturn. With a public discourse dominated for many years by free market ideas and individualism, public interest principles, and the schema of the licence fee in particular, have gradually lost support among the general public too. The backing provided by the German Constitutional Court has temporarily eased the pressure on PSB, but it is clear that it cannot in the long term be a substitute for a broad societal consensus.

German PSBs' challenges and dilemmas are particularly highlighted in the case of their position in the digital and online world. As a result of a compromise reached with the EC, the German federal states agreed on new regulations for PSB. From the point of view of PSB the positive aspects are twofold: the inter-state Treaty on broadcasting for the first time acknowledges the right of PSB to offer online content which is not directly related to any radio or television programme; and online activity has been recognised as being part of PSB's remit.

On the other hand, the new regulation only gradually departs from the concept of 'programme-related' PSB online content. It defines 'telemedia' as a new area for the activities of PSB, but it falls short of establishing online as a platform equal to broadcasting. Intensive lobbying from the print media sector in particular resulted in detailed limits on the scope and duration of content PSBs are allowed to offer online. The new three-steps test for large parts of existing and future online content contains a risk for PSB of disclosing plans and prolonging procedures in such a way that private competitors may gain an advantage.

In the public debate in Germany it has been argued that the new regulation may enhance transparency of the online activities of PSB. It may also strengthen the position of the governing bodies entrusted with the three-steps test, thereby adding to the legitimacy of the system of PSB as a whole. Critics, however, point to the restrictions laid on PSB online as a 'loss in diversity' (Lilienthal, 2009) and a missed opportunity: the question of whether PSB is a real or imagined threat for the online activities of the printed press has overshadowed serious consideration of what kind of service the public can expect from PSB online. As in other countries, the remit of PSB in Germany has traditionally been based upon the idea of broadcasting as a special area which justifies public intervention. Transferring the PSB remit to the online world is therefore no trivial undertaking. Private lobbyists and the European Commission have so far been remarkably successful in demarcating the internet as a field first and foremost for private enterprise. The German Länder have tried to stem the tide by upholding the principles of the Amsterdam Protocol against the dominance of European competition policy. But domestically they have been unwilling or unable to seriously address the fundamental issue of the benefits – and risks – of a transition from public service broadcasting to public service media.

In comparison to other national regulatory frameworks, the German tradition of the inter-state Treaty on broadcasting allows for fairly frequent (normally every two years) updating of the rules for public service and private broadcasting. This apparently facilitates fairly detailed adjustments of the regulation in response to short-term or mid-term market developments and changes in the political field (for example, at European level). Because it puts its emphasis on a 'step-by-step' approach, the system is obviously less well suited to address more fundamental issues, for instance the role of PSB in the internet sphere. As with other such issues in media policy before, it is therefore not unlikely that it will be the Constitutional Court – instead of media policy – which eventually will have to clarify the position of PSB in a changing media landscape.

References

AGF/GfK (2009a) *Entwicklung der TV-Empfangsebenen*, www.agf.de/fsforschung/methoden/empfangsebenen, accessed 17 February 2009.

AGF/GfK (2009b) *Digitalisierungsgrad*, www.agf.de/daten/zuschauermarkt/digitaltvgrad, accessed 17 February 2009.

AGOF (2008) *Berichtsband – Teil 2 zur internet facts 2008-III, December 2008*, www.agof.de/if-2008-iii-teil-2-vermarkter.download.4a056da10d72ab7574b23f034e0d46dd.pdf, accessed 25 February 2009.

Basisdaten (2008) *Daten zur Mediensituation in Deutschland 2008* (Frankfurt).

BBC (2008) *Annual Report and Accounts 2007/08: the BBC Executive's Review and Assessment* (London: British Broadcasting Corporation).

BMWA (Bundesministerium für Wirtschaft und Arbeit) (2000) *Startszenario 2000 – Aufbruch in eine neue Fernsehwelt* (Berlin).
Bundesverfassungsgericht (1986) *Bundesverfassungsgerichtsentscheidung, BverfGE 73, 118 – NiedersachsenUrteil.*
Bundesverfassungsgericht (1991) *Bundesverfassungsgerichtsentscheidung, BVerfGE 83, 238 – NRW-Urteil.*
Bundesverfassungsgericht (2007) *Bundesverfassungsgerichtsentscheidung, 1 BvR 2270/05*, 11 September 2007, www.bverfg.de/entscheidungen/rs20070911_1bvr227005.html, accessed 25 February 2009.
Darschin, W. and H. Gerhard (2003) 'Tendenzen im Zuschauerverhalten. Fernsehgewohnheiten und Fernsehreichweiten im Jahr 2002', *Media Perspektiven*, 4: 158–66.
European Communities (1997) 'Protocol on the System of Public Broadcasting in the Member States', Protocols to the Treaty of Amsterdam amending the Treaty on European Union, *Official Journal* of the European Communities, C 340/109, 10 November (Brussels: European Communities).
Eimeren, B. van and B. Frees (2008) 'Internetverbreitung: Größter Zuwachs bei Silver-Surfern. Ergebnisse der ARD/ZDF-Onlinestudie 2008', *Media Perspektiven*, 7: 330–44.
Grundgesetz (1949) *Grundgesetz für die Bundesrepublik Deutschland*, 23 May (BGBl. I S. 1).
InfoMonitor (2008) *InfoMonitor Jahresüberblick 2008*, www.politik-digital.de/infomonitor-jahresrueckblick-2008-wirtschaftskrise-obama-und-olympia-waren-topthemen, accessed 3 March 2009.
KEF (2007) *16. Bericht, Dezember 2007*, www.kef-online.de/inhalte/bericht16/kef_16bericht.pdf, accessed 3 March 2009.
KEK (2008) *Elfter Jahresbericht*, www.kek-online.de/Inhalte/jahresbericht_07–08.pdf, accessed 12 February 2009.
Knappmann, L. and R. Hönighaus (2008) 'Rundfunkvertrag dient EU als Vorlage', *Financial Times Deutschland*, 24 October.
Krüger, U. M. and T. Zapf-Schramm (2008) 'Programmanalyse von ARD/Das Erste, ZDF, RTL, Sat.1 und ProSieben Sparten, Sendungsformen und Inhalte im deutschen Fernsehangebot 2007', *Media Perspektiven*, 4: 166–89.
Lilienthal, V. (2009) 'Vielfaltsverluste, gesetzlich normiert. Der neue Rundfunkstaatsvertrag, notifiziert von Brüssel', *epd medien*, 86 (29 October): 3–6.
Maier, M., G. Ruhrmann and K. Stengel (2009) *Der Wert von Nachrichten im deutschen Fernsehen. Inhaltsanalyse von TV-Nachrichten im Jahr 2007*, www.lfm-nrw.de/downloads/nachrichtenanalyse_1992–2007.pdf, accessed 3 March 2009.
Ofcom (2008a) *The Communications Market 2008. Part 3 – Television*, www.ofcom.org.uk/research/cm/cmr08/tv/tv.pdf, accessed 3 March 2009.
Ofcom (2008b) *Digital Television Update – 2008 Q3*, www.ofcom.org.uk/research/tv/reports/dtv/dtu_2008_03/q3_2008.pdf, accessed 3 March 2009.
Peters, B. (2009) 'Der "Drei-StufenTest". Die Zukunft der öffentlich-rechtlichen Onlineangebote', *Kommunikation & Recht*, 12(1): 26–39.
Rundfunkstaatsvertrag (2007) *Neunter Staatsvertrag zur Änderung rundfunkrechtlicher Staatsverträge (Neunter Rundfunkänderungsstaatsvertrag)*, in force since 1 March 2007.
Rundfunkstaatsvertrag (2008) *Zwölfter Staatsvertrag zur Änderung rundfunkrechtlicher Staatsverträge*, signed on 18 December 2008.
Überall-TV (2009) www.ueberall-tv.de, accessed 15 February 2009.
Zubayr, C. and H. Gerhard (2009) 'Tendenzen im Zuschauerverhalten. Fernsehgewohnheiten und Fernsehreichweiten im Jahr 2008', *Media Perspektiven*, 3: 98–112.

14

Public Service Communication in Italy: Challenges and Opportunities

Cinzia Padovani

Introduction

RAI (Radioaudizioni Italiane), the Italian Public Service Broadcaster (PSB), has a distinctive history, which, in some respects, sets it apart from other European PSBs. Its mixed funding structure of advertising and licence fee, and its regulatory framework and governance, which have fostered a close relationship between the government of the day and the leadership of the corporation, are some of the most important elements defining RAI's history. A practice known as *lottizzazione* (the allocation of positions of power inside the broadcaster according to a quota system based on political affiliation) and the concentration of media and political power in the hands of Silvio Berlusconi, majority owner of Fininvest/Mediaset[1] (the main commercial broadcaster) and three times prime minister (1994–5; 2001–6; 2008–present) add layers of complexity to this history. In an environment characterised by a large number of parties and a conflict-based democracy (Hallin and Mancini, 2004), and poisoned by decades of an unresolved conflict of interests between Berlusconi's private holdings and his public office,[2] questions arise as to the conditions of the country's PSB, the future of its Public Service Media (PSM) and, more generally, about the significance of the Italian experience for broader concerns about the health of contemporary democracies (Ginsborg, 2005: 185–90).

In order to explore the conditions of public service communication in Italy, the first part of this chapter focuses on the developments that have occurred in RAI during the decade between the late 1990s and the late 2000s. This was an important period for the development of the Italian TV sector as technological advancements (satellite, pay-TV, and the launch of digital terrestrial television – DTT) and new legislation, including the media reform law number 112 of 4 May 2004 (Italian Government, 2004a), paved the way for important changes in the media industry. During this decade, which was politically dominated by Silvio Berlusconi, the public broadcaster suffered the consequences of legislative and financial circumstances that significantly impacted its performance and ability to prepare for the transition to the digital era.

From this historical analysis some questions arise. Given the intense social transformation that occurred during this decade due, in part, to the intensification of immigration from Northern Africa and Eastern Europe, as well as to the process of European expansion and unification, how can RAI (still the largest cultural industry in the country) continue to be an integral part of Italian society? How can it embrace the challenges that new technologies present to the traditional broadcasting model of communication? How can the PSB tap into the needs of an increasingly diverse society?

From a single medium and nationally based broadcaster, in order to stay competitive and socially relevant RAI must evolve into a public service multimedia organisation, serving a new multicultural society. In order to look at these transformations, the second part of this chapter explores RAI's plans with regard to the transition to a multimedia public service corporation. In an attempt to examine how RAI has negotiated its historical role as a nation-based broadcaster with its obligations to contribute to a new multicultural society and to the formation of a European public sphere, the third and final part of the chapter discusses the PSB's coverage of European Union (EU)-related news.

Conditions of PSB

Regulatory framework

RAI remained a monopoly from 1954 (when TV broadcasting began) until the mid-1970s, when private radio and television stations started their transmission. By the mid-1980s, the TV sector had been consolidated in a duopoly consisting of the PSB and the Berlusconi channels. In the absence of effective anti-trust regulations, the duopoly continued for decades: in 2007, the two broadcasters were still commanding 82.3 per cent of national audiences and more than 84 per cent of TV revenues (AGCOM, 2008: 6).

Ad hoc legislation contributed to making this situation possible. Law 223 of 1990 legitimised the concentration in the market by establishing that no broadcaster could control more than 25 per cent (or three) of all available national channels (Italian Government, 1990: Article 15, para 4). Three was exactly the number that RAI and Mediaset each controlled. Law 249 of 1997 apparently sought to curb the duopoly by lowering the anti-trust limits to 20 per cent of national channels (Italian Government, 1997: Article 2, para 6). However, the legislator failed to establish a deadline as to when each broadcaster would have to comply with those limits.

Law 112 of 2004, one of Berlusconi's most controversial, broadened the scope of the advertising market for the calculation of anti-trust limits. Once again, both Mediaset and RAI were able to keep all their terrestrial channels.

In order to promote a 'fast and easy' transition to DTT, the 2004 law imposed on RAI the task of providing coverage of at least 50 per cent of the population by 1 January 2004, and 70 per cent by 1 January 2005 (Italian Government, 2004a: Article 25, para 2). However, the legislator made no provisions for the allocation of the necessary financial resources. Instead of ensuring appropriate support, the law warned that the licence fee should be used exclusively for 'tasks of general public service interest' and that 'no other forms of public financing would be allowed' (Article 6, paras 5 and 6). The law also reasserted a cap on advertising revenues that had been imposed on the public broadcaster fourteen years earlier by the 1990 law (Article 17, para 2, o).

The centre-left government of Romano Prodi, inaugurated in May 2006, gave priority to designing new legislation that would allow more operators to enter the market, free RAI from the influence of politicians and, as a result, promote more pluralism of information. In October 2006, the government passed a reform bill, which imposed stricter anti-trust limitations on analogue broadcasting and the migration to other platforms of any terrestrial channel in breach of those limits. In early January 2007, then Communications Minister Paolo Gentiloni also proposed to reform the PSB. In an effort to free RAI from political interference, it was suggested that the public broadcaster be governed by an independent foundation, whose task would be to 'nominate [RAI's top leadership and], defend [its] autonomy' (Gentiloni in Natale, 2007). The reform of the corporation also foresaw a drastic change in its funding structure: in order for the public broadcaster to become less dependent upon advertising revenues, two of RAI's channels would become completely public and financed entirely by the licence fee, whereas its third channel would turn fully commercial.

Neither the bill nor the proposal to reform RAI made it through the phase of parliamentary discussions. They both encountered fierce resistance, not only from Silvio Berlusconi (at the time leader of the opposition), but also from those who supported the PSB. For instance, according to the Rifondazione Comunista party, Gentiloni's proposal was intended to reduce 'the space of public television' (Bellucci quoted in Natale, 2007), and therefore could not be supported.

With the fall of the Prodi government in January 2008 all parliamentary discussions were interrupted. The return of Berlusconi as head of government in May 2008 further undermined any hope for a serious discussion about reforming the TV sector in Italy.

Educate, inform and entertain?

Since its early days, the financial arrangement of the PSB consisted of a mixed system of advertising revenues and the licence fee, an arrangement established by Decree Law number 655 of 1924. However, since competition for audiences and advertising revenues began in the early 1980s, this mixed source of funding became a reason for criticism. As citizens pay the licence fee, populist critics argue that the broadcaster should only offer distinctive public service programmes. On the other side, if RAI wants to continue to sell airtime to advertisers, then it should turn into a pure commercial broadcaster. Not surprisingly, Silvio Berlusconi has often echoed those arguments, publicly expressing his frustration with the PSB's programming and its funding structure. '[T]he function of private commercial TV and that of public TV – the media-mogul-turned-politician pointed out shortly after becoming Prime Minister for the third time – should be completely separated' (quoted in Natale, 2008). The role of commercial TV should be to 'entertain first, second to inform, and third to educate ... [while that of] public TV should be exactly the opposite: to educate, inform, and maybe, entertain' (ibid.). The issue is indeed one that speaks to the core of public service, its definition and mission in society. Should RAI continue to compete with commercial operators? Or should it focus on education and information, leaving entertainment to the market? Should advertising be taken away, so that a more pure but weaker public broadcaster emerges? And if that were the case, is the government ready to push for a higher licence fee and provide more public funds to compensate for the loss of revenues?

Funding

The point is that RAI is heavily dependent on advertising revenues, with nearly 40 per cent of all its resources coming from this source during the period 2001–7 (see Figure 14.1).

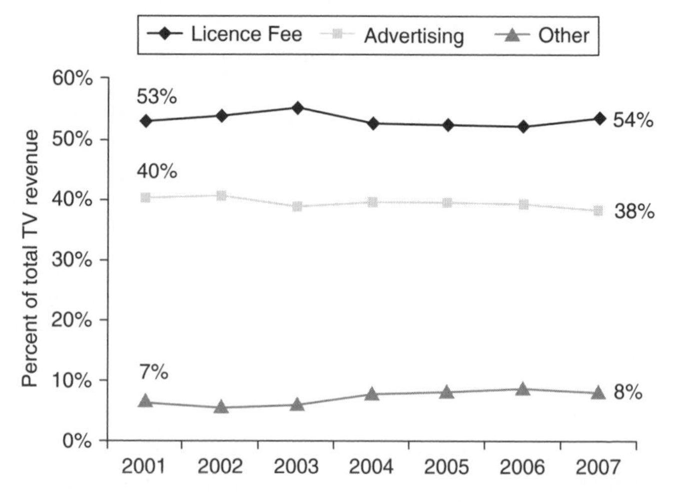

Figure 14.1 Source of RAI's revenue by year (2001–7)
Source: RAI Group (2002, 2005, 2006, 2007).

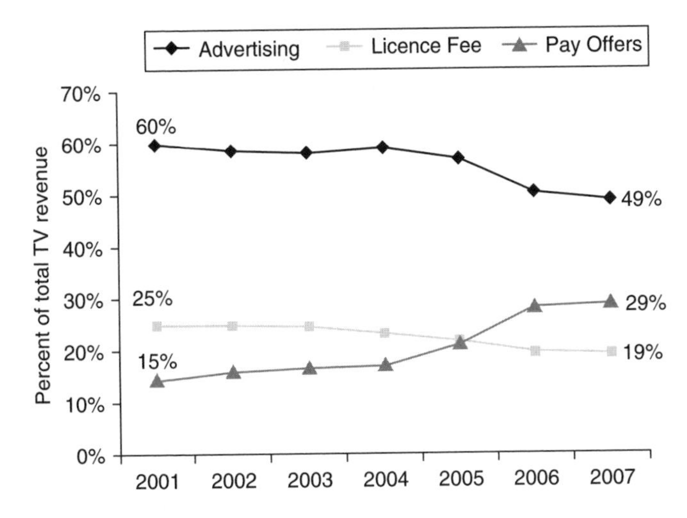

Figure 14.2 Source of TV revenues in Italy by year (2001–7)
Source: AGCOM (2005, 2008).

It should be noted that, between 2001 and 2007, the structure of RAI's revenues appears static, sclerotic if compared to the rest of the TV sector. Indeed, over this period, revenues from advertising for the industry as a whole dropped from 60 to 49 per cent, while those from pay offers increased from 15 to 29 per cent. The contribution of the licence fee to total revenues for the whole sector went down from 25 per cent in 2001 to only 19 per cent in 2007 (see Figure 14.2).

Although RAI remains the industry leader with TV revenues coming in at 2.7 billion euros for 2007, its per annum growth rate of only 2.5 per cent over the 2001–7 period is significantly lower than the overall growth of the industry (7.5 per cent) and considerably less than Mediaset's growth of 4.4 per cent (see Table 14.1). Meanwhile, thanks to increasing consumer demand for pay-TV services, Sky Italia has grown at a remarkable rate with total revenues increasing to 2.35 billion euros in 2007.

Some observers have noticed that Mediaset's better performance (compared to RAI) might have been due, at least in part, to the broadcaster's 'benefi[cial] connection to ... Berlusconi' (Meichtry, 2009). Analysts have underlined, for instance, a direct correlation between Berlusconi's elections to the top political office with an increment of advertising money going to Mediaset channels. In this regard, Paolo Stucchi (head of advertising agency MindShare Italia), pointed out that in 2003 'the share of investment in Publitalia channels increased versus other media ... [and Berlusconi's] share shifted by around 3 to 4 per cent' (quoted in Esposito, 2008). Publitalia, part of the Fininvest holding company, is Italy's biggest advertising agency.

Table 14.1 TV revenues and growth (2001–7) by broadcaster in Italy

Broadcaster	*Revenue (billions of euros)*		*Growth (percentage)*	
	2001	*2007*	*2001–2007*	*per annum*
RAI	2.33	2.7	15.9%	2.5%
Mediaset	1.85	2.4	29.7%	4.4%
Sky	0.82	2.35	187%	19.2%
Other	0.39	0.87	123%	14.3%
All broadcasters	*5.39*	*8.32*	*54.4%*	*7.5%*

Source: AGCOM (2005, 2008).

Low licence fee

Although the public broadcaster commands higher audience shares than other European PSBs, its licence fee remains one of the lowest. In 2007 the level of RAI's licence fee was 104 euros, compared to 287 euros for the Swiss PSB, 221 euros for the Swedish PSB, 204 euros for the German public broadcasters, 201 euros for the BBC, and 117 euros for France Télévisions (RAI Group, 2007).

Moreover, since the early 2000s, the licence fee has not been increased to compensate for cost of living and remained flat for two consecutive years in 2005 and 2006. Once Berlusconi became prime minister in May 2008, he promised to lower the licence fee even further as part of his plans to cut taxes. Although Berlusconi rejects accusations that his private interests colour his decisions with regard to the public broadcaster, 'limiting RAI's revenue stream, regardless of the reason, [inevitably] underscores the inherent impropriety in Berlusconi drafting policy that affects the pub-caster' (Vivarelli, 2008).

High audiences

Overall, audience shares for generalist TV have been diminishing since the early 2000s and RAI's audience has diminished at a faster pace than that of Mediaset. Nevertheless, in 2008 the public broadcaster still commanded the largest average daily audience: over 41 per cent in the whole day and more than 43 per cent in prime time (see Table 14.2).

According to media analyst Francesco Siliato (2009), RAI's high ratings are due mainly to the fact that the PSB 'still maintains an institutional credibility. One that Mediaset never had.' Above all, this is the strength of public broadcasting in Italy.

From duopoly to monopoly

This institutional credibility continues notwithstanding the many problems and the political struggles that surround RAI. For instance, when Berlusconi

Table 14.2 Change in RAI and Mediaset's whole day (24 hours) and prime time (8.30–10.30 p.m.) audience shares (2001–8)

	Whole day (%)			*Prime time (%)*		
	2001	*2008*	*Change*	*2001*	*2008*	*Change*
RAI Networks						
RAI1	23.9	21.8	−2.1	23.7	22.7	−1.0
RAI2	13.7	10.6	−3.1	13.5	10.7	−2.8
RAI3	9.7	9.1	−0.6	10.4	10.1	−0.4
All	*47.3*	*41.5*	*−5.8*	*47.6*	*43.4*	*−4.1*
Mediaset networks						
Canale5	23.5	20.3	−3.2	24.1	20.7	−3.4
Italia1	10.3	10.8	0.5	10.9	10.3	−0.6
Rete4	9.3	8.3	−1.0	8.1	8.6	0.5
All	*43.1*	*39.4*	*−3.7*	*43.1*	*39.6*	*−3.5*
All other networks	9.61	19.09	9.5	9.37	17.02	7.7

Source: RAI Group (2002); Studio Frasi, 2009.

became prime minister in 2001, key appointments inside the corporation were assigned to people from his own entourage, or close to his ideological positions: starting with Antonio Baldassarre, a former Italian Constitutional Court president, supporter of Berlusconi and member of the National Alliance party (an important component of the centre-right government coalition), who was appointed head of RAI's Board of Directors in 2002 (the Board is the one that chooses the heads of each channel and the directors of news programming). In the same year, Agostino Saccà, an outspoken supporter of Berlusconi's party Forza Italia, was named Director General of the corporation. Heads of news programmes were also replaced by people close to the prime minister (Padovani, 2005: 149–53). By the end of his second year in office, the media-mogul-turned-politician had secured control, either directly or indirectly (via proxies and relatives), of 'more than 50 per cent of Italy's advertising market' (Kramer, 2003: 95). '[T]he duopoly – as opposition leader Enrico Boselli emphasized – ha[d now] become a monopoly' (Zecchinelli, 2002).

When Berlusconi became prime minister again in May 2008, the mandate of the existing Board of Directors expired. It took months of negotiations to agree on new appointments, including that of RAI's president and of the president of RAI's Parliamentary Commission, the committee that appoints the majority of the members of RAI's Board of Directors. Obviously, long periods of vacuum and high turnover of its leadership are not good for the public broadcaster: between 1999 and 2009 RAI had nine different presidents and six director generals.

Programming

At times, critics have accused the public broadcaster for its flashy shows on prime time, part of what analysts have defined, since the 1990s, as RAI's *counter scheduling strategy* (Siliato, 2001). The result of such strategy, driven by the 'total hegemony of advertising over programmes' (Bellucci quoted in Natale, 2007), is the homogenisation of content, and a public broadcaster that is, at times, hardly distinguishable from the commercial competitor.

Attempts have been made to change this. For instance, the service contract between RAI and the Italian state dictates that at least 65 per cent of the PSB's overall programming must qualify as public service, which includes sports, children programming, information, public affairs and culture (Ministry of Communications, 2007: Article 4, para 2). In general, the broadcaster has been able to comply with these guidelines by loading up its third channel, RAI3, with public service programmes. By doing so, RAI has had plenty of room for 'entertainment' and 'non-European films and series' (the types of programmes that do not fall into the public service category) on its flagship channels. Furthermore, although the service contract establishes that public service programming should also be aired during the most lucrative time slots, there is very little of it on prime time, either on RAI1 or RAI2. Indeed, these time slots are often filled with game shows and Hollywood films.

Nevertheless, the quality of public service programmes is usually very high, with RAI3 being one of the best channels, with its excellent public affairs shows, and Italian and European fiction. This quality has to be cultivated also in light of the transition to a multimedia public service corporation (PSM), where RAI, like other PSBs, will have to establish a distinctive brand in order to be successful (EBU, 2006). RAI should build upon its credibility as the institution of Italian broadcasting and aim at 'bringing today's users on to digital channels ... while trying to reach new audiences in the process' (Di Stefano, 2007).

The transition from PSB to PSM

RAI's most pressing requirement in the so-called new media sector is to expand its DTT offering. By early 2004, the DTT division was placed under the New Media structure, which also comprised RAINet, RAISat, and RAIClick. Besides its three generalist channels, RAI's DTT offering includes Rai News 24, RaiEdu (in simulcast with the corresponding satellite broadcasts), RaiDoc (specifically created for DTT), RAI Utile (one of the first interactive DTT channels, shut down in 2008 due to poor ratings), Rai Sport, RAI Gulp (for children and teens) and RAI Quattro.

In an effort to expand RAI's digital offering, the 2007 service contract points out that the public broadcaster must provide a 'proper strategy for utilising its editorial production and audiovisual rights' on new platforms (Ministry of Communications, 2007: Article 6, para 1). RAI should also 'improve and update the service that it offers on its websites, in order to promote visibility of [its own] content, especially its broadcasting content' (Article 6, para 2). In order to complete its multimedia offering, the public broadcaster should 'make available ... on the RAI.it portal, [all] radio and television content that is part of its broadcasting offer' (Article 6, para 3, b); 'create specific content for new platforms' (Article 6, para 3, e); 'offer visible spaces for users to communicate and discuss' and 'provide the opportunity for users to publish ... self-produced content' (Article 6, para 3, f and g).

In view of these guidelines, the public broadcaster acknowledges that it 'will need to face important changes in technology ... as regards the younger, more advanced viewing public' (RAI Group, 2006: 9). Indeed, various initiatives have been promoted based on the acknowledgement that, in the 'television industry, it is becoming essential to turn to new platforms, such as DTT, IPTV, WebTV, and mobile TV' (ibid.).

Considerable progress, for instance, has been made with regard to RAI's main internet portal, RAI.it, which, since 2006, has been offering access to live streaming of radio and TV programmes, as well as podcasting and video on demand of selected programmes. In the original draft of the 2007 service contract, then Communications Minister Gentiloni promised that RAI.it would allow users to download all content produced with public funds within the Creative Commons licensing agreement. Unfortunately, that provision was eliminated from the final version of the contract, and RAI has yet to clear the rights of much of its premium programming for use on the internet, or other platforms.

Some progress has also been made with regard to providing more interactive possibilities for internet users, although there is still a long way to go before the promises of full access and interactivity can be realised. For instance, there are few opportunities for users to upload their videos (only eight programmes, the majority of which are reality TV and game shows, let people do that). A noticeable exception is Tg1, the most prestigious news bulletin on Italian TV, the only news programme that allows users to send in their videos. A selection of videos is posted on Tg1's website.

A missing philosophy

Although technological progress is being pursued, what seems to be missing is a philosophy guiding the transition from a public service broadcaster to a public service media corporation. Given its historical position as the institution of Italian broadcasting, RAI should take the lead rather than let the market dictate the quality of evolution in the new media technologies. However, it is the language of the market that permeates the public broadcaster's commitment in the multimedia environment. For instance, instead of lobbying for an increase of public funds, the hope is that the broadcaster's multi-channel capabilities will enable 'RAI to attract additional advertising revenues and other new revenues' (RAI Group, 2006: 9). Rather than looking for ways to engage and promote the formation of new publics, RAI's interest in new media seems to be dictated by the desire to 'increase ... *customer* loyalty' (RAI Group, 2007: 70, emphasis added).

A philosophy guiding the transition should emphasise elements of continuity with the foundational ethos of PSB, while at the same time underlying a paradigmatic shift, which should include the need to rethink ways to reach users and engage the 'public as partner' (Lowe and Bardoel, 2008: 20–2). Obviously one should refrain from believing that new technologies might solve the decades-old problems that have undermined the core philosophy of PSB institutions (Tracey, 1998). In fact, as Splichal (2008: 253) underlines, 'these technological driven changes do not radically challenge the historical roots of public service as an ideal', and cannot lend them legitimacy either.

Certainly, technological opportunities present a chance for RAI to renew its role. But whether or not the public broadcaster will be able to position itself as a provider of quality content in the multimedia environment will depend on whether it finds ways to be socially relevant. A renewed commitment in this sense is necessary: rethinking the role of public service should mean taking into consideration the profound social changes that are transforming the fabric of Italian society. As Richeri (2008) underlines, we should not begin 'from the new technologies, or from the need to compete with private broadcasters', rather we should think about which 'components of the new society [might still] need public TV'. In order to stay relevant, PSM will need to tap into those new components and into their communicative needs.

European news on RAI

Changes in the social and cultural configuration of Italian society, alongside the process of European unification and enlargement, should prompt the broadcaster to negotiate its own identity as a nation-based broadcaster with programming and services addressing these social and cultural changes. Indeed, the 2007 service contract dedicates a whole article to the international requirements of RAI: this includes RAI's obligation to promote Italian culture, history and language abroad; to provide coverage of news and information related to the EU; and to collaborate with European institutions on the realisation of programmes and other cultural initiatives (Ministry of Communications, 2007: Article 9, paras 3 and 5).

However, coverage of European news has never been prevalent on RAI's schedules, mainly due to the fact that such output has never attracted large audiences. Moreover, coverage of EU news has diminished since the early 2000s, when public support for European integration began to fade due to the process of EU expansion on its Eastern borders and the introduction of the euro (Della Porta and Caiani, 2006: 84). The second Berlusconi government

(2001–6) was also a period of uneasy relationships between the Italian government and European institutions, when there was 'even more [of an] interest to resist the influence of the EU in Italian politics' (Quaglia and Radaelli, 2007: 924). Since then, as Roberto Santaniello of the European Committee in Rome points out, EU institutions have come to be perceived as 'constrictive' rather than as sources of 'opportunities' (Santaniello, 2009).

Some of the themes that have defined the process of Europeanisation of public discourse in Italy, including the 'prevalence of national agendas even in relation to purely EU news; and the prevalence of intergovernmental and elite-driven images of the EU' (Della Porta and Caiani, 2006: 79), are confirmed in the findings of the Observatory of Pavia on the coverage of European news in Italian media. Stefano Mosti, the president of the Observatory, reported that the EU was often presented as an 'economic and monetary union [and that] priority [was] given to political institutional subjects linked to the economic and financial sphere' (Mosti in Esposto, 2009). Mosti also found that of all the airtime dedicated to news of political and economic relevance, only 3 per cent covered news related to the EU – of that, more than 60 per cent was aired on RAI channels and only 10 per cent on Mediaset channels. Not surprisingly, RAI3 was the channel with the highest percentage of EU-related news (Mosti, 2008).

Whereas thematic channels and new communication technologies are the media where young users seem to be more likely to engage in the formation of some shared European identity (Presidency of the Council of Ministers, 2007), RAI does not seem to be interested in proposing forums for discussing European-related issues on its multimedia platforms. Indeed, some have given up on the idea that RAI will ever cover topics related to the EU satisfactorily: experts are convinced that local radio stations and thematic channels will be more likely to do so. The path to explore, according to Santaniello (2009), is 'going local', and create synergies of interest between local realities and European issues.

Conclusion

The decade from the late 1990s to the late 2000s has proven to be very important for the development of public service communication in Italy. This was a time characterised by rapid technological developments, the transition from broadcasting to multimedia, and the introduction of legislation to secure the dominance of Berlusconi's group. It was also a decade when the media-mogul-turned-politician governed the country almost continuously, with only brief intervals of centre-left governments. A low licence fee, a 'dwindling advertising market' and years of unsympathetic government have forced the public broadcaster to 'make do with less' (Vivarelli, 2008). The impact of the so-called 'Berlusconi factor' during much of this decade cannot be underestimated.

However, amidst great difficulties – some of which are typical of PSBs in other countries (increased competition, decreased legitimacy, technological challenges), while others are related to the specific problems facing public broadcasting in Italy (frequent changes of leadership, concentration of media and political power in the hands of the head of government, an asphyctic duopoly) – RAI has been able to mantain its 'institutional credibility'. The challenge will be to carry this legacy into the future. This certainly cannot be done without engaging the communicative needs of a new multicultural society. The lack of coverage of EU-related issues seems to be a missed opportunity for RAI to engage in the discussions surrounding the process of EU unification and the formation of a critical European public sphere.

Acknowledgement

I would like to express my gratitude to Dr Rodney Richards for his help with data analysis.

Notes

1. Fininvest is a holding company created in 1978 by Silvio Berlusconi; Mediaset, Fininvest's TV channel, was established in 1996.
2. The existing conflict of interest regulation, passed by the Berlusconi government in 2004 (Law number 215, 20 July), forbids those holding public office from having executive positions in private corporations, and states that private entrepreneurs must relinquish their positions and duties to a designated individual if elected to public offices (Italian Government, 2004b: Article 2, paras 1 and 2). As a result, Berlusconi has been able to maintain ownership of Fininvest, although he had to give up his executive position inside the corporation.

References

AGCOM (2005) *Annual Report on Activities Carried Out and Work Programme*, 30 June (Naples: Autorità per le Garanzie nelle Comunicazioni), http://www2.agcom.it/rel_05/eng/rel_05_eng.pdf, accessed 17 March 2009.

AGCOM (2008) *Annual Report on Activities Carried Out and Work Programme*, 15 July (Naples: Autorità per le Garanzie nelle Comunicazioni), http://www2.agcom.it/rel_08/eng/rel_08_eng.pdf, accessed 17 March 2009.

Della Porta, D. and M. Caiani (2006) 'The Europeanization of Public Discourse in Italy: a Top-Down Process?', *European Union Politics*, 7(1): 77–112.

Di Stefano, T. (2007) 'Il Digitale Terrestre' (interview with the author), Rome, 27 November.

EBU (2006) *Public Service Media in the Digital Age*, 6 June (Geneva: European Broadcasting Union).

Esposito, M. (2008) 'Political Return Paves Way for TV Power-Grab Ploy', *Campaign*, 7 March, Lexis Nexis.

Esposto, E. (2009) 'Verso le Europee 2009. Le elezioni europee e i media locali', home page, http://www.osservatorio.it, accessed 9 March 2009.

Ginsborg, P. (2005) *Silvio Berlusconi: Television, Power and Patrimony* (London and New York: Verso).

Hallin, D. C. and P. Mancini (2004) *Comparing Media Systems: Three Models of Media and Politics* (Cambridge: Cambridge University Press).

Italian Government (1990) Law number 223, 6 August, 'Disciplina del sistema radiotelevisivo pubblico e privato' (Rome: Italian Government), http://www.agcom.it/L_naz/L223_90.htm, accessed 19 April 2008.

Italian Government (1997) Law number 249, 31 July, 'Istituzione dell'Autorità per le garanzie nelle comunicazioni e norme sui sistemi delle telecomunicazioni e radiotelevisivo' (Rome: Italian Government), http://www.agcom.it/L_naz/L_249.htm, accessed 21 April 2008.

Italian Government (2004a) Law number 112, 3 May, 'Norme di principio in materia di assetto del sistema radiotelevisivo e della RAI- Radiotelevisione italiana S.p.A., nonché delega al Governo per l'emanazione del testo unico della radiotelevisione' (Rome: Italian Government), http://www.agcom.it/L_naz/L_112_04.htm, accessed 10 May 2008.

Italian Government (2004b) Law number 215, 20 July, 'Norme in materia di risoluzione dei conflitti di interessi' (Rome: Italian Government), http://www2.agcom.it/L_naz/l_215_04.htm, accessed 24 May 2009.

Kramer, J. (2003) 'Letter from Europe "All He Surveys"', *The New Yorker*, 10 November, pp. 94–105.

Lowe, G. F. and J. Bardoel (2008) 'From Public Service Broadcasting to Public Service Media: the Core Challenge', in G. F. Lowe and J. Bardoel (eds), *From Public Service Broadcasting to Public Service Media* (Göteborg: NORDICOM), pp. 9–26.

Meichtry, S. (2009) 'Mediaset Shakes Up the Italian TV Market', *The Wall Street Journal*, 3 March.

Ministry of Communications (2007) 'Contratto Nazionale di Servizio tra il ministero delle comunicazioni e la RAI Radioaudizioni Italiane S.P.A.', 5 April (Rome: Ministero delle Comunicazioni).

Mosti, S. (2008) 'L'Europa nei media italiani I quadrimestre 2008', paper presented at the conference Davvero l'Europa non sa comunicare? L'Europa e i media italiani, 25 June (Milan: Office of the European Parliament).

Natale, R. (2007) 'Rai: a chi piace la riforma Gentiloni?', *Key4Biz*, 11 January, http://www.key4biz.it, accessed 1 June 2009.

Natale, R. (2008) 'Rai: Berlusconi auspica un ritorno al ruolo del servizio pubblico', *Key4Biz*, 9 June, http://www.key4biz.it, accessed 25 November 2008.

Padovani, C. (2005) *Fatal Attraction: Public Television and Politics in Italy* (Boulder: Rowman & Littlefield).

Presidency of the Council of Ministers (2007) 'Comunicare l'Europa Scarsa informazione dai media, meglio internet: i risultati della consultazione italiana sul libro bianco sulla comunicazione', 19 May (Rome: Presidency of the Council of Ministers, Dipartimento per le Politiche Comunitarie), http://www.politichecomunitarie.it/, accessed 10 June 2009.

Quaglia, L. and C. M. Radaelli (2007) 'Italian Politics and the European Union', *West European Politics*, 30(4): 924–43.

RAI Group (2002, 2005, 2006, 2007) *Reports and Financial Statements*, 31 December (Rome: RAI Group).

Richeri, G. (2008) 'Come ripensare il servizio pubblico radiotelevisivo?' 1 July (Laretenet).

Santaniello, R. (2009) 'EU and the Role of Italian Media' (interview with the author), Rome, 21 February.

Siliato, F. (2001) 'Television' (interview with the author), Milan, 3 October.

Siliato, F. (2009) 'Ascolti RAI' (electronic correspondence with the author), 6–9 June.
Splichal, S. (2008) 'Does History Matter? Grasping the Idea of Public Service Media at its Roots', in G. F. Lowe and J. Bardoel (eds), *From Public Service Broadcasting to Public Service Media* (Göteborg: NORDICOM), pp. 237–56.
Studio Frasi (2009) 'Ascolti 2008' (electronic correspondence with the author), 3 June.
Tracey, M. (1998) *The Decline and Fall of Public Service Broadcasting* (Oxford: Oxford University Press).
Vivarelli, N. (2008) 'Berlusconi Fuels RAI-Mediaset Rivalry', *Variety*, 27 June, Lexis Nexis.
Zecchinelli, C. (2002) 'Berlusconi Crony Wins Top Post at Pubcaster', *Variety*, 14 March, Lexis Nexis.

15

Spanish Public Service Media on the Verge of a New Era

Bienvenido León

Introduction

In the last decade, Spanish national public media have faced a dramatic economic and identity crisis, deriving from a lack of a clear definition of their role in Spanish society. But the new regulation promulgated in 2006 opened the way to a new era, which will be discussed in this chapter. The work starts by explaining the transformation process and the main characteristics of the public media institution, in the light of the new regulation. It then moves on to discussing whether the performance of Public Service Media (PSM) has been effective in the digital environment under the new strategy that has been adopted. Finally, it analyses the output of the television offerings in order to determine whether the new arrangements contributed to a better fulfilment of the public service role.

Reorganisation of PSM in Spain

Spanish national PSM are integrated in the RTVE group (Radiotelevisión española), which includes television (TVE, with the generalist channels TVE1 and La 2), radio (RNE), the rtve.es website, a training institution (IORTV), a symphonic orchestra (Orquesta Sinfónica de Radiotelevisión Española) and a record label (RTVE). RNE was launched in 1937 whereas TVE began television broadcasts in 1956. TVE maintained a monopoly in the Spanish television market until 1986, when the first regional public channel appeared. When commercial television began in 1990, the resulting multiplication of channels led to an unprecedented commercialisation of advertising-funded TVE programming. This inadequate funding model resulted in a profound economic crisis, with the group reaching in 2006 an accumulated debt of over 7.8 billion euros (*El País*, 2006), which the state had to take on.

Besides the economic crisis, Spanish public television has been frequently accused of lack of independence from the government, with criticisms originating from the political parties of the opposition, trade unions and viewers' associations. The lack of a clear definition of the public TV remit, alongside the lack of independence from political control have been among the most negative facets of the Spanish audio-visual panorama (Telefónica, 2006: 338). Indeed, as Bustamante (2006) recalls, since 1975, when democracy was established in Spain, governments of all political colours have used TVE to promote their own interests. The lack of independence and the frequent changes in management also impacted on the capability to construct a coherent strategic plan for the corporation.

After winning the 2004 election, the Socialist government set the reform of public media as a priority. A few months later an independent commission of experts was formed tasked with the analysis of the situation and proposing solutions. After nine months of discussions, in February 2005, the commission completed the report on the reform of PSB, which concluded that Spanish public broadcasting was 'obsolete and deficient', and pointed towards the commercialisation of programming and an unfeasible financial situation, as the two major problems. On the one hand, TVE's programming, as a whole, had come close to that of its commercial competitors, which blurred its image as a public service provider. On the other hand, the huge debt accumulated along the years was difficult to maintain (Consejo para la reforma de los medios de comunicación de titularidad del Estado, 2005).

Based on the report of the commission of experts, in May 2006 the Parliament approved a revision of the regulation concerning PSB. The aim of the new law was to provide a framework that would guarantee PSB's 'independence, neutrality and objectivity, providing organisation structures and a funding model that would allow it to fulfil its public service mission with efficiency, quality and public recognition' (Ley 17, 2006). The new RTVE is a state-owned corporation with 'special autonomy', ruled by a Board of Director made up of twelve members who are elected for a period of six years by the Parliament (eight by the Congress and four by the Senate). The Parliament also appoints the president of the Board and of the corporation. The law provides for the independence of the corporation from the government by establishing that the members of the Board and the president must be appointed by a majority of two-thirds of the Congress or the Senate. Luis Fernández, appointed president of the new RTVE in January 2007, was, for the first time in the history of the institution, elected after agreement of the main political parties.

The independence of the corporation is ensured by the setting up of three bodies for external control, at the Parliament, the Court of Accounts (Tribunal de Cuentas) and an independent audio-visual authority (Consejo Audiovisual), which has yet to be created. As an independent authority, Consejo Audiovisual could play a crucial role (as is the case in other countries) by guaranteeing that citizens' rights are not infringed. In addition, the Councils of Information (Consejos de Informativos), created in July 2008 and elected for a period of two years, are formed as the internal bodies through which journalists can oversee editorial autonomy. TVE's Council consists of thirteen members, whereas that of RNE has nine.

The new regulation provides for a fresh financial model. Based on the suggestions of the experts' commission, the current financial arrangement consists of public subsidies (which account for 45–50 per cent), advertising (40 per cent) and commercial activities (10–15 per cent). At the beginning of this new era, the state once again assumed the huge accumulated debt of the previous years and signed a new agreement to provide subsidies, so that the corporation could start as a 'sane, strong and competitive' institution (Atarés, 2006). The 2007 budget of the corporation was 1.2 billion euros, which reflected a reduction of 22.32 per cent, compared to 2006. The state subsidies accounted for 433 million euros (35.8 per cent) (RTVE, 2008a). In 2009, reductions have continued, coming down to a total budget of 1.1 billion euros, 50.6 per cent of which consisted of public subsidies (RTVE, 2008b). But compared with its counterparts in the UK, Germany and Italy, funding for the Spanish PSB is insufficient to cover operational and programming costs (Bustamante, 2008b: 188). For example, the new law foresees a reduction of advertising. In 2009 TVE will include a maximum of ten minutes of advertisements per hour (instead of twelve minutes allowed by European regulation for public and commercial broadcasters). It has to be said, however, that this reduction is much smaller than the one that has occurred in France since January 2009, where public channels do not carry any advertising from 8 p.m. until 6 a.m. (*El Mundo*, 2009a).

Meanwhile, a new strategic plan was adopted, in order to improve productivity and efficiency. The most dramatic part of the new strategy was a severe reduction of staff, a process started in 2006. Over 4150 employees (almost 50 per cent of the permanent staff) from 50 to 65 years old were given incentives for early redundancy or were forced to retire. The new public corporation started operation in January 2007 with a staff of 6400 workers, of whom 5900 were permanent and 500 temporary (Gómez, 2007). The current number of staff is clearly smaller than that of PSBs in the UK (over 20,000), Germany (over 20,000), France (over 15,000) or Italy (over 10,000). The plan, however, has been criticised on the grounds of wasting an experienced generation of professionals who were still productive (*El País*, 2009).

Public service for the 'information society'

In some European countries, public broadcasters have played a leading role in the development of the so-called 'information society' (Iosifidis, 2005), but in Spain, the report of the commission for the reform of PSB claims that, until 2005, official policies have deprived RTVE from playing a leading role in the development of digital services (Consejo para la reforma de los medios de comunicación de titularidad del Estado, 2005: 116). Consequently, the new law of 2006 explicitly mentions that PSB should contribute to the development of the 'information society', and establishes that public media must develop new interactive services that 'enhance or complete their programming offer and bring public administration and citizens closer' (Ley 17, 2006).

In fact, RTVE has only modestly contributed to the development of digital services in the country. As far as digital terrestrial television (DTT) is concerned, since 2005 TVE owns five of the eighteen frequencies allocated by the government. These are used to broadcast the two generalist channels TVE1 and La 2, which are also available in an analogue format, besides four thematic channels: Canal 24 horas (news), Clan TVE (children), sharing frequency with Canal 50 años (archive programming), and Teledeporte (sports). But all these channels were already available either as analogue broadcast offerings or through cable or satellite platforms and therefore none was specifically created for DTT, as a contribution to develop this new technology.

The multiplication of channels boosted by digital technology poses some new challenges and opportunities for PSB, including opening the way for creating content to meet minority needs as well as distinctive universal content, as suggested by the European Broadcasting Union (EBU, 2002). RTVE seems to have adopted a mixed strategy that combines the above: on the one hand, the group maintains two generalist channels (TVE1 and La 2) that carry popular output and, on the other, it offers some niche channels addressed to specific target audiences (for example, Clan TVE, for children; and Teledeporte, for sports fans). In this context the public group tries to offer programmes of various genres, to satisfy diverse types of audiences, as we will discuss in the next section of this chapter. In theory, this would have been a valid strategy, similar to what other European PSBs have adopted. In practice, though, economic difficulties have not allowed TVE to create an attractive set of free-to-view channels that could be of interest to those citizens not willing to pay for television content.

In terms of ratings, the Spanish audience seems to be uninterested in these digital offerings: in December 2008, the audience share of all DTT channels in Spain was 21.9 per cent of all television viewing. As Table 15.1 shows, among them, RTVE's channels held 20 per cent of the DTT share. TVE1 was in the third position, behind commercial broadcasters Antena 3 and Telecinco. Among the thematic channels, Clan TVE was second, trailing Antena.Neox, part of Antena 3 group (Mundoplus.tv, 2009).

Although initially one multiplex of the DTT spectrum was allocated to RTVE, the regulation foresees that after the 'analogue switch-off', scheduled to occur in March 2010, RTVE will have two full multiplexes, which will allow for eight channels (Real Decreto 944, 2005). One of the channels that will be included in the new offer is 'Canal cultura', a cultural niche channel which has been created as a joint venture with the Spanish Ministry of Culture and

Table 15.1 DTT audience shares of Spanish channels (December 2008) (%)

Antena 3	13.9
Telecinco	13.4
TVE1	12.4
La Sexta	8.9
Cuatro	7.4
Antena.Neox	3.5
Clan TVE	3.0
Disney Channel	3.0
La 2	2.8
Antena.Nova	2.1
FDF	1.1
Telecinco 2	1.0
Teledeporte	1.0
Intereconomía	0.9
24 horas	0.8
Sony TV en VEO	0.8
CNN+	0.7
40 Latino	0.6
Veo	0.5
Hogar 10	0.4

Source: Mundoplus.tv.

is devoted to artistic content of any kind (literature, cinema, theatre, music, and so on) with the aim of spreading Spanish culture abroad (*El Mundo*, 2007).

Further, TVE's offering of interactive services is very limited. Since July 2006 it has an Electronic Programming Guide (EPG) of the group channels. The interactive offer also includes a digital teletext, traffic reports, weather and stock-exchange information, as well as an employment service (Mundoplus.tv, 2006). Since March 2007 it has been made possible to fulfil some forms for tax declaration through the television set (RTVE, 2007).

As Hujanen (2004: 133) points out, the launch of DTT accelerates the transformation of PSBs from production-oriented institutions to programming-oriented industries. In the new environment, 'public service television as a unique, original production will be replaced by *public service as a brand*, developed and maintained by strategic programming choices and marketing campaigns'. In other words, PSM need to increase their programme output to ensure a sufficient presence in the new digital media. According to its president, Luis Fernández, RTVE embraced the new notion 'more than one decade late, but the moment to react has come' (RTVE, 2008c). However, the backwardness of RTVE in providing online services is evident, compared to other European broadcasters. In June 2007 the online reach of the group websites was only 4.9 per cent of the Spanish population, the lowest percentage of all EBU members (EBU, 2007: 3).

In this context, it is not surprising that, in May 2008, the public group started an online venture (rtve.es), which was presented as an ambitious plan to upload all RTVE content on the internet, including the huge library of TVE and RNE with over one million hours of programmes. Similar initiatives of archive digitisation have been undertaken by the BBC and various national PSBs in the Netherlands, France and Norway, thereby creating some unease among commercial channels. In the UK, for instance, the BBC initiative to upload its archive online and offer it for free has created controversy and opposition from commercial groups who argued that this might create unfair competition. But contrary to the British case, RTVE's initiative has not raised a similar controversy, for Spanish commercial companies are not planning to offer any such services.

Among the most outstanding elements of rtve.es, there is a news service (produced by a separate newsroom), as well as video and audio on-demand services. Users can access most television and radio programmes, including a four-minute version of television newscasts and an ample offering of archive programmes. Staffed with forty journalists and thirty technicians, it offers blogs of the best-known presenters of the group (RTVE, 2008d).

Experts foresee a dramatic increase of television consumption through mobile phones and personal digital assistants (PDAs) within the next few years (EBU, 2007) and the Spanish PSB is trying to position itself in this emerging market. For example, during the Beijing Olympic Games, in the summer of 2008, RTVE offered, for the first time, television broadcasts for mobile devices. It included live video of TVE1, La 2 and Teledeporte, as well as some summaries of the most important events in a special format. But no further initiatives have been taken and it might be fair to say that RTVE is also behind in this area compared with other Northern European PSBs.

To sum up, RTVE's position in the digital environment is ambivalent. On the one hand, the thematic channel offerings on DTT do not seem to attract a significant number of citizens, while the interactive services available are very limited. But, on the other hand, the ambitious launch of the website, as well as the recently created new services for mobile gadgets, can be regarded as the first outcome of a new strategy, which could lead the public group to strengthen its position in this new environment.

Programming

Although the new legal framework provides motives for more plural and objective PSB news programmes (Bustamante, 2008a: 24, for example, points out that in this new phase there is 'visible independence of the information'), in general terms, the new era of RTVE has not been accompanied by a subsequent change in its television output. Along with the implementation of the

Table 15.2 TVE1 and La 2 programming by genre as % of broadcast time

	2004		*2007*	
	TVE1	*La 2*	*TVE1*	*La 2*
News and current affairs	35.3	10.5	42.5	9.5
Fiction	24.2	14.7	24.5	34.4
Entertainment	17.6	13.2	15.3	10.8
Cultural programmes	8.1	18.0	9.4	20.5
Sports	5.0	15.6	3.1	16.6
Quiz shows and game shows	2.2	2.7	4.1	1.6
Other	7.6	25.3	1.1	6.6
Total	100	100	100	100

Source: Author's analysis based on data from RTVE (2005) and UTECA (2008).

above-mentioned structural and strategic changes, the managers of the group have opted for an increase in the offering of public service programming and asserted that achieving high audience ratings should not be a goal in itself. However, maintaining advertising as a fundamental source of revenue makes it very difficult to implement this policy. The concern is that the need to achieve high ratings to meet advertisers' needs will result – as it has done in the past – in airing populist programming which often does not fit into the public service mission, especially in peak hours.

Table 15.2 shows that in the years 2004–7 both generalist channels TVE1 and La 2 kept their programming strategy unchanged. The first pillar of TVE1 is 'news and current affairs', which has even increased in the period under review, while the second most popular genre ('fiction') remains unchanged as far as the flagship channel TVE1 is concerned. La 2 remains a cultural and sports channel, although the amount of fiction has increased dramatically between 2004 and 2007. The data indeed reflect the profile of a generalist channel (TVE1), in which the search for high audience ratings is of paramount importance. Compared to commercial channels, TVE1's programming offers more informational content but approximately the same level of fiction. However, according to some critics, these differences are not substantial enough to differentiate between public and private channels, and may delegitimise RTVE's public service role in the audio-visual market (Artero, 2008: 120). La 2 differs more from its commercial competitors, with a higher percentage of cultural programmes, although it keeps similar levels of fiction (UTECA, 2008). This confirms the earlier statement about the mixed strategy of the group, in which generalist channels try to compete in various areas of programming.

In addition, the union of commercial channels (UTECA) has criticised TVE for spending a huge amount of money on acquiring American films. UTECA

Table 15.3 TVE 1 prime-time programming – number of programmes and (%) by genre (2004–8)

	2004	*2008*
Fiction series	13 (26.0)	9 (16.9)
News and current affairs	5 (10.0)	9 (16.9)
Magazines	3 (6.0)	1 (1.9)
Sports	2 (4.0)	2 (3.8)
Game shows and quiz shows	4 (8.0)	4 (7.6)
Reality shows and reality games	0 (0.0)	2 (3.8)
Documentaries	3 (6.0)	3 (5.7)
Feature films	7 (14.0)	14 (26.4)
Talk shows	5 (10.0)	8 (15.1)
Music5	5 (10.0)	0 (0.0)
Humour	2 (4.0)	1 (1.9)
Other1	1 (2.0)	0 (0.0)
Total	100.0	100.0

Source: Author's analysis.

considers that public television should promote the national audio-visual industry rather than buying expensive content produced outside Europe (*El Mundo*, 2009b). Theorising, one could argue that the mission of PSM is linked to diversity, as a way of serving a wide audience (Blumler, 1992). A relevant indicator is the number of different types of programmes (genres) of a channel or a programming slot within a period of time, also known as 'vertical diversity' (Ishikawa, 1996; Litman, 1979; McQuail, 1992). Since TVE1's prime-time slot reaches larger audiences than other slots, it is especially relevant to know if programming diversity here has been increased as the PSB enters a new phase. Therefore we have analysed the number of programmes of each genre that TVE1 has broadcast on prime time in 2004 and 2008.[1] The results are portrayed in Table 15.3. Based on the data presented in this table, we have calculated the index of diversity for this slot. We have used the method of Dominick and Pearce (1976: 73), which reflects 'the extent to which a few categories dominate prime-time'.[2]

The index of diversity of TVE1 prime time is lower in 2008 (39.8) than in 2004 (50) (see note 2 for more details). This reduction is due to the fact that the channel has increased substantially the number of feature films and, at the same time, has maintained a relatively high percentage of fiction series and current affairs programmes. In spite of the reduction in the index of diversity, prime time for TVE1 relies on similar types of programmes in both years, as illustrated by the percentages shown in Table 15.2. This seems to confirm that economic profitability remains one of the main criteria for programming choices. In this regard, the new era of the public corporation has in fact meant a step backwards in the fulfilment of its public service mission.

Another important way of serving a diverse audience are the programmes addressed to specific interest groups, including minorities. La 2 airs programmes for different religious groups, like *Últimas preguntas*, *Testimonio*, *El día del Señor* and *Pueblo de Dios* (Catholics), *Buenas noticias TV* (Evangelists) and *Shalom* (Jews), but it has to be said that similar programmes were already on air before RTVE's restructuring.

In sum, as far as television programming is concerned, the new phase has not meant a better fulfilment of RTVE's mission as a public broadcaster. Compared to the commercial channels, there are some differences between the output of TVE1 and particularly La 2. However, TVE1's output – which is of capital importance, considering it reaches large audiences – remains very similar to that of the commercial competitors, and it has even reduced its diversity on prime-time slots.

Conclusion

Spanish national PSM are in the midst of a process of change that may lead to notable advances in the fulfilment of their public service mission in a modern society. The new legislation provides an appropriate framework to establish a new organisation that can effectively serve the citizens in the new millennium. Of special importance among the new positive regulatory provisions are those addressed to ensure independence from the government and pluralism, as well as the new strategic organisational plan.

In the new era, some important steps have been taken to ensure a more plural content. The mechanisms established by the new regulation for the election of the Board of Directors and the president of the corporation are of particular importance. The setting up of bodies for internal and external control can also play a crucial role, although it is still early days to assess their effectiveness. The new mechanisms for political and economic control are crucial to the survival of the group, and the imminent creation of the independent external authority (Consejo Audiovisual) would complete the set of necessary mechanisms.

However, some shadows have appeared in the current panorama of RTVE. The backwardness in the development of multi-platform content and services for the digital environment is especially relevant. In this regard, the recent creation of the new website rtve.es seems appropriate, as the internet will play a fundamental role in the distribution of content in the digital age, but it is not enough and more online initiatives should be considered. It is crucial that RTVE does not repeat the mistakes of the past and takes on a leading role in the development of the digital society.

In general terms, the programming strategy adopted by TVE in the new phase seems adequate, since it combines two generalist channels (TVE1 and

La 2), addressed to large audiences, with some niche channels for special interest groups. However, it is advisable that TVE1 modifies its programming grill (especially on prime time), so that it differentiates itself more clearly from the commercial channels, thereby increasing diversity. To this effect, the new financial model may be positive for the transformation of the public group, since it will reduce dependence on advertising income. However, there is a danger that public funds will not compensate this reduction, and the total budget may be gradually reduced, to the detriment of the quality of the output.

In the near future, a strong public service group can play an important role in the Spanish media ecology. It can work as a counterweight to commercial media, in order to ensure the provision of a wider range of quality content that would help to spread cultural values addressed to an audience of citizens rather than an audience of consumers. This is a crucial moment, as public media could either reproduce the mistakes of the past, or adopt new strategies that could effectively allow them to fulfil a public service mission.

Notes

1. We have selected a sample of four weeks per year. 2004: 9–24 January, 12–18 April, 12–18 July, 8–14 November. 2008: 19–25 January, 14–20 April, 14–20 July, 10–16 November. Data were obtained from programming guides. Prime time in Spain goes from 10 p.m. to 1 a.m. Total number of programmes codified = 103.
2. The index is calculated by summing the percentages of the top three categories (genres) and subtracting this sum from 100. A low percentage indicates a restricted range of choices for the audience. Following this method, we can establish the following indexes of diversity for TVE1's prime time: In 2004, the main genres are 'fiction series' (26 per cent), 'feature films' (14 per cent) and 'news and current affairs' (10 per cent). Therefore, the index of diversity is: $100 - (26 + 14 + 10) = 50$. In 2008, the main genres are 'feature films' (26.4 per cent), 'fiction series' (16.9 per cent) and 'news and current affairs' (16.9 per cent). Therefore, the index of diversity is: $100 - (26.4 + 16.9 + 16.9) = 39.8$.

References

Artero, J. P. (2008) 'Los contenidos de las cadenas', in UTECA, *La televisión en España. Informe 2008* (Barcelona: CIEC-Ediciones Deusto), pp. 81–122.

Atarés, M. L. (2006) 'Moraleda afirma que la oferta de Sepi para RTVE es "mayor que la de Izar"', *El Mundo*, 27 June, www.elmundo.es, accessed 10 January 2009.

Blumler, J. (ed.) (1992) *Television and Public Interest: Vulnerable Values in West European Broadcasting* (London: Sage).

Bustamante, E. (2006) 'Hacia un servicio público democrático', in Telefónica, *Tendencias 2006. Medios de comunicación* (Madrid: Telefónica), pp. 357–62.

Bustamante, E. (2008a) 'La televisión en la era digital. El debate español en la encrucijada', in B. León (ed.), *Transformar la televisión. Otra televisión es posible* (Seville: Comunicación social ediciones), pp. 22–31.

Bustamante, E. (2008b) 'Public Service in the Digital Age: Opportunities and Threats in a Diverse Europe', in I. Fernández and M. de Moragas (eds), *Communication and Cultural Policies in Europe* (Barcelona: Generalitat de Catalunya-Cátedra Unesco de Comunicació), pp. 186–215.

Consejo para la reforma de los medios de comunicación de titularidad del Estado (2005) *Informe para la reforma de los medios de comunicación de titularidad del Estado* (Madrid: Ministerio de la presidencia), www.mpr.es, accessed 20 October 2008.

Dominick, J. R. and M. C. Pearce (1976) 'Trends in Network Prime Time Programming, 1953–1974', *Journal of Communication*, 26: 70–88.

EBU (2002) *Media with a Purpose: Public Service Broadcasting in the Digital Era* (Geneva: European Broadcasting Union).

EBU (2007) *Broadcasters and the Internet: Executive Summary* (Geneva: European Broadcasting Union).

EBU (2008) *Mobile Content: Any Time, Anywhere in Any Format?* (Geneva: European Broadcasting Union), www.ebu.ch, accessed 10 January 2009.

El Mundo (2007) 'TVE crea un canal cultural con el Ministerio de Cultura', 12 October, www.elmundo.es, accessed 14 January 2009.

El Mundo (2009a) 'Desaparece la publicidad nocturna de las cadenas públicas francesas', 5 January, www.elmundo.es, accessed 10 January 2009.

El Mundo (2009b) 'Uteca critica que TVE y Forta hayan gastado 10 millones en cine americano', 11 January, www.elmundo.es, accessed 12 January 2009.

El País (2006) 'RTVE cerrará el año con una deuda acumulada de 7.811 millones', 22 September, accessed 22 October 2008.

El País (2009) 'Rosa María Calaf. Ahora tengo que devolver algo de lo aprendido', 8 January, www.elpais.com, accessed 20 January 2009.

Gómez, R. G. (2007) 'La nueva vida de RTVE', *El País*, 15 January, www.elpais.com, accessed 10 November 2008.

Hujanen, T. (2004) 'Public Service Strategy in Digital Television: From Schedule to Content', *Journal of Media Practice*, 4: 133–53.

Iosifidis, P. (2005) 'Digital Switchover and the Role of the New BBC Services in Digital Television Take-up', *Convergence: The International Journal of Research into New Media Technologies*, 11(3): 57–74.

Ishikawa, S. (ed.) (1996) *Quality Assesment of Television* (Luton: University of Luton Press).

León, B. (2007) 'Commercialisation and Programming Strategies of European Public Television: a Comparative Study of Purpose, Genres and Diversity', *Observatorio (OBS*) Journal*, 2: 81–102.

Ley 17 (2006) de 5 de junio, de la radio y la televisión de titularidad estatal (Madrid: Boletín Oficial del Estado), www.boe.es, accessed 15 October 2008.

Litman, B. (1979) 'The Television Networks, Competition and Program Diversity', *Journal of Broadcasting*, 23: 393–409.

McQuail, D. (1992) *Media Performance: Mass Communication and the Public Interest* (London: Sage).

Mundoplus.tv (2006) 'RTVE presenta sus aplicaciones interactivas en TDT', www.mundoplus.tv, accessed 7 November 2006.

Mundoplus.tv (2009) 'Audiencias TDT diciembre', www.mundoplus.tv, accessed 15 January 2009.

Real Decreto 944 (2005) 29 July (Madrid: Boletín oficial del Estado), www.boe.es, accessed 10 January 2009.

RTVE (2005) 'Informe 2004', www.rtve.es, accessed 20 January 2009.

RTVE (2007) 'Desde este jueves se puede solicitar el borrador de la Renta a través de la TDT de RTVE', 1 March.

RTVE (2008a) 'RTVE cierra 2007 con un resultado positivo de 18,4 millones de euros, antes de impuestos', 12 March.

RTVE (2008b) 'RTVE prevé reducir sus gastos en 2009 en casi 99 millones de euros respecto al presupuesto de 2008'.
RTVE (2008c) 'Luis Fernández: RTVE quiere entrar con paso firme en la tercera era de la televisión', 27 March.
RTVE (2008d) 'Un canal de noticias y programas a la carta, primeras grandes ofertas de rtve.es', 20 May.
Telefónica (2006) *Medios de comunicación. Tendencias 2006* (Madrid: Telefónica).
UTECA (2008) *La televisión en España. Informe 2008* (Barcelona: CIEC-Ediciones Deusto).

16

Squeezed and Uneasy: PSM in Small States – Limited Media Governance Options in Austria and Switzerland

Josef Trappel

Introduction

This chapter concentrates on ownership issues, legal provisions, market performance and the public debate on Public Service Media (PSM) in Austria and Switzerland. In both countries there is a unique situation as television offerings are strongly influenced by non-national players. Large neighbours influence the development of public and private broadcasting by setting standards and absorbing considerable amounts of viewers' time and advertisers' money – with no adequate balance or reciprocity. Smaller states appear vulnerable. Despite this challenge – or because of it – both public service organisations in Austria and Switzerland have developed a highly sophisticated model and understanding of the notion of public service. In Switzerland, the formal owner of the public service broadcaster SRG SSR is a civil society-rooted membership association, having the last say on strategic decisions. In Austria, the ORF has recently worked out and published a media governance report on the role and conduct of PSM in society. The chapter argues that expectations of ownership diversity and programming autonomy in smaller states such as Austria and Switzerland are unrealistic. Thus, media governance requirements concentrate on the conduct of PSM organisations.

Background

Austria and Switzerland share several characteristics: both countries qualify as small states, according to any classification discussed in literature so far (Meier and Trappel, 1992; Puppis, 2009: 235; Siegert, 2006; Trappel, 1991a). At least three criteria are used for these classifications: geographical size, population and economic performance. Austria's territory covers 84,000 km^2, it is inhabited by 8.4 million people and its GDP per capita reached 30.800 euros in 2007. Switzerland covers 42,300 km^2, has a population of 7.7 million and its GDP per capita of 34.200 euros in 2007 exceeded that of Austria (Eurostat, 2009). In contrast to geographical size, population size and economic performance are important indicators for media markets. Population size is important for the potential volume of revenues from media subscriptions, while economic performance is a crucial indicator for advertising revenues.

Both countries share the national language with neighbours that are much larger in size and much stronger in economic terms, including audio-visual production. While Austria shares its single nationwide spoken language with Germany (and, of course, Switzerland), Switzerland is divided into four language areas, three of which are shared with much larger neighbours, namely France, Germany and Italy. The fourth language, Rhaeto-Romanic, is spoken only by a small minority.

Austria and Switzerland manage to maintain a high number of television services for their respective population: in Austria, three national television channels (two public, one private) and five national radio channels (four public, one private) are available by air, cable and satellite. Local and regional services complement the national infrastructure. In Switzerland, each of the three larger language areas is supplied with two public television channels and at least three public radio networks. Private operators cover smaller areas within these language territories. It is obvious from these few basic facts that in both countries the PSM organisation is the dominant player in the field of radio and television (see Table 16.1).

Starting from this observation, the following text will discuss three main claims with respect to the position and perspective of PSM:

- Austria and Switzerland represent small states. This fact determines to a large extent the mass media landscape. Inevitably, media ownership concentration is a dominating feature. Both countries are strictly limited to a small number of media companies controlling large portions of the respective media markets (both audience markets and advertising markets). PSM organisations play a dominating role in the audio-visual media markets. Ownership diversity in broadcasting markets is an unrealistic objective for Austria and Switzerland. Private radio and television operators are (and are likely to remain) marginalised for the foreseeable future.
- High levels of media ownership concentration require an elevated level of responsibility by the dominating actors. Generally, the stronger media companies dominate their respective markets, the higher their public accountability becomes – irrespective of whether they like it or not. PSM in Austria and Switzerland are dominating forces in their radio, television and (partly) in the online news markets. Therefore, a high degree of social responsibility and public accountability are to be expected. In both countries, PSM undertake efforts to live up to these expectations.
- The available space for governance action is limited in Austria and Switzerland. Both countries are highly dependent on policy decisions taken by

non-national forces, such as the European Union (EU), as well as neighbouring countries. National media governance is thus restricted to a rather small area. These restrictions equally apply to the national governmental level as well as to the corporate governance level. Programming policy depends as much on decisions taken by non-national competitors as broadcasting policy on decisions taken by the non-national institutions.

Social science research on the characteristics of small states in the mass media field has been on the research agenda for some twenty years (Siegert, 2006: 191). In the late 1980s and early 1990s there were some efforts to examine the specific development space for small states in Europe (Kleinsteuber, 1990; Meier and Trappel, 1992; Trappel, 1991b). Small states in Western Europe were vaguely but sufficiently defined as smaller than the big five European countries (France, Germany, Italy, Spain and the UK) and larger than the dwarf states like Andorra, Liechtenstein and Luxemburg. The focus was on small states in Western Europe. The following common features have been listed (Meier and Trappel, 1992; Trappel, 1991a):

- Shortage of resources: small states lack financial/capital resources. Investment into high-risk business such as film production is rarely executed in small states. Large markets seem to provide better opportunities for a positive return on investment than products developed for small states. If – in exceptional cases – large amounts of media capital are available in small states there is a tendency to invest it abroad in larger markets with better perspectives to exploit economies of scale and scope. Moreover, small states suffer from a shortage of talent, well-trained media professionals, attractive carrier options and so on. It is simply much more likely for talented actors, musicians and media managers to follow a carrier in larger states than to restrict themselves to the fewer options in small ones.
- Small market size: both advertising revenues and direct revenues from viewers/readers/users are limited by the size of the market. Income from these two main sources is low in absolute terms compared to larger countries. The cost of producing television programmes and newspapers, however, is almost as high as in larger countries, given the economic characteristics of high fixed cost (capital cost for setting up studios and printing presses, labour cost for first copy) and low marginal cost (additional copies and/or viewers/listeners/users) (Doyle, 2002: 9).
- Dependence: smaller states depend to a large degree on decisions taken by actors outside the country with little power to influence preparatory deliberations. Television programme formats are unlikely to be developed in response to audience needs of small states but rather with a view to mass markets. Anglo-Saxon, French and German formats are developed for English, French and German language markets, ignoring specificities of small states within these language areas, such as Austria, Belgium and

Ireland. Moreover, transnational media policy decisions rarely consider the needs of smaller states. The history of the Television Without Frontiers Directive, now the Audiovisual Media Services Directive (Humphreys, 2008), provides ample evidence. It is rarely the small states' media companies that profit from the removal of trade barriers in the audio-visual field, as the main transnational media companies in Europe are located in one of the larger states (such as Bertelsmann/RTL in Germany, Canal Plus in France, and so on).

- Vulnerability: there is no balance in relations between larger and smaller states, in particular in those territories that share the same language. Media companies in smaller states are much more likely to be taken over by companies based in larger states than vice versa. In terms of media consumption, media products from smaller states are generally less consumed in large language areas than products originating in larger countries. This imbalance is simply due to the fact that television programmes from larger states are readily available in small states sharing the same language while programmes from smaller states normally are not available in larger ones.
- Corporatism: small states tend to organise their media policy within relatively closed circles. Included in these circles are the dominating social forces such as ruling political parties, large trade unions and vested industrial interests (in particular, the advertising industry). Excluded are civil society actors and smaller interest groups such as journalist unions, representatives of third sector media actors and the viewing, listening and reading public at large.

In early 2009, a Special Issue of the *International Communication Gazette* was published – and the main findings mirror what we stated almost twenty years ago. In the introduction of this Special Issue, one of the editors enlisted the same characteristics of small states as exist now: shortage of resources, small markets, dependence and vulnerability (Puppis, 2009: 10–11). Apparently, not much has changed over the past two decades with respect to small states' media policy and performance. Such a conclusion, however, would be misleading and deserves closer attention.

Traces of commercialism in small media landscapes

During the last twenty years or so the broadcasting landscape has been transformed from what was called a public service monopoly to a dual system with public and private operators sharing the markets for audiences' attention and advertisers' money. In both countries under scrutiny, broadcasting laws were passed that allowed for continued exploitation of advertising markets by both public and private television operators. From a media policy point of view, it was and is evident that the cost per household to finance the operation of public broadcasting entirely out of licence fee returns is unviable and simply too costly. When new broadcasting legislation was introduced in Austria (2001) and Switzerland (1991 and 2006), there was broad political consensus that public broadcasting as an institution is important and desirable. Advertising revenues should be used to lower the financial burden on each television household obliged to pay licence fees for PSM. The case was slightly different in radio, where Swiss legislation leaves the entire radio advertising market to private operators and excludes the public service radio organisation from any advertising income. In Austria, just one of the public radio channels is free of advertising. The most popular radio channels compete for advertisers' money with their private competitors.

Broadcasting ownership as an issue of media policy was strongly influenced in both countries by developments in other European territories. Austria and Switzerland observed with close attention the results of the introduction of private operators in continental Europe, in particular in neighbouring states. Austria's early contribution to the opening of Germany's television markets to commercial operators was the export of management capacity. Germany's market leading commercial television station RTL (owned by Bertelsmann) has been managed since its inception in 1984 by Austrians, who were formerly employed by the national PSB in Vienna (Helmut Thoma 1984–98, Gerhard Zeiler since 1998). The observation results were not really convincing: Germany, France and Italy experienced changes in television programming quality and a shift of audiences to the new private commercial operators. All these new television stations were (and still are) readily available in Austria and Switzerland by terrestrial means, cable or satellite. This availability lowered the political pressure to provide for legislation in Austria and Switzerland as well, allowing for private operators to distribute national commercial television. It was obvious what kind of programme was to be expected from market deregulation. The early days of commercial television broadcasting were filled with abundance of cheap programming purchased from international programme markets, with equally cheaply produced talk shows and other little welcomed cost effective programming.

It took Austria until 2001 to legally allow for private commercial television operators. The 2001 law on broadcasting finally provided for the opening of the market and introduced a licensing system for any would-be television broadcaster. In 2002, the first national commercial television station went on air, funded by Austrian banks and a German media investor. At that time, Austria's newspaper publishers were paralysed by the burst of the new economy euphoria. Otherwise, they would have been expected to conquer the new medium from the outset. In contrast, they collectively shied away from another uncertain investment – given the costly experience with the internet and their failure to implement a working business model for online news.

Swiss legislation was more than ten years ahead. After allowing for private radio stations on an experimental basis in 1983, the then groundbreaking broadcasting law of 1991 opened the local and regional level to private television and radio initiative and reserved the national coverage to the public operator (Meier, 2004). National coverage in the Swiss context always refers to broadcasting within one language area, for example, German, French or Italian.

What followed in Switzerland and Austria was a short period of enthusiasm, and a lasting period of disillusion. Public service television (and radio) reacted to the new market entrants, adopted new programme concepts and ultimately remained highly popular despite all efforts by commercial operators to offer an alternative. In Switzerland, private radio prospered, mainly due to the fact that the market leading public radio channels were excluded from advertising. Ambitious commercial radio projects with frequent newscasts and local and regional news coverage turned out to be too expensive to sustain, while purely entertainment radio was profitable. This situation motivated several newspaper publishers to invest in commercial radio. Currently, and with few exceptions, commercial radio is controlled by newspaper publishers in most parts of Switzerland.

Encouraged by the commercial success of mainstream radio, newspaper publishers prepared for the launch of commercial television in the late 1990s. Finally, in 1999, one of the leading publishing companies in Zurich, Tamedia, launched its own television channel (TV3) – and failed spectacularly. Just two years later, in December 2001, the costly experiment was terminated and all further plans to launch a commercial television channel for one of the three language areas were discouraged by this experience. The Swiss market proved to be too small for both a fully-fledged public service television broadcaster and a commercial competitor.

The much shorter experience of commercial media in Austria is somewhat similar. Commercial radio and television was finally licensed in 2001. Radio went quickly into the hands of those newspaper publishers who dominate their regional market. By far the largest national newspaper, *Kronen Zeitung*, was granted after several years the one (and only) national terrestrial radio licence (Radio Kronehit).

Despite the long and fierce political battle to allow for private commercial television, the industry did not manage to set up a viable alternative to the fairly successful television programmes offered by the public service operator. In 2002, one national licence was granted to contender ATV, a consortium of Austrian banks and German media capital (Herbert Kloiber). Since 2008, and after the exit of the Austrian banks, ATV is entirely owned by Mr Kloiber, who is a successful film rights owner and trader, based in Munich, Germany. Despite all efforts and ambitions to develop a viable alternative to the PSB, the market success of ATV is strictly limited. ATV reached a market

Table 16.1 Market shares of TV channels in Austria and Switzerland (%) (2008)

Austria		*Switzerland (German)*		*Switzerland (French)*		*Switzerland (Italian)*	
ORF1	25.1	SF1	23.6	TSR1	22.9	TSI1	24.6
ORF2	16.8	SF2	8.3	TFI1	13.7	Canale5 5	11.6
Sat.1	7.3	RTL	6.9	M6	9.2	RAI1	10.4
RTL	5.6	Sat.1	6.2	France 2	8.1	ITAL 1	7.7
Pro Sieben	4.8	ARD	5.4	TSR2	7.0	RAI2	7.2
ZDF	4.3	ZDF	4.9	France 3	6.7	TSI2	6.6
VOX	3.9	Pro Sieben	4.5	RTL9	2.9	Rete4	6.2
ARD	3.8	VOX	3.7	ARTE	2.6	RAI3	4.3
ATV	3.0	Super RTL	3.1	TNT	1.8	SF 1	1.9
Kabel1	2.5	ORF1	2.9	TV5	1.6	private	1.7
RTL2	2.5	RTL2	2.9	private	0.8	La7	1.3
3SAT	1.7	Kabel1	2.5	SF1	0.8	TSR1	0.9
BFS	1.7	ORF2	1.8	TSI1	0.2	others	15.5
Super RTL	1.6	3plus	1.5	others	21.4		
Puls4	1.2	SF Info	1.5				
others	14.2	others	20.3				
National public	43.6		33.4		33.5		34.0
National commercial	4.2		1.5		0.8		1.7
Foreign	52.2		65.1		65.7		64.3

Note: Austria: population 12+, Switzerland population 3+
Source: GfK-Teletest (2009); Publicadata (2009).

share of just 3 per cent in 2008, compared to 25.1 per cent for ORF1 and 16.8 for ORF2 (GfK Teletest 2009, adult 12+).

Table 16.1 shows market shares of the leading television channels in Austria and the three main language areas in Switzerland. It is obvious that the public channels ORF1 and ORF2 (Austria), SF1 and SF2 (Switzerland German), TSR1 and TSR2 (Switzerland French) and TSI1 and TSI2 (Switzerland Italian) are by far the most successful operators in their respective markets. Their national commercial competitors have not managed to challenge their leading position. In Austria, ATV reaches 3 per cent of the population, the second private commercial channel Puls4 broadcasts in Vienna and reaches 1.2 per cent. In Switzerland, private commercial television is even smaller.

But Table 16.1 can also be read differently. Taken together, the national television broadcasters do not manage to attract even half of the audiences' attention. In Austria, foreign channels cover 52 per cent of the television market, in Switzerland almost two-thirds of the market in all three language areas. This illustrates the strong impact foreign television channels have on these two countries – and it helps to explain why there is so little space for private commercial broadcasters. The audience in Austria and Switzerland has ample access to commercial television from Germany, France and Italy. Most of these foreign commercial channels have opened so-called advertising windows for Switzerland. Attractive advertising blocks are supplied with Austrian or Swiss advertising, generating substantial revenue. In 2007, non-national television advertising reached 442 million CHF in the German part of Switzerland, out of a total television advertising expenditure of 1.041 million CHF in that part of Switzerland (Publisuisse Media Focus, 2008: 11).

National commercial operators therefore face competition in both relevant markets: they need to compete for the audience with the respective national public service television broadcaster and foreign public and commercial broadcasters. They are also forced to compete for advertising money with the national PSB and the foreign commercial competitors. Finally, there is even competition in the upstream programme markets: many broadcasting rights are attributed not to states but to language areas. Austrian and Swiss broadcasters have to negotiate programming contracts with rights holders in France, Germany and Italy – and not with rights holders in the UK or the US, because the broadcasting rights were attributed to language areas in Europe, not to nation-states. There is no evidence that this situation is likely to change. The conclusions to be drawn from the experience in Austria and Switzerland are:

- These two markets are unable to sustain public and private television operators addressing the entire population with comprehensive television programming, although there might be niche markets for specific audience segments. Television is in clear contrast to radio and the newspaper press, the latter being driven by commercial actors.
- Transnational flows of television do not favour smaller countries within larger language areas. More than half of the audiences' attention goes to foreign television programmes. Consequently, this share of the audience market is sold by foreign television operators to national advertisers. Further, the share of the advertising market is not invested into national programme production or any other form of national content.
- It is therefore unlikely that a potential ban on advertising for PSB would change the economics of commercial television broadcasting. Advertisements are placed where the largest return in terms of audience attention can be expected and documented. The logic is quite simple: low market shares attract little advertising money.

After almost two decades of experience with television broadcasting competition in Switzerland and about ten years of similar experience in Austria, a disappointing conclusion must be drawn. In both countries, relevant market players such as newspaper publishers invested into commercial broadcasting, but ownership diversity has not increased. In Switzerland, newspaper publishers control the commercial radio markets and the regional television broadcasting markets but have failed to conquer the national (language) television markets. In Austria, successful commercial radio stations are in the hands of newspaper publishers who abstain from television investment, given the discouraging market performance of the one and only contender in the national market. Competition between a large PSB and commercial broadcasters resulted in considerable increased media concentration. There are very few, if any, new independent television operators providing for ownership diversity. Instead, higher levels of ownership concentration at the local and regional areas are evident in both countries and this in fact supports my claim above that ownership diversity in the broadcasting markets is an unrealistic objective for Austria and Switzerland.

High expectations of PSM's accountability

Two main consequences follow from the observation that ownership diversity in broadcasting in smaller states is unfeasible. The first and most far-reaching consequence concerns the conduct of the PSM, for if there is little or no private competition in television broadcasting then the demand of public responsibility is higher than under conditions of more or less workable competition. The second consequence concerns public policy issues. If ownership diversity is unrealistic, accountability of the few (the only) television broadcaster increases. 'There seems to be a growing consensus that increasing "power" or, better, "influence" of the media has to be counterbalanced by greater media transparency and accountability' (Bardoel and d'Haenens, 2004: 10–11). Thus, all four quadrants of McQuail's governance model need to be taken into consideration when analysing the consequences of the lack of external television diversity. McQuail (2005: 234–5) defines governance as 'the overall set of laws, regulations, rules and conventions which serve the purpose of control in the general interest, including that of media industries' and his model distinguishes four fields: *formal* and *external* governance refers to law and regulation, *formal* and *internal* governance refers to the conduct of management, *informal* and *external* governance is made up by market forces as well as lobby groups and public opinion, and finally, *informal* and *internal* governance refers to professionalism (McQuail, 2003: 98). PSM in small states face the challenge of living up to all these governance requirements. An application of McQuail's model is presented in Table 16.2.

In response to their privileged market position but also in their own interest, PSM are required to document their governance efforts. One possibility is to publish a 'governance report' in order to enhance the public debate about the role and function of PSM in contemporary democratic societies. Such a 'governance report' could contain information about internal rules and procedures on how editorial decisions have been taken, illustrated

Table 16.2 Public service media governance

external	governance reports	public criticism and public discourse
internal	management and journalism culture	internal codes of conduct
	formal	**informal**

Source: Adapted model from McQuail (2005: 235).

by cases from day-to-day practice. Moreover, such a report would be the place to describe efforts in gender equality, minority treatment and quality management. A 'governance report' would refer to all other fields in the adapted model (see Table 16.2). In that respect, transparency on management and journalism conduct as well as any forms of internal codes of conduct would be made publicly available and subject to public deliberations.

Contemporary thinking about accountability of public service organisations in the media field is strongly influenced by the debate originating in the UK. The periodic BBC Charter review processes provide for fresh thoughts on PSM. The most recent renewal was shaped by the BBC itself and the publication of its 'manifesto' *Building Public Value* (BBC, 2004). In his reflections upon the process of reorganising the UK governance structure of public broadcasting Richard Collins sheds some light on the notion of accountability. He underlines 'the insight that public services are generally delivered better when delivered through collaboration between users and producers' and if public services respond 'to a growing demand by users for involvement and participation' (Collins, 2007: 173). Although the BBC and the overall governance setting in the UK cannot – and should not – be applied to (or compared with) small states such as Austria and Switzerland, the general governance orientation towards more involvement of the civil society in the conduct of PSM is similar.

In an effort to launch a public debate on the accountable and responsible performance of the ORF the Austrian PSM organisation published a governance report in late 2008, entitled *Wert über Gebühr* (literally translated as 'excessive value' which does not reflect the German play on words) (ORF, 2008). This report describes the main tasks and obligations of the ORF and gives a voice to prominent ORF journalists and responsible ORF editors who describe in their own words their professional attitudes. Although weak in critical observations (much weaker than the self-critical BBC 'manifesto') and lacking external views on the ORF's performance, the report is a first step in enhancing transparency on conduct issues in Austria's dominating PSM.

The Swiss public service media organisation SRG SSR lives up to the increased expectations on media governance and civic participation by referring to its unique organisational statute. SRG SSR is formally organised as a membership association. Citizens from all over the country may become members in this association which formally owns public radio and television. Each language region has its own board, elected by the association's members. The assembly of delegates, composed of elected members from the regions, then nominates seven members of the executive board, with two additional members nominated by the Swiss federal government. Membership associations meet regularly and have the right to invite and interrogate all directors including the Director-General of SRG SSR. Under this organisational structure, a close link to the Swiss population is established which gives a voice to those interested in the governance of Swiss PSM. In 2009, a reorganisation was planned to reduce the power of the membership associations, though at the time of writing the final decision had yet to be taken.

In both countries, PSM have started to establish closer links with their respective national populations – in Austria clearly in response to a growing demand for legitimacy of PSM; in Switzerland these links were built into the understanding of PSM from the outset. The absence of powerful national competitors in the television market together with the increasing presence of foreign channels in the consumption patterns of a growing portion of the audience require PSM to increase their efforts in contemporary media governance in the public interest.

Restricted media regulation options

Media governance efforts undertaken by PSM are even more important given the limited media policy options available to national governments. Over the last two decades or so media policy in Europe has been steered and dominated by the interests of large states or – more subtly – by the interests of large media corporations. The most basic concept of 'Television Without Frontiers', as laid down by the EU's Directive and the Council of Europe's Convention in 1989, never favoured the interests of smaller states or smaller media companies. There are not many television companies with transborder activities originating in small states. One of the few exceptions has been SBS (Scandinavian Broadcasting System), which was, however, taken over in 2006 by the German ProSiebenSat.1 group.

Even Switzerland, a non-EU member state, has accepted the so-called audio-visual *acquis communautaire* in its broadcasting legislation in order to benefit from membership in the EU MEDIA programme. Transfrontier television has certainly helped to increase the presence and consumption of non-national television programming in smaller states, but might have been less successful in larger states. The permanent presence of foreign television programmes in smaller territories limits the policy options even further. Restrictions in the public interest are virtually impossible to impose as long as incoming television programmes are not covered by such rules.

Although European legislation allows for stricter rules for broadcasters under national jurisdiction, it is unrealistic to implement stricter rules in a situation where foreign channels are overwhelmingly present.

Similarly, television programming decisions taken by national public service (and private commercial) broadcasters are strongly influenced by formats chosen by broadcasters in neighbouring countries. Although programming autonomy in smaller states is formally untouched, it is unrealistic for broadcasters in smaller states to ignore formats that are successful in neighbouring countries, such as *Pop Idol, Dancing Star, Deal or No Deal* or *Big Brother*. Even if such programme formats do not fit the public service remit, smaller public broadcasters in large language areas have no choice but to follow up with these formats, otherwise they risk losing substantial parts of their audience to foreign competitors.

Conclusion

Much has changed in small states' PSM over the past twenty years. Dependency and vulnerability have increased in response to transnational media policy and to programme policy decisions taken by larger television broadcasters in same-language neighbouring countries. Austria and Switzerland are two showcases that provide evidence of unrealistic ambitions to maintain programming autonomy on the one hand and ownership diversity on the other. The consequence of this observation is the maintenance of a strong PSM organisation with increased standards for transparency, accountability and responsibility. 'Public demands for attention to corporate governance have led a number of the leading media companies to create and disseminate guidelines and policies about governance in recent years, but their existence is still not common among media firms' (Picard, 2005: 5). PSM might become early birds, according to Picard's observation.

Given the limited options for media regulation, media governance measures in the public interest have to be enforced. In the absence of workable national competition PSM are requested to provide measures, such as the publication of governance reports, to foster and initiate public discourse and deliberation on the role of the media in democratic societies.

References

Bardoel, J. and L. d'Haenens (2004) 'Media Responsibility and Accountability: New Conceptualizations and Practices', *Communications*, 29(1): 5–25.

BBC (2004) *Building Public Value: Renewing the BBC for a Digital World* (London: British Broadcasting Corporation).

Collins, R. (2007) 'The BBC and "Public Value"', *Medien & Kommunikationswissenschaft*, 55(2): 164–84.

Doyle, G. (2002) *Understanding Media Economics* (London: Sage).

Eurostat (2009) 'Gross Domestic Product at Market Prices', http://epp.eurostat.ec.europa.eu/portal/page?_pageid=1090,30070682,1090_33076576&_dad=portal&_schema=PORTAL, accessed 31 March 2009.

GfK Teletest (2009) 'TV-Marktanteile 2008', http://mediaresearch.orf.at/index2.htm?fernsehen/fernsehen_ma.htm, accessed 31 March 2009.

Humphreys, P. (2008) 'The Principal Axes of the European Union's Audiovisual Policy', in I. Fernández Alonso and M. De Moragas I Spà (eds), *Communication and Cultural Policies in Europe* (Barcelona: Generalitat de Catalunya), pp. 152–82.

Kleinsteuber, H. (1990) 'Kleinstaatliche Medienpolitik und gemeinsamer Markt', *Medien Journal*, 14(2): 97–111.

McQuail, D. (2003) *Media Accountability and Freedom of Publication* (Oxford: Oxford University Press).

McQuail, D. (2005) *McQuail's Mass Communication Theory* (5th edn) (London: Sage).

Meier, W. A. (2004) 'Switzerland', in M. Kelly, G. Mazzoleni and D. McQuail (eds), *The Media in Europe: the Euromedia Handbook* (London: Sage), pp. 249–61.

Meier, W. A. and J. Trappel (1992) 'Small States in the Shade of Giants', in K. Siune and W. Truetzschler (eds), *Dynamics of Media Politics: Broadcast and Electronic Media in Western Europe* (London: Sage), pp. 129–42.

ORF (Österreichischer Rundfunk) (ed.) (2008) *Wert über Gebühr. Public Value Bericht* (Vienna: ORF), http://zukunft.orf.at/show_content.php?hid=22, accessed 26 March 2009.

Picard, R. (2005) 'Corporate Governance: Issues and Challenges', in R. Picard (ed.), *Corporate Governance of Media Companies* (Sweden: Jönköping), pp. 1–10.

Publicadata (2009) 'Marktanteile 2, Semester 2008', http://www.publicadata.ch/de/tv/tv-daten/2008.html, accessed 31 March 2009.

Publisuisse Media Forum (2008) *Geschäftsbericht*, Annual Report (Bern).

Puppis, M. (2009) 'Introduction: Media Regulation in Small States', *The International Communication Gazette*, 71(1–2): 7–17.

Siegert, G. (2006) 'The Role of Small Countries in Media Competition in Europe', in J. Heinrich and G. G. Kopper (eds), *Media Economics in Europe* (Berlin: Vistas), pp. 191–210.

Trappel, J. (1991a) 'Born Losers or Flexible Adjustment? The Media Policy Dilemma of Small States', *European Journal of Communication*, 6(3): 355–71.

Trappel, J. (1991b) *Medien Macht Markt. Medienpolitik westeuropäischer Kleinstaaten* (Vienna, St. Johann: Österreichischer Kunst- und Kulturverlag).

17
The 'State' of 'Public' Broadcasting in Greece

Stylianos Papathanassopoulos

Introduction

Greek broadcasting underwent spectacular change in the late 1980s. From a broadcasting environment with two state TV channels and four state radio stations in the mid-1990s, it now comprises 135 private TV channels and 890 private radio stations, 23 national and 135 local newspapers as well as 800 magazines. However, all private national and local TV stations technically speaking operate without an official licence to broadcast since the state has not yet awarded licences in an official manner.

The rapid and haphazard deregulation of Greek broadcasting has not only led to an overcrowded and commercialised TV universe but also caused major problems to the public broadcaster, Elliniki Radiofonia Teleorasi – ERT (Hellenic Broadcasting Corporation). In effect, ERT seems to be one of the few public broadcasters in Europe that have suffered widely from the introduction of private TV. This chapter analyses how a public broadcaster of low esteem in the eyes of the citizens/viewers of its country has tried to react in recent years on both the analogue and the digital fronts.

Broadcasting and the state in Greece

All public broadcasting systems are to some degree subject to political influence (Etzioni-Halevy, 1987), and disputes over the independence of public broadcasting are common in the history of European media. In the case of Greece, ERT even now after two decades of deregulation and the break-up of its broadcasting monopoly is still called by many a 'state' rather than 'public' broadcaster. This wide public perception and the reason for its problems stem from ERT's one-time role as a mouthpiece of government propaganda. To understand this, one must examine the relationship between the state and the media in Greece.

Broadcasting has had a symbiotic relationship with the political upheavals of the country – both radio and television broadcasting were introduced under dictatorships in modern Greece's troubled history. Radio was formed in the late 1930s under the Metaxas dictatorship and television in the mid-1960s under the Colonels (1967–74) (Papathanassopoulos, 1989: 29–35). Consequently, both radio and television have been regarded as 'arms of the state'. Moreover, the whole debate about the electronic state media in Greece before deregulation of the sector was focused on governmental control and interference in television programmes. This condition has become part of post-dictatorship ritualised politics and since Parliament was re-established in 1974, the Conservatives and Socialists have dominated the political scene accusing each other of too much governmental control over state broadcasting media (Papathanassopoulos, 1990).

This situation has largely arisen from the political tensions in Greek society since World War Two. These tensions, combined with the absence of a strong civil society, have made the state an autonomous and dominant factor in Greek society. The state is not only relatively autonomous but also has an 'over-extended' character. Mouzelis (1980: 261–4) points out that this situation has been associated with a weak, atrophied civil society where the state has to take on an additional politico-ideological function. This makes the system less self-regulatory than nations with developed capitalism such as exists in Britain or in the US. Thus, the state has to intervene and adopt a *dirigiste* attitude because it has to 'fill the gaps' in various sectors of the society and economy. Mouzelis (1980: 263) notes that because of the persistence of patronage politics, even bourgeois parties and interest groups are articulated within the state machinery in a clientist/personalistic manner. This led the state to promote the interests of particular types of capital (such as shipowners and developers) rather than the interests of capital as a whole (Mouzelis, 1987; Tsoukalas, 1986). Therefore, lack of self-regulation makes the state intervene also in the politico-ideological sphere and thus diffuse its repressive mechanisms. The fact that the state plays a decisive role in the formation of the Greek economy and policy illustrates the state's relative autonomy from society.

Looking at the mass communication sector in general, the strong state (not only in Greece) in its role as a rule-maker defines the extent of the relative autonomy it is willing to grant to the media. Even in the case of the press, which enjoys a liberal regime, the state defines press autonomy. This can easily be seen in the press laws or in cases of declared national emergency where the state reserves the right to reduce press autonomy. In a more indirect but equally effective way, the state acts to enforce these formal rules, as well as to implement the unwritten rules of power politics, by using a wide range of means of intervention that are at its disposal. One very effective means that the state uses to enforce written press limits is to provide sizeable financial aid to the press, on which individual enterprises become dependent since they cannot cover their production costs. Other means that are rarely used are courts, censorship, suspension of publication, and so on (Papathanassopoulos, 1990: 338–9).

In the case of broadcasting, as noted the state not only intervenes but is the active agent. Greek broadcasting was established, as in most European countries, as a state monopoly which remained after the restoration of Parliament. According to the constitution of 1975, 'radio and television will be under the direct control of the state' (Alivizatos, 1986; Dagtoglou, 1989). Although 'direct control' did not necessarily mean 'state monopoly', state monopoly was justified on the grounds of limited frequencies being available, as well as the need to provide full coverage for such a mountainous country with its many islands. Therefore, the state became the sole agent of the broadcast media. The government manipulation of state TV news output is a suitable example of the *dirigiste* role of the state, since it has traditionally reflected and reinforced government views and policies (Papathanassopoulos, 1990).

As a result, ministerial censorship was common practice and state control greater than was usual elsewhere. The general pattern of the Greek state broadcasting media was (and still is) that a transfer of political power will be followed by an equivalent changeover in the state media institutions' executives. The outcome, especially in the past, was news and editorial judgements of particular importance in close agreement, if not identical, to the government announcements on a whole range of policies and decisions. Thus, it is not surprising that the responsible posts in state broadcasting have come and gone with great frequency, and when the major political parties, New Democracy (Conservatives) and PASOK (Socialists), come to power they usually adopt a policy they strongly criticised when they were in opposition. In the era of the dominance of private television such a practice is rather absurd. However, the political affiliation of the executives of the public broadcaster is self-evident as all parties in the opposition still accuse the government of the day over its control of the news output. In this sense, it could be said that PSB has never really existed in Greece. The troubled political history of the country formed a state rather than a public broadcaster. To understand this, one has to note that the licence fee has never been collected directly from the TV households, but from the very beginning through electricity bills. In this sense there was never a licence fee in a Western sense. By and large, in Greece the public broadcaster was unable to function according to the public service regulations compared with those in Britain or Scandinavia or other Northern European countries. As Hallin and Mancini (2004: 106–7) note, 'it is probably significant that democracy was restored ... at a time when the welfare state was on the defensive in Europe, and global forces of neoliberalism were strong'. In other words, the deregulatory deluge of the 1980s found the state broadcaster unprepared and weak.

The effects of deregulation

The deregulation of Greek broadcasting, as in other European countries, was more than the removal of certain rules and regulations. Greece, as an EU member state, was also influenced by the Community's policies and the European political environment (Iosifidis, 2007: 77; Papathanassopoulos, 1990: 391–4). As is known, the 1980s were the era of broadcasting deregulation in almost all European countries. This environment provided strong motives to domestic Greek forces with neoliberal ideologies to press for the removal of obstacles to the introduction of market forces in the sector (Papathanassopoulos, 2006).

But when politics become the determinant factor in shaping the reorganisation of the broadcasting scene, it is bound to produce less-than-ideal results and many side effects. Some of them have been:

- The rapid and disproportionate increase in the number of channels in a small country with just 3.3 million households.
- The increase of media cross-ownership since the speed with which the publishers and other business interests moved into the broadcasting landscape was impressive. In fact, leading politicians and analysts have been concerned over how easily and quickly the media sector could be concentrated in the hands of a few influential media magnates (Papathanassopoulos, 2004: 67). This is probably the main reason why politicians want to have the upper hand with the public broadcaster since they feel vulnerable in a confrontation with the vested interests which at the same time own the mainstream private media of the country.
- Successive governments have shown a characteristic inability to intervene and regulate. It is no coincidence that every time the government announces its willingness to grant official TV licences, general elections come to interrupt the procedure. In effect, the procedure for granting operating licences to broadcasting stations has been an unresolved issue from the very first days of the introduction of commercial television. In this unregulated field, all private local and national TV channels are 'illegal', without operating licences, using television frequencies that are state property.
- Unregulated and indebted television channels degrade notions of quality and freedom of speech. It has been argued that only a strong PSB could 'show' the way to quality in such a commercialised and anarchic environment (Panagiotopoulou, 2006). But the political parties, while climbing on and off the commercial bandwagon, gave no real thought as to how to renovate the public sector and redefine the concept and mission of PSB.

The effects of deregulation on ERT

ERT's history is identified with the history of Greek broadcasting, but the emergence of private stations has been disastrous for the public broadcaster. ERT has sharply declined in ratings and advertising revenues, which resulted in large advertising losses. Nowadays, 80 per cent of ERT's funds

Table 17.1 The evolution of ET1 and NET* market share (%) (1989–2007)

Channels	*1989***	*1990*	*1993*	*1996*	*1997****	*2004*	*2007*
ET1	37.3	19.7	7.9	4.8	6.1	8.7	9.6
NET	24.3	8.7	5.3	3.5	3.4	4.2	3.8
ET3	n.a.	n.a.	n.a.	n.a.	1.1	2.0	2.5

* NET was formerly known as ET-2
** Before private TV
*** After the restructure of ERT in 1997
Source: Author's analysis based on data from AGB Hellas Media Research.

derive from the licence fee, while 20 per cent come from advertising expenditure. In effect, all ERT's three channels have witnessed a steady erosion of market share since private TV launched in late 1989 (see Table 17.1). ERT's management and the government realised that the public broadcaster could no longer justify its presence in the system. ERT was too bureaucratic, it was in debt (its accumulated debt was 112 million euros by 1997), its programming was uncompetitive and its news output lacked credibility. Moreover, since 1989 politicians had been unable to approve any of the numerous plans for the public broadcaster's salvation.

Since 1987, ERT has consisted of:

- Two national coverage channels (ET1, NET) based in Athens and a third semi-national channel (ET3) based in Thessaloniki.
- Since 2006, ERT operates three digital terrestrial channels (Prisma Plus, Sport Plus, and Cine 1) and the round-the-clock free international satellite channel ERT World.
- Hellenic Radio (ERA) with five national radio programmes (NET, Second, Kosmos, ERA3, ERA/Sport) and an international sixth programme (ERA5), the 'Voice of Greece' which targets Greeks abroad and regional programmes.
- It publishes the weekly TV/radio listings magazine *Radioteleorasis* and has a strong presence on the web (www.ert.gr) and in musical ensembles.

A long-due restructuring

ERT's management has been aiming to turn a new page in its troubled history. Since the turn of the century, the PSB's managing directors aimed, with the government's blessing, to restructure the corporation. The reorganisation of ERT has been directed along two paths: first, on the organisational structure of the broadcaster and second, renovating ERT channels' profiles and eventually their public image.

Table 17.2 ERT (ET1 and NET) programming mix by genre (% of air time)

Genres	*1995/96*	*1996/97*	*2005/06*	*2006/07*
Series	14.6	13.0	6.0	8.0
Films	15.1	15.0	8.0	8.0
Light entertainment	1.1	1.0	2.0	2.0
Arts/culture	13.3	10.9	6.1	7.0
News/ information	33.4	37.1	57.0	58.0
Children's programmes	11.3	10.8	7.0	6.0
Sports	7.5	8.4	10.0	8.0
Other	3.7	3.8	5.0	5.0

Source: Author's analysis based on data from AGB Hellas Media Research.

Profile and organisational restructuring

In 1997, ERT's management, in line with the government's wishes, changed the face of the state broadcaster in order to reapproach the Greek public. In effect the first channel, ET1, has become a general quality entertainment channel and has adopted a family entertainment profile. Its programming consists of motion pictures, telefilms, Greek series (in the last three TV seasons, it has produced twenty-seven new TV Greek series with well-known Greek actors and directors), children's programming as well as international sporting events such as the Olympic Games, World and European soccer championships and European final four basketball championships.

The second channel, formerly known as ET2 does not exist any longer, for it has been relaunched and dubbed NET (Nea Elliniki Teleorasi – New Hellenic Television). It is mainly a 24-hour information channel with news bulletins, information programmes, talk shows, documentaries and live soccer games. As is expected, soccer is a popular draw and ERT, especially NET, has considerable involvement in both national and international coverage. In addition to exclusive rights to the home games of popular Greek teams such as Panathinaikos and Olympiakos, it can air thirteen live Championship League matches per season. Other programmes include the local version of international formats such as *Who Wants to be a Millionaire?* as well as content from the History Channel. US imports include *The Bold and the Beautiful, The Young and the Restless, CSI: NY* and *Desperate Housewives* (see Table 17.2).

ET3, as noted, is fairly independent from the main corporation and it also forms its programming independently from the other two channels of ERT. In effect, it is a generalist channel giving emphasis on news and quality programmes while its focus is on northern Greece. By and large, the changes have been welcomed by the audience and this can be seen in the TV ratings shown in Table 17.1. Since 1997, ERT's strategy has aimed both to increase its profile in the Greek market and develop its digital terrestrial services in order to get a competitive advantage in the digital era (see below).

Both aims have upset the powerful media magnates who want their private TV channels to always have the upper hand in the Greek media and in the final analysis the economy, society and polity since there is a strong tendency in Greece for media to be controlled by private interests with political alliances and ambitions who seek to use their media properties for political ends. In effect, industrialists with interests in shipping, travel, construction, telecommunications and the oil industry also have a strong media presence (Papathanassopoulos, 1999).

Management restructuring

Since the late 1990s, ERT's management has aimed to reduce labour costs by applying a system of voluntary retirement of some of its personnel. It should be noted that in 2002 ERT's management aimed to retire 1062 of its personnel through a redundancy plan. This plan was considered because 76 per cent of ERT's revenues went to payroll and only 24 per cent to production and to the upgrading of the technical infrastructure. By following this path, ERT's management considered that, on the one hand, it would reduce one of the major financial burdens of the company, and on the other, by saving money it would have resources to invest in programming and respond to technological developments. It also decided to reduce the number of external collaborators and increase the productivity of the existing personnel, but even in these cases it did not achieve impressive financial savings since in recent years, it has had to increase their number to fulfil its new ventures.

ERT management aimed to turn a new page in its troubled history, but the financial problems did not disappear. The public broadcaster is still 'in the red'; for the period from 1 January to 30 August 2008, ERT's revenue was 283 million euros, while its debts increased to 293 million euros (*Paron*, 2009). This can be attributed to two factors: first, the acquisition of expensive premium programming such as live football games (for example, for the live broadcasts in the period 2006–8 of the two most popular Greek football clubs, ERT paid 12 million euros for Olympiakos and 11 million euros for Panathinaikos); second, ERT still employs a considerable number of personnel (approximately 2800), which is quite high considering the Greek market size.

Digital terrestrial television (DTT)

While there is no digital or analogue cable TV service in Greece, DTT seems to be the next priority of the country due in part to the recommendation of the EU to its member states to switch from analogue to digital TV by 2012 (see Iosifidis, 2006). The government aims to undertake the integration of the Greek television industry into DTT through the public broadcaster (Papathanassopoulos, 2007b).

As in many other European countries (Iosifidis, 2007) ERT has acted as a pioneer introducing DTT exclusive TV services to the Greek public. The digital channels are being broadcast free-to-air and are funded exclusively from ERT's budget as they carry no advertisements. ERT's digital terrestrial offerings are only available in Athens, Thessaloniki and a handful of other major cities. ERT also plans to launch a second multiplex which will broadcast the current analogue channels (ET1, NET and ET3) and the Parliament Channel. Meanwhile it is planning to launch an interactive information service for citizens called info+ (Papathanassopoulos and Papavasilopoulos, 2009).

Further, ERT is active as a network operator and according to the new law (3592/2007 'New Act on Concentration and Licensing of Media Undertakings' which was passed by the Greek Parliament late in 2007), commercial analogue TV broadcasters are encouraged to collaborate with ERT in forming a single multiplex operator company that will act as the network operator for the whole Greek digital terrestrial platform. As usual, the venture was politicised. One of the main points of contention for opposition parties over ERT's new digital subsidiary was that ERT Digital would be a mixed public–private company, with the state retaining a 51 per cent stake. The opposition parties charged that this signalled the beginning of the end of the public nature of the public broadcaster since private investors would participate in its capital.

Moreover, the law provides that 15 per cent of the taxes earmarked for ERT go to the new public–private digital company and allows the ERT board to provide material resources to the new company. ERT's union employees (POSPERT) conducted a series of work stoppages to protest the bill as a threat to the public character of ERT, bringing newscasts to a temporary standstill. ESIEA, the Union of Athens Dailies' Journalists, and the Greek Federation of Labour supported POSPERT's protest. The government responded that ERT Digital was created by the previous Socialist government, which also envisaged the entire privatisation of ERT Digital.

On the other hand, the private terrestrial broadcasters accuse the government of giving the 'green light' to the public broadcaster to enter the digital terrestrial landscape and they are left only with promises. In effect, the first law that deals with the issue of digital TV (independently of platform such as satellite, cable, terrestrial or IPTV) is Law 3592/2007 which makes a clear distinction between platform, or multiplex operator (sometimes it is called network operator) and content provider. The platform or multiplex operator is under a general licence regime, provided that the undertaking/company is registered by the Hellenic Telecommunications and Post Commission (EETT). The Ministry of Transport and Communications and the Ministry of Press and the Media are responsible for establishing the digital frequencies map and plan for the relevant assignments and allotments. The law makes it possible for licensed television stations to digitally transmit their analogue TV programmes using frequencies that are to be allocated for the period up until the digital switchover. The majority of those frequencies are being used for analogue TV broadcasting by local TV stations but the frequencies will be cleared so as to be available only for digital terrestrial TV broadcasting. The procedure for licensing DTT consortia is to be handled through a Presidential Decree.

The law is following the direction of the French regulatory framework for DTT as the frequency is being allocated to each channel editor and not to the platform, multiplex or technical operator. Contrary to the French law, in Greece a financial charge will be levied for spectrum usage. The new law does not provide for a special authority (organisation or body) with competence to settle issues relating to the switchover process, nor does it propose a timetable for this process. However, it is widely believed that Greece will be ready for the switchover within the period 2012–15. This delay is mainly due to the indecision of the private broadcasters, especially within the current financial crisis. These private consortia have adopted a 'wait-and-see' policy in case they can be supported by the government at a later stage.

According to the law, the responsibilities of the Ministry of Transport and Communications and the Ministry of Press and the Media are to establish the regulatory framework for the licensing procedure; create the frequency map and establish the technical requirements; and grant the licences. Currently, the two ministries have created a provisional frequency map where the whole country is being divided into fourteen broader services.

Conclusion

In contrast to the past two decades, it seems that the government aims, at least in principle, on the one hand to regulate the emerging digital landscape (with resource allocation and licences, for example) in order to avoid the chaotic situation of the analogue frequencies. On the other hand, with the Law 3592 of 2007, which also tries to deal with the issue of media concentration, the government attempts to address the dominance of the private broadcasters, leaving aside useless regulations such as that a company or a person cannot own more that 25 per cent of a TV station. Nevertheless, the players from the private sector are following a 'wait-and-see' policy since the development of digital television comes in an age of global financial and social crisis. However, with DTT the government may find a good reason, at last, to regulate the UHF frequencies.

Since the analogue terrestrial television landscape is still unregulated and any attempt to 'bring order' in the chaotic analogue UHF frequencies would cause major problems – mainly for those who would be excluded from getting licences – only the move to the digital 'level' would perhaps 'liberate' the audio-visual system from the anarchy that has operated for the last two decades.

But even in this case there have been objections from the private terrestrial broadcasters. As noted, they accuse the government of giving the 'green light' to the public broadcaster to enter the digital terrestrial landscape, and they are left with mere promises that the government will help them to adjust in the digital environment. They also argue that the government's action means the public broadcaster will receive a competitive advantage in the digital era, while the private broadcasters still hold no licences (either analogue or digital), although they are ready to use the digital frequencies.

It seems clear that, as in the past, centres of power compete within the vast bureaucracy, which means the application of tactical advantages, manoeuvres and conflict of interests. The media in general and their owners – businessmen in particular, in their attempt to promote their vested interests – accuse the government and attack the public broadcaster. It is not a coincidence that the majority of the political world considers that a strong public broadcaster could better serve democracy rather than the private channels. In the era of digital explosion and media convergence, political, social and economic inefficiencies and the related insecurity provide the media with what they need, i.e. to do whatever they want due to the lack of a strong government and political leadership. If politicians do not make difficult decisions, regardless of the political cost, the situation will largely remain the same, as in the analogue era.

Nevertheless, one has to admit that the Greek broadcasting system has been surprisingly adaptable and flexible in the face of new developments. To understand this, one must remember that this system has worked under Western democratic rule for thirty years now, and suddenly had to face all the upheavals that other Western broadcasting systems have taken years to deal with. This situation is related to the fact that Greece, like other Southern European states, entered late into 'modernity' and has neither a strong civil society nor a strong market. In Greece parliamentarianism was established in the absence of a strong civil society, and the media were used as vehicles for negotiating with and pressuring the government of the day, rather than representing the public discourse of society. The overextended state, meanwhile, is considered by the private interests as the place to create business (for example, public works), more than the market, which has remained underdeveloped. It is in this context that one can understand why the power of the media has increased so considerably, but not the power of journalists and of course not that of the public broadcaster.

References

Alivizatos, N. (1986) *State and Broadcasting: the Regulatory Dimension* (Athens: Sakkoulas) (in Greek).

Dagtoglou, P. D. (1989) *Broadcasting and Constitution* (Athens: Sakkoulas) (in Greek).

Etzioni-Halevy, E. (1987) *National Broadcasting under Siege: a Comparative Study of Australia, Britain, Israel, and West Germany* (Basingstoke: Macmillan).

Hallin, D. and P. Mancini (2004) *Comparing Media Systems: Three Models of Media and Politics* (Cambridge: Cambridge University Press).

Iosifidis, P. (2006) 'Digital Switchover in Europe', *The International Communication Gazette*, 68(3): 249–68.

Iosifidis, P. (2007) 'Public Television in Small European Countries: Challenges and Strategies', *International Journal of Media and Cultural Politics*, 3(1): 65–87.

Mouzelis, N. (1980) 'Capitalism and Development of the Greek State', in R. Scase (ed.), *The State in Western Europe* (London: Croom Helm).

Mouzelis, N. (1987) 'Continuities and Discontinuities in Greek Politics: From Elefterios Venizelos to Andreas Papandreou', in K. Featherstone and D. K. Katsoudas (eds), *Political Change in Greece: Before and After the Colonels* (London: Croom Helm).

Panagiotopoulou, R. (2006) 'Culture on Television', *Zitimata Epikoinonias*, 4: 41–56 (in Greek).

Papathanassopoulos, S. (1989) 'Greece: Nothing is More Permanent than the Provisional', *Intermedia*, 17(2): 29–35.

Papathanassopoulos, S. (1990) 'Broadcasting, Politics and the State in Socialist Greece', *Media, Culture and Society*, 12(3): 338–97.

Papathanassopoulos, S. (1999) 'The Effects of Media Commercialization on Journalism and Politics in Greece', *The Communication Review*, 3(4): 379–402.

Papathanassopoulos, S. (2001) 'Media Commercialization and Journalism in Greece', *European Journal of Communication*, 16(4): 505–21.

Papathanassopoulos, S. (2002) *European Television in the Digital Age: Issues, Dynamics and Realities* (London and New York: Polity & Blackwell).

Papathanassopoulos, S. (2004) *Politics and the Media* (Athens: Kastaniotis) (in Greek).

Papathanassopoulos, S. (2006) *Television in the 21st century* (Athens: Kastaniotis) (in Greek).

Papathanassopoulos, S. (2007a) 'The Mediterranean/Polarized Pluralist Media Model Countries: Introduction', in G. Terzis (ed.), *European Media Governance: National and Regional Dimension* (Bristol, UK: Intellect), pp. 191–200.

Papathanassopoulos, S. (2007b) 'The Development of Digital Television in Greece', *Javnost/The Public*, 14(1): 93–108.

Papathanassopoulos, S. and K. Papavasilopoulos (2009) 'The Status of Digital Television in Greece', in W. Van den Broeck and J. Pierson (eds), *Digital Television in Europe* (Brussels: VUB Press), pp. 91–100.

Paron (2009) 'Damages for ERT', 18 January 2009, http://www.paron.gr (in Greek), accessed 21 February 2009.

Tsoukalas, K. (1986) *State, Society, Labour* (Athens: Themelio) (in Greek).

18

Public Service Broadcasting in Poland: Between Politics and Market

Paweł Stępka

Introduction

There are many reasons to study in-depth the system of Public Service Broadcasting (PSB) in Poland, not least because the Polish broadcasting market is the largest among the new European Union (EU) member states. While market size has not 'protected' PSBs from the typical (for this region) difficulties, that is, pressure exerted by the main political forces and ineffectiveness of the funding mechanism, Polish PSBs enjoy a fairly strong position, which makes this system unique among other Central-Eastern European countries. Both PSB organisations Polish Television (TVP) and Polish Radio (PR) still maintain a high audience share – a combined share of 46.8 per cent for TVP and 25.2 per cent for PR in 2007 (KRRiT, 2008a: 107–12). But the licence fee collection system appears inadequate and is characterised by one of the highest evasion rates in the EU.

This chapter starts by unravelling the historical development of this system starting from 1993 when PSBs were officially established in order to mark the long-lasting transition process from 'state' to 'public' broadcasting. It then moves on to describing the current Polish PSB system with a special focus on its political independence and funding mechanism. The strong market position enjoyed by the PSBs paradoxically makes them vulnerable to political influence and is simultaneously one of the reasons why PSB is still high on the agenda of the main political parties. Meanwhile the large number of licence fee evaders makes PSBs dependent more on commercial income, which in turn affects programming choices and results in commercially oriented output. A content analysis is therefore conducted and an assessment is made of PSB programming distinctiveness.

Furthermore this chapter discusses the cultural role of Polish PSBs which are meant to serve and promote national culture and identity on the one hand, and preserve European values and heritage on the other. This latter task became especially important after the accession of Poland to the EU in May 2004.

The chapter does not, however, touch upon the substance of the bill under preparation in March 2009, though this controversial measure is likely to change the PSB system radically with its provisions to abolish the licence fee, introduce the so-called 'public mission fund' which would be annually defined by the Parliament and operated by the regulatory authority (KRRiT), change the PSB governance model, and introduce programming licences which would more accurately define public tasks for PSBs and other commercial broadcasters. Precisely because of these controversies, as well as differences of opinions between political parties, it is doubtful whether it will ever be adopted (Poland, 2009).

Organisation of PSBs in Poland

After the collapse of the authoritarian system in Poland in 1989 a long-lasting and dynamic process of political and economic transformation began with the aim of creating a democratic system and a free-market economy following Western Europe's example. This basically involved an attempt to copy and import a PSB system from other established democratic regimes, a procedure that was dubbed in the Parliamentary Assembly of the Council of Europe report as 'transplantation' of those institutions into systems with an insufficiently developed political and organisational culture (Council of Europe, 2004). The idea of importing a Western-style PSB system to Poland coincided with an equally challenging process of building civil society.

It is worth emphasising that the so-called 'socialisation' of Polish mass media was proposed by the democratic opposition during the Polish Round Table talks (6 February–5 April 1989).[1] Eventually PSBs in Poland (Polish Television – TVP and Polish Radio – PR) were established on the basis of the Broadcasting Act of 29 December 1992 (Poland, 1992), which was one of the longest debated bills in Polish Parliament (Ociepka, 2003: 142). The new entities started operations on 1 January 1994, at the time when the first commercial licence was granted to private terrestrial broadcaster Polsat. Currently TVP operates three nationwide channels (TVP1, TVP2 and TVP Info), TV Polonia, which is targeted at Polish diasporas, the high definition channel TVP HD, and three satellite thematic channels (TVP Culture, TVP History and TV Sport). PR operates four nationwide programme services (PR1, PR2, PR3 and Radio Euro which in May 2008 replaced Radio Bis), Radio Parliament, Radio dla Zagranicy (Polish Radio External Service) and seventeen regional radio stations.

The law provided for the general structure of PSB, determining the membership of the companies and the method of members' appointment, and defining the companies' bodies as follows: Management Board; Supervisory Board; and General Assembly. The Management Board of the PSB companies consists of one to five members appointed for a four-year term of office by a Supervisory Board, which is composed of five to nine members, but the final number of members in both Boards rests with the regulatory authority National Broadcasting Council (KRRiT). A representative in the Supervisory Board is at the disposal of the Minister of Treasury, whereas the General Assembly is constituted by the Minister of Treasury (Poland, 1992). Further, fifteen-member Programme Committees have been formed and tasked with the evaluation of programming quality. The structure of these committees is determined by the KRRiT which appoints ten members recommended by parliamentary groups and five members with a record of expertise in culture and media (Poland, 1992).

Does the above-mentioned legal framework ensure political independence of PSBs? Taking into consideration the Council of Europe standards determined in Recommendation Number R (96) 10 on the guarantee of the independence of PSB (Council of Europe, 1996), we can assume that the Broadcasting Act includes various mechanisms that would strengthen the independence of PSBs, including the Supervisory Boards and the Programme Committees whose members are appointed by the independent regulatory agency. The Minister of Treasury can neither issue orders to the Management Board on programming policy, nor change public broadcasting companies' statutes without the consent of KRRiT. This may give the impression that PSB in Poland enjoys a good degree of political independence, but a number of experts have rightly expressed doubts about it (Jaskiernia, 2006: 172; Ociepka, 2003: 145–6). Despite PSBs' political independence on paper, political parties do interfere in the organisation of public broadcasters, for the Sejm (the lower chamber of Parliament), the Senate and the president responsible for the choice of the KRRiT's members are quite often guided by political criteria (Jakubowicz, 2007: 225), which impacts upon the process of selecting the members of the Supervisory Boards.

Another example reinforcing the argument that PSB companies are in fact politically dependent is that any change of government is typically associated with a change of the management of PSB. Besides, it should be noted that the Management Boards are collective and reflect a political compromise reached within KRRiT. That in exceptional cases can lead to a situation which would considerably hinder an efficient management, as was the case in December 2008 with the TVP when a conflict within the Supervisory Board resulted in the suspension of three members of the Board. This kind of political interference in the PSB system has led some experts to compare this system with the Mediterranean model of 'polarised pluralism' in which political institutions exert some influence on PSBs (Dobek-Ostrowska, 2007: 54).

Funding mechanism

There has been some concern over the commercialisation of PSB programming and this is intimately connected with the inefficiency of the funding mechanism which is based on two main sources: licence fee and commercial income (Poland, 1992, 2005). The licence fee is collected by all households that own radio and television sets (Poland, 2005). In 2009 households that owned a radio set had to pay an annual fee of 58.20 PLN (about 13 euros) and those owning both a radio and a TV set paid 186.70 PLN (42 euros) (KRRiT, 2008c). Although radio and TV set ownership does not depart from the European average, the actual income generated is not sufficient to allow public media to produce quality public service programmes. This can be attributed to a number of factors. Firstly, large groups such as the disabled and senior citizens over 75 years of age are exempted from paying the licence fee (Poland, 2005). Secondly, the system is exceptionally leaky and with about 50 per cent of households evading payment in fact results in the highest evasion rate among the EU member states. According to the regulator (KRRiT, 2008a) by the end of December 2007 only 43.5 per cent of households had registered radio and TV sets and paid the licence fee in a timely manner. It is striking that even public organisations evade the licence fee, for only 5 per cent of them have registered radio and/or television sets. As a result, the amount of the licence fee collected is said to be one of the lowest in Europe (Woźniak et al., 2007).

Actual revenues from the licence fee fell in the period 1994–2007 by 25.3 per cent to 887 million PLN (201 million euros) (KRRiT, 2008a: 123). The collected resources are distributed by the KRRiT to the PSB entities as follows: on average 60 per cent is allocated to PR and the remaining 40 per cent to TVP. Despite transferring the largest amount to PR, radio broadcasters feel most severely the decrease of the licence fee income due to their greater degree of dependency on public funds (in 2007 about 72 per cent of their budget). In the case of the TVP, the share of the licence fee dropped to just 24.6 per cent in 2007 from 28.3 per cent in 2005, while in the same period advertising income remained unchanged at around 55 per cent (see Table 18.1).

Table 18.1 Sources of income for Polish PSB companies (%) (2005–7)

Sources	*TVP*			*Polish Radio*		
Year	*2005*	*2006*	*2007*	*2005*	*2006*	*2007*
Licence fee	28.3	28.3	24.6	74.3	72.1	72.0
Advertisement	55.3	57.0	55.1	14.1	14.4	14.5
Sponsoring	5.5	5.7	5.2	–	–	–
Financial revenues	2.7	1.5	1.9	0.8	1.0	1.0
Other sources	8.2	7.4	13.2	10.8	12.5	12.5

Source: KRRiT (2008b: 11).

The distinctiveness of PSB content

The public service remit, as defined in Article 21.1 of the Broadcasting Act, obliges TVP and PR to offer to 'the entire society and its individual groups diversified programme services in the areas of information, journalism, culture, entertainment, education and sports which shall be pluralistic, impartial, well balanced, independent and innovative, marked by high quality and integrity of broadcast' (Poland, 1992). However, PSB tasks have not been specified in other documents, known in some European countries as 'agreements' or contracts such as those in Denmark, Italy or the UK (Jakubowicz, 2008: 160–6). Therefore the Polish system can be described as an 'autonomy model' in which PSBs interpret for themselves the scope of their public service remit (Murawska-Najmiec, 2004: 1). It is obvious that under this system PSBs are allowed to conduct a relatively flexible programming policy that lacks independent and rigorous evaluation of their programming output (Jakubowicz, 2007: 229).

But how distinctive is PSB output? To answer this question we attempted to compare the programming of the public channels with that of commercial rivals. In the case of TVP, we focused on the output of the two main programme services (TVP1 and TVP2), as well as TVP Info and two theme satellite programmes (TVP Culture and TVP History) and compared it with the programming output of the main commercial broadcasters TVN, Polsat and Puls. Table 18.2 shows that the most popular programming genres on

Table 18.2 Programming output of Polish public and commercial broadcasters (%) (2007)

	TVP					*Commercial Broadcasters*		
	TVP1	*TVP2*	*TVP Info*	*TVP Culture*	*TVP History*	*TVN*	*Polsat*	*Puls*
News	6.80	4.40	44.50	0.80	0.00	2.30	3.20	0.90
Current affairs	11.90	4.60	24.30	10.20	6.60	15.80	1.60	13.00
Education	3.40	0.10	0.70	0.00	58.60	0.00	0.00	1.00
Life style	0.70	5.00	0.20	0.00	0.00	2.40	0.20	1.90
Religion	2.00	0.60	0.70	0.03	0.00	0.00	0.20	8.40
Documentaries	7.00	9.00	9.50	18.30	16.80	0.00	0.90	6.80
Films	44.40	42.20	0.10	38.10	9.10	28.90	42.00	43.40
Dramas	7.00	0.10	0.00	3.40	0.00	0.00	0.00	0.00
Music	2.30	3.10	0.10	21.60	0.00	0.30	0.80	0.30
Entertainment	3.80	9.30	0.00	1.90	0.02	24.20	25.20	1.40
Sport	2.60	3.30	7.50	0.00	0.00	1.20	4.50	0.30
Self-promotion	3.20	2.90	4.90	4.90	8.80	4.90	4.80	7.00
Advertisements and sponsoring	11.10	13.40	7.50	0.70	0.00	20.10	16.60	15.60

Source: KRRiT (2008b).

both the main public channels TVP1, TVP2 and the commercial channels are films and advertisements. In the case of TVP2, the third most popular category is entertainment including talk shows, games or sitcoms while in the case of TVP1 the third most popular genre is current affairs. Where the programming of public and private channels differs is in the provision of education, drama, theatre, documentaries and information programmes, areas in which TVP scores higher compared with the commercial operators.

As far as the themed satellite public services are concerned, these have a completely different programming structure. According to Table 18.2, TVP Info offers a high proportion of news and current affairs, whereas TVP Culture offers documentaries, dramas, music and films and TVP History focuses mainly on educational broadcasts and documentaries. In these cases, the dominance of public service genres such as information, documentary and education programmes is more visible, but it should be noted that with less than 0.5 per cent audience share each, these thematic services are far less popular than TVP1 (23.2 per cent share) and TVP2 (18 per cent) (KRRiT, 2008b).

To sum up, a higher percentage of public service categories is only observable when one compares the programming of commercial channels with the programming of thematic public channels, but a so-called 'programming convergence' is noticeable when it comes to comparing the output of the main public channels TVP1 and TVP2 with the output of their commercial counterparts. This can be attributed to a shortage of public funds which forces the main public channels to compete with commercial broadcasters for advertising income (commercial sources constitute about 60 per cent of the total TVP budget). In this context public broadcasters generally avoid embarking upon risky and experimental programming that would potentially alienate advertisers.

PSB and new technologies

The 1992 Broadcasting Act provides for a wide definition of the public service remit, thus making it possible for PSBs to expand to new technologies and offer not just *programmes*, but also *services* within the scope of information, journalism, culture, entertainment, education and sport (Jakubowicz, 2007: 246). On the basis of this provision PSBs are allowed to offer new media services such as interactive websites, on-demand services (catch-up TV) and experiment with mobile TV services and the YouTube channel. Apart from its official website (tvp.pl) TVP is involved in interactive television projects accessible on the internet (ITVP). ITVP was in fact one of the first internet-based platforms in Poland which offered on-demand services and the possibility of watching episodes of popular series. In 2008 ITVP was merged with the official website and turned more interactive. Consumers also have an opportunity to use a number of internet-based theme services devoted to film, information, sport and knowledge. A significant part of such content is also available on TVP's YouTube channel. Apart from providing a good quantity of audio-visual public service content on the internet, TVP's channels are also available on other new platforms. In 2008 TVP together with other broadcasters and mobile operators took part in the DVB-H pilot project, which paved the way for the introduction of mobile TV in Poland. The mobile package includes the five TVP programme services TVP1, TVP2, TVP Info, TVP History and TVP Sport.

Moreover, TVP channels are available in cable packages and on the three competing digital satellite platforms Polsat Cyfrowy, Cyfra Plus and 'n'. This is important in terms of consumer reach since digital terrestrial television (DTT) in the country has not started yet due to the lack of a clear and coherent governmental strategy. Despite this delay the analogue terrestrial switch-off is foreseen in 2015 (European Commission, 2009). Nowadays the digitalisation process is well under way, but it is shaped primarily by the commercial sector (terrestrial, cable and satellite operators), whereas the public sector is lagging behind. The lack of governmental backing has pushed the TVP to consider the possibility of establishing its own digital satellite platform in order to broaden its digital reach and compete on equal terms with commercial broadcasters which have set up such platforms. However, no binding decision has been taken yet in this respect.

Polish culture and minority needs

Alongside incorporating new technologies, the scope of the public service remit should be expanded to reflect the complex character of society and its peculiarities (Council of Europe, 2006). It therefore follows that PSBs should appeal to different types of minorities in order to meet the expectations of a multicultural society. The 2006 Council of Europe resolution also indicated that PSBs should contribute to the 'greater appreciation and dissemination' of European cultural heritage (ibid.: 35–8), implying that PSB programming should not focus solely on a given society and its culture but also 'bring Europe' to the citizens. It can be inferred then that European PSBs have a threefold task: (a) to protect and develop values and culture specific to a given society; (b) take into account the needs of national and ethnic minorities; and (c) reflect the cultural heritage of other European nations. These tasks are realised by Polish public broadcasters, for the first two are explicitly mentioned in Article 21.1 of the Broadcasting Act (Poland, 1992), which envisages the obligations imposed on PSBs to meet the needs of the whole society as well as different social groups, regardless of geographic, ethnic and linguistic criteria. The Act enumerates among others the following obligations: 'dissemination of the Polish language', and 'paying due regard to the needs of national and ethnic minorities and communities speaking regional languages including broadcasting news programmes in the languages of the national and ethnic minorities and in regional languages'.

But all this should be considered in terms of the fact that Polish society is one of the most homogeneous in Europe. According to the 2002 Polish census, an overwhelming majority of 96 per cent of surveyed people declared Polish nationality and a mere 1.23 per cent declared other nationality (Główny Urząd Statystyczny, 2002). The uniform character of Polish society is fostered by a common religion since nearly 89 per cent of the population are followers of the Roman Catholic Church. Therefore in spite of joining the European Union, challenges connected with reflecting a multicultural society appear at present not to concern Poland so much as other European countries.

This peculiarity of the Polish society also impacts on decisions concerning the amount of programmes produced in the Polish language. According to Article 15 of the Broadcasting Act all TV broadcasters are legally obliged to reserve at least 33 per cent of their quarterly transmission time to programmes originally produced in the Polish language while they are also obliged to reserve at least 33 per cent of their quarterly transmission time to vocal-musical compositions for compositions performed in the Polish language (Poland, 1992). These quotas, which are realised by a considerable surplus by most public TV and radio operators, are aimed at promoting Polish culture and language, and supporting a domestic audio-visual and phonographic sector (KRRiT, 2008a: 80–1).

These quotas also apply to commercial broadcasters, though the obligation of 'taking into consideration the needs of national and ethnic minorities and societies using a local language' (Article 21a of the Broadcasting Act) is imposed only on PSBs (Poland, 1992). This latter obligation is primarily fulfilled by regional public radio stations as well as the TVP regional branches, but nationwide PSBs also dedicate a certain amount of their output to national and ethnic minorities. In 2006, nationwide public radio stations broadcasted in total 420 hours of programming in Polish dedicated to minorities but in 2007 this fell to 50.2 hours. This visible decrease has been compensated by regional companies. In 2006 regional public stations' share of programming dedicated to minorities was 0.9 per cent (about 1373 hours) (KRRiT, 2007: 63) but in 2007 this increased to 1.1 per cent (1657 hours) (KRRiT, 2008a: 76).

In the case of TVP Info, programmes addressed to the national and ethnic minorities occupied on average only 0.3 per cent (about 334 hours) of air time (KRRiT, 2008a: 83). Some minority groups such as the Ukrainian national minority complained to the KRRiT about the PSB's insufficient coverage of national and ethnic minorities (KRRiT, 2008a: 32). However, given the lack of a so-called 'contract' between PSBs and the government it is difficult to evaluate unambiguously whether PSBs fulfil their obligations towards minorities.

PSBs and European cultural heritage

With the beginning of the lengthy process of preparing Poland for European Union membership, EU matters were intensely accentuated by PSBs. However, the issue of public media reflecting a European cultural heritage is not clearly defined in the Broadcasting Act, for the Act envisages merely that public radio and television broadcasting should 'provide reliable information about the vast diversity of events and processes taking place in Poland and abroad' (Article 21 of Act 1).

But what is the share of European informational and current affairs programmes in the Polish broadcasters' daily schedule? These types of programmes are evident in the schedule of the four nationwide public radio stations, which placed particular emphasis on special events such as Poland's accession to the EU, discussion on the EU constitution, and Poland's accession to the Schengen Area. In the period 2004–7 coverage of European affairs in the areas of information and public affairs was adequate and remained almost unchanged with 6421 minutes in 2004 and 5935 minutes in 2007. The somewhat higher coverage of 2004 could be attributed to the interest generated by Poland's accession to the EU in May that year and the first European Parliament elections taking place in Poland in June of that year.

PSBs are also often used as a platform for managing social campaigns, funded either by EU or governmental means, aimed at promoting knowledge of the EU. In the case of Polish Radio, such campaigns have been aired mainly through Programmes 1 (PR1) and 3 (PR3) and concerned, for example, the issue of EU subsidies. In the case of public television campaigns conducted these have been about promoting the ideals and programmes of the Community. An example was a campaign commissioned in 2008 by the Ministry of Agriculture and Rural Development of the Republic of Poland which concerned European funds' 'innovative economy'.

For public television, the important goal of bringing European culture to the Polish society is realised through the so-called European quotas. Among European works, European films occupy a significant space in TVP's programming schedule and viewed this way are a carrier of European values. Table 18.3 shows that the amount of films in the offering of the two most popular TVP channels TVP1 and TVP2 constitute about 40 per cent of the total output, with European works in 2007 totalling 11.1 per cent in TVP1 (just behind American films at 17.6 per cent) and 4.4 per cent in TVP2. In the period 2004–7 the quantity of European films in TVP1's offering slightly increased, perhaps due to the Polish accession to the EU, whereas in the same period the percentage of European films aired in the second channel TVP2 appeared to decline. In any case, the presence of both Polish and European films in the daily schedule of public channels appears to be fairly strong when compared with the presence of American films.

Table 18.3 Films on TVP1 and TVP2 (%) (2004–7)

	TVP1				*TVP2*			
	2004	*2005*	*2006*	*2007*	*2004*	*2005*	*2006*	*2007*
Polish films	10.88	10.79	14.0	13.7	18.54	21.98	20.5	16.6
European films	9.9	7.8	8.5	11.1	3.35	5.8	5.8	4.4
American films	13.74	20.16	16.0	17.6	7.31	12.7	15.0	16.8
Films on offer	**42.5**	**40.3**	**43.2**	**44.4**	**34.8**	**42.3**	**43.2**	**42.2**

Source: KRRiT.

Finally public media also air entertainment programmes which reflect the culture of other European countries. One example was the talk show *Europa da się lubić* aired by TVP2 in the period from 2003 to 2008 (144 episodes in total) which aimed to familiarise Polish viewers with the customs of various European nations by inviting migrants living in Poland to discuss their experiences.

Conclusion

This chapter has shown that the PSB ideals, transferred to Poland in the early 1990s, have not taken root in the country yet, due to the difficulties (inherent in most Central-Eastern European territories) in transforming the political system and the political culture, as well as the lack of a sufficiently developed civil society. The position of PSBs is further weakened by the lack of a clear media policy on the part of successive governments which have failed to define sufficiently the role and remit of national public media. The primary concern of the political parties appears to be exerting influence on PSBs, rather than ensuring their political independence. The apparent ongoing political pressure on PSBs adds to the continuation of a negative image of these broadcasters, which are still perceived by many citizens as 'state', rather than 'public' media.

The lack of a coherent vision of PSBs' remit is also reflected in the inadequate funding mechanism which does not provide sufficient public funds and makes those broadcasters, especially TVP, dependent upon commercial sources. Not surprisingly, the main consequence is increased commercialisation of the programme offering of the main public channels TVP1 and TVP2. Alongside political dependence, programming commercialisation seems to be an equally important factor which undermines the idea of PSB in Polish society. Nevertheless as has been shown above the content of some, albeit less popular, PSB channels, is still distinctive from the programming offer of commercial broadcasters. By serving different social groups, including various types of minorities, PSBs play a unique social role in an increasingly commercially dominated audio-visual market.

It should also be underlined that PSBs enjoy a fairly strong market position, especially TVP, which is among the most influential public TV channels in Europe, not least because of its large market size. However, such a strong market position makes these broadcasters even more attractive to politicians who perceive PSBs as a powerful means of influencing public opinion. The degree of PSBs' political independence as well as their economic viability may further deteriorate if the above-mentioned bill, currently being prepared by the ruling parties and supported by the Social-Democratic party, enters into force in its current version. The proposed amendments could result in deepening the funding crisis of PSBs and increasing their financial dependence on commercial means, which in turn could adversely affect the programme offering of the main TVP channels. Meanwhile Polish Radio and especially its regional companies would almost certainly cease to exist without sufficient public funding.

Note

1. The Round Table refers to negotiations conducted from 6 February to 5 April 1989 by representatives of the Polish People's Republic authorities, the opposition and the Church, as a result of which a political transformation began in Poland.

References

Council of Europe (1996) Recommendation Number R (96) 10 on the guarantee of the independence of public service broadcasting (Strasbourg: Council of Europe).

Council of Europe (2004) 'Public Service Broadcasting', Report of the Committee on Culture, Science and Education, 12 January 2004, Doc. 10029 (Strasbourg: Council of Europe).

Council of Europe (2006) '4th European Ministerial Conference on Mass Media Policy', Prague (Czech Republic), 7–8 December, 1994, in Council of Europe, *European Ministerial Conferences on Mass Media Policy: Text Adopted* (Strasbourg: Council of Europe).

Dobek-Ostrowska, B (2007) 'Współczesne systemy medialne: zewnętrzne ograniczenia rozwoju', in B. Dobek-Ostrowska (ed.), *Media masowe na świecie. Modele systemów medialnych i ich dynamika rozwojowa* (Wrocław: Wydawnictwo Uniwersytetu Wrocławskiego).

European Commission (2009) 'EU Member States on Course for Analogue Terrestrial TV Switch-off', IP/09/266 (Brussels: European Commission).

Główny Urząd Statystyczny (2002) *Narodowy Spis Powszechny Ludności i Mieszkań 2002* (Warsaw: GUS).

Jakubowicz, K. (2007) *Media publiczne. Początek końca czy nowy początek* (Warsaw: Wydawnictwa Akademickie i Profesjonalne).

Jakubowicz, K. (2008) *Polityka medialna a media elektroniczne* (Warsaw: Wydawnictwa Akademickie i Profesjonalne).

Jaskiernia, A. (2006) *Publiczne media elektroniczne w Europie* (Warsaw: Oficyna wydawnicza Aspra JR).

KRRiT (2005) *Obrona lokalności i demokracji lokalnej – Strategia działania Krajowej Rady Radiofonii i Telewizji na rzecz ochrony lokalnego charakteru i pluralizmu oferty programowej w lokalnych mediach elektronicznych* (Warsaw: KRRiT).

KRRiT (2007) *Sprawozdanie Krajowej Rady Radiofonii i Telewizji za rok 2006* (Warsaw: KRRiT).

KRRiT (2008a) *Sprawozdanie Krajowej Rady Radiofonii i Telewizji za rok 2007* (Warsaw: KRRiT).

KRRiT (2008b) *Informacja o podstawowych problemach radiofonii i telewizji* (Warsaw: KRRiT).

KRRiT (2008c) *Rozporządzenie KRRiT z dnia 29 kwietnia 2008 r. w sprawie wysokości opłat abonamentowych za używanie odbiorników radiofonicznych i telewizyjnych oraz zniżek za ich uiszczanie z góry za okres dłuższy niż jeden miesiąc w 2009 r* (Warsaw: KRRiT).

Murawska-Najmiec, E. (2004) *Informacja o sposobie określania zadań programowych nadawców publicznych oraz o sposobach ich finansowania i rozliczania/kontroli na podstawie wybranych przykładów nadawców telewizyjnych*, Analiza Biura KRRiT no. 8/2004 (Warsaw: KRRiT).

Ociepka, B. (2003) *Dla kogo telewizja? Model publiczny w postkomunistycznej Europie Środkowej* (Wrocław: Wydawnictwo Uniwersytetu Wrocławskiego).

Poland (1992) Broadcasting Act of 29 December 1992.

Poland (2005) Licence Fee Act of 21 April 2005.

Poland (2009), Projekt ustawy o zadaniach publicznych w dziedzinie usług medialnych, 18 March 2009.

Woźniak, A., P. Stępka, M. Borkowska and E. Murawska-Najmiec (2007) *System poboru opłat abonamentowych w państwach europejskich*, Analiza Biura KRRiT no. 5/2007, December (Warsaw: KRRiT).

19
From 'State Broadcasting' to 'Public Service Media' in Hungary

Márk Lengyel

Introduction: the formation of the Hungarian system of public service broadcasting

The role and the status of public service media in Hungary cannot be fully understood unless we delve into its history. Until 1989–90 the term 'public service' stood exclusively for *state* broadcasting and for *broadcasting* itself. The term Public Service Broadcasting (PSB) gained real meaning only after the transition from the former communist regime to democracy.

The architects of the new democratic framework attributed special importance to the role of the media. As a consequence an explicitly new provision referring to PSBs was added to the constitution of 1989 among other provisions, with the aim of establishing a democratic playing field for political parties. According to this, 'a majority of two-thirds of the votes of the Members of Parliament present is required to pass a Law on the supervision of public radio, television and the public news agency, as well as the appointment of the directors thereof' (Constitution, §61 (4)).

For a period of six years following the introduction of this constitutional arrangement the political consensus did not reach the level necessary to pass an Act that would have regulated inter alia the system of Hungarian PSB. Several decisions on issues relating to PSB that the Constitutional Court passed during this period are worth noting. In these decisions the Court urged the Parliament to adopt the necessary legislation and gave guidance especially on the issues of institutional and financial independence. The period 1990–6 was characterised not just by the absence of proper legislation but also by fierce political struggles over the governance of PSBs.

Act I of 1996 on radio and television broadcasting (Broadcasting Act) was expected to end this 'media war', as this period is commonly called by analysts. It defined the legal framework for the operation of PSBs and meanwhile paved the way for the introduction of commercial broadcasting. In effect, 1997 witnessed the launch of two national commercial television channels (TV2 aired by MTM-SBS and M-RTL) which soon became very popular in the

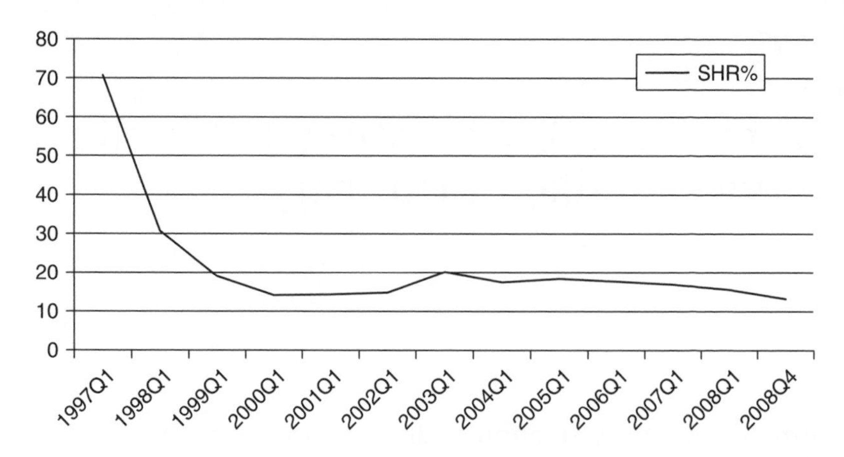

Figure 19.1 Hungarian public TV channels' audience shares (1997–2008) (quarterly average)
Source: AGB Nielsen, Hungary (2009).

country. As elsewhere, their emergence proved to be a shock for the public broadcasters (MTV and Duna Tv.) which lost a significant audience share (see Figure 19.1).

Such audience share loss was not merely extremely fast but also proved fatal. PSBs tried to respond to the challenge in different ways. Duna Tv. and the second channel of MTV moved towards providing distinctive content, consciously sacrificing high ratings. In contrast the main channel of MTV tried to increase its audience, frequently at the expense of public service programming. If we use Christian Nissen's categories (Nissen, 2006a: 75), in the first case we can identify the symptoms of the 'heroic hara-kiri' and, in the latter case, the 'rash kamikaze'. Hungarian public television has never really recovered from this initial shock of the advent of the dual media system. Public radio also experienced similar trends, but it managed to keep a much stronger position in terms of both distinctiveness of content and reach.

Since the adoption of the 1996 Broadcasting Act the rules governing matters relating to PSB have remained unchanged. Although the Act intended to transform the Hungarian broadcasting landscape into a modern dual broadcasting system, in practice it proved to be an inadequate tool to eliminate the political tensions around public service organisations. Today it is widely acknowledged among Hungarian analysts and decision-makers that the Broadcasting Act has become outdated by the rapid development of the media sector and it needs to be revised with special regard to the rules relating to PSB. However, finding legal solutions that would meet the requirement of approval by a qualified majority in Parliament is a task still ahead.

Overview of the present system of PSB

The current structure of the Hungarian PSB comprises three companies:

- Magyar Televízió Zrt. (Hungarian Television; MTV), a successor of the former state broadcasting company.
- Magyar Rádió Zrt. (Hungarian Radio; MR), also a successor of the same state broadcasting organisation.
- Duna Televízió Zrt. (Duna Television; Duna Tv.), founded in 1992 with the objective of serving the needs of Hungarian nationals living abroad.

These broadcasters offer the following channels:

- MTV is present in the audio-visual landscape with two national channels: m1 is a national programme service addressing the general audience via terrestrial means and m2 is a satellite channel with a cultural profile.
- Duna Televízió provides a satellite channel under the name Duna Tv. In 2005 Duna Televízió also began to provide an additional public service satellite channel called Autonómia. This channel is officially devoted to 'the presentation of national identity and cultural diversity'.
- MR provides three national radio channels on terrestrial networks: MR1 Kossuth – a general news and talk channel, MR2 Petőfi – a light entertainment channel, and MR3 Bartók – a cultural channel. These are complemented by a parliamentary channel (MR5), a number of other local, regional programmes (MR6) and a channel for ethnic minorities (MR4).

However, these PSB companies are not the only providers of public service content, for there are significant public service tasks assigned to the national terrestrial commercial radio and TV broadcasters. According to the Hungarian law these broadcasters can provide their programmes on the basis of a broadcasting contract concluded with the National Radio and Television Commission (ORTT), the national regulatory authority for the media. Such contracts issued with the four national terrestrial commercial broadcasters (MTM-SBS, M-RTL, Sláger Rádió and Danubius Rádió) contain obligations to provide a significant proportion of public service programmes. The profile of these programmes is also defined in accordance with the categories set out in the Broadcasting Act.

The definition of the public service remit

The 1996 Broadcasting Act lays down the basic criteria for the PSB remit. This remit is defined essentially by the category of 'public service programme item'. According to the Broadcasting Act the general aim of such programme items is to 'serve the informational, cultural, civic and everyday needs of the (local, regional and national) audience in the area of reception of the broadcaster'. This refers in particular to programmes of a cultural, scientific or educational nature, presentation of church and religious activities, children's programmes, or news (Broadcasting Act, 1996, §2 (19)).

From the legal point of view the main task of public service television is to provide a 'public service programme' (or programmes) that constitutes 'a programme in which public service programme items play a decisive role, and which regularly informs the listeners and viewers living in the area of reception of the broadcaster about issues deserving the attention of the public' (Broadcasting Act, 1996, §2 (18)). Beyond this basic obligation the Broadcasting Act also formulates several additional programming requirements, including the obligation to provide regular, comprehensive, unbiased and accurate news; the fostering of the values of universal and national cultural heritage, and promoting cultural diversity; providing programmes for minors, serving their physical, psychological and moral development; and providing assistance to people with disabilities (Broadcasting Act, 1996, §23 (4)).

As mentioned above, the public service remit, as defined by the Broadcasting Act, concerns exclusively the broadcasting of 'public service programme items' in radio or television programmes. The remit does not make any provisions for programme production or on-demand services. To this effect, the system of public service content requirements which the law has established is very static and any future changes in the public service mission would require an amendment to the Act or the voluntary commitment of broadcasters. Furthermore, the definition of public service mission in the Broadcasting Act is essentially reduced to a declaration. As is typical for declarations, the definition provides neither any details for the types and quality of programmes that would count towards fulfilling the public service mission, nor any penalties that would result from the failure to meet qualitative programme requirements.

A further questionable characteristic of this way of defining the public service content is that essentially the same commitments apply to both PSBs and commercial broadcasters. It is striking that the public service nature of programming is much more scrutinised by the ORTT in the case of the public service obligations of the commercial broadcasters than in the case of the PSBs themselves. While the public service tasks carried out by the national terrestrial commercial televisions are subject to monitoring by the regulator twice a year there is no similar exercise concerning public service television at all. As a consequence, the definition of 'public service programme item', which is at the core of the idea of public service, is defined for the jurisprudence almost entirely on the basis of the practice of commercial broadcasters.

Developing the public service remit

When analysing the remit of the Hungarian PSBs one realises that it has not been developed to a great extent by the regulator. It seems that at the time of establishing the system of PSBs the issues of governance and funding took regulatory priority over the definition of the public remit, which took second place. This regulatory peculiarity has remained unchanged in the past decade, for until 2008 the regulator had not launched any further initiative to define the public service remit in order to keep up with recent technological developments in the media sector. This inactivity leaves a series of questions for PSBs: What is their room for manoeuvre in building up their concepts of public service content and how freely can they develop their brands? Are they free to launch new services on their own initiative?

The first occasion when a practical decision on the extent of the public service remit was needed occurred in 2005, when MTV announced the launch of its third channel m3, intended to be a thematic satellite news channel covering parliamentary affairs. Shortly after MTV's announcement, Duna Televízió also applied to the regulatory authority for the launch of an additional public service satellite channel called Autonómia. Since PSBs benefit from 'must carry' obligations the launch of these new programme services triggered legal debates with cable operators, which are obliged to distribute all the channels of the PSBs free of charge and therefore challenged the decisions of the ORTT to approve m3 and Autonómia. The dispute concerned the interpretation of the Broadcasting Act, and in particular the question posed by cable operators as to whether the Act allowed PSBs to launch new satellite services without an explicit mandate. Furthermore the lack of a specific public service mandate for launching new services also raised concerns under the state aid rules of EC competition law. In 2008 the competent courts rejected the appeal of the cable operators, but they did so merely on procedural grounds.

A similar issue emerged in 2007 when MR, the public service radio, decided to change the profile of its second national channel. Traditionally the station provided a programme consisting of a mixture of light entertainment, sports and news, but it sought to become a channel of contemporary light music. Following the change it managed to boost its audience ratings significantly especially by attracting younger listeners. However, the programming shift triggered debates as to whether the new output corresponds with the standards of public service content as specified by the Broadcasting Act. At the end of 2008 the ORTT delivered a decision that MR breaches its obligations because the programmes of MR2 Petőfi cannot be regarded as public service programmes (ORTT decision 242/2008 (I.30)). The MR appealed against this decision, but the case is still pending.

The role PSBs should play in the process of digital switchover has received even less attention. The presence of the PSBs in the digital terrestrial services, launched at the end of 2008, is granted by the general 'must carry' rules. However, broadcasting the programmes of the public service items in HD quality still remains a controversial issue. These examples show that the lack of a clear definition of the public service remit by the regulator may result in confusion: the broadcasters have no real freedom to interpret this remit on their own and the lack of legal certainty prevents them from embarking upon new initiatives.

The emergence of new platforms of content delivery and new types of content poses significant challenges for PSBs. It has been suggested that PSBs should extend their remit to new services (Schulz, 2005: 9). Recommendations of the Council of Europe also emphasise the importance of PSBs' responses given to the challenges of the development of the information society. According to one of the recent recommendations of the Committee of Ministers, 'member states should ensure that the public service remit is extended to cover provision of appropriate content also via new communication platforms' (Council of Europe, 2007). Consistent with this approach the Council of Europe (ibid.) concluded that the term 'Public Service Media' better reflects the role of such service providers than the traditional notion of 'Public Service Broadcasting'. But in Hungary the perception of PSB still dominates over the concept of PSM. However, it is also clear for the national regulator that in response to the general trends of development of the media it should make provisions for the transformation of the country's PSBs into real PSM providers by properly adjusting the public service remit to the needs of the information society and extending the scope of this remit to new media.

The organisational structure of Hungarian PSB

The provisions in the Broadcasting Act governing the institutional aspects of PSB build to a large extent on various decisions that the Constitutional Court had issued during the years 1992–5 and in which the independence of the PSBs as well as their institutional and financial guarantees were a central focus. In these standard-setting decisions the Constitutional Court established that the need for the uncompromised independence of PSBs derives from the basic right to freedom of expression. The Court emphasised several times that the independence requirement applies not only with regard to the government but also to the broadcasters' relations with Parliament, the political parties and with the state as such (Constitutional Court decision 37/1992, VI. 10, AB, III. 3).

On these grounds, the Broadcasting Act provides in essence the following institutional structure for the PSBs (see Figure 19.2). It can be seen that the owners of the PSBs are the respective public foundations (Broadcasting Act, 1996, §§64–5). The rationale for the setting up of public foundations between the Parliament and the PSBs was to secure the independence of the latter.

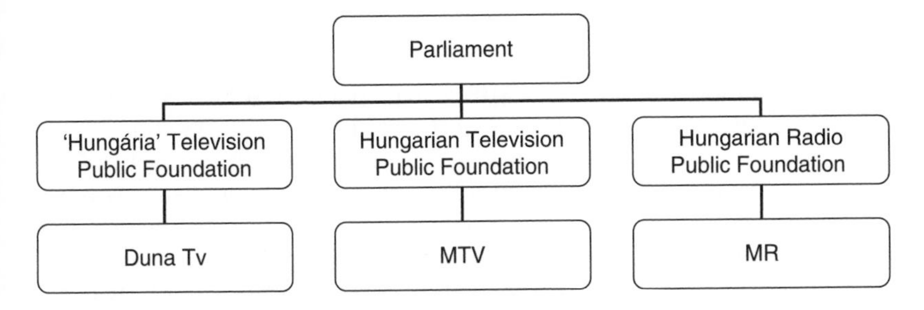

Figure 19.2 The institutional structure of the PSBs in Hungary
Source: Broadcasting Act, 1996.

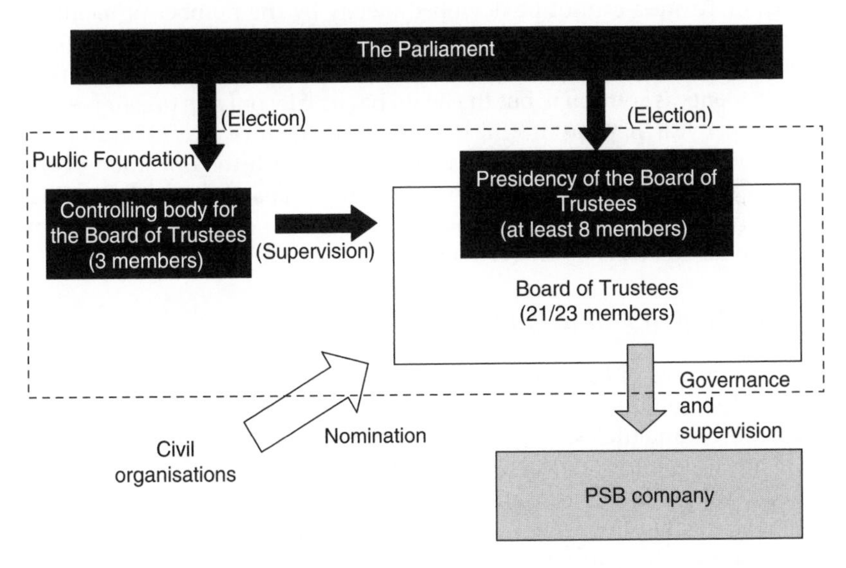

Figure 19.3 The organisational structure of the public foundations in Hungary
Source: Broadcasting Act, 1996.

The organisational structure of the public foundations can be represented as shown in Figure 19.3.

The Boards of Trustees of these public foundations are composed of two different elements. The Presidency has the role of representing the state in the governance of the PSB and therefore mirrors the composition of the Parliament to a certain extent. In contrast, the body of the Board of Trustees is expected to represent society. The 'ordinary' members of the Board are nominated by civil associations (for example, ethnic minority groups, churches, human rights organisations, trade unions, professional organisations representing journalists, women, children and the youth, people with disabilities, and so on).

The main rationale for involving the general public in the decision-making process relating to the performance of the public service tasks is the expectation that civil members can exercise genuine social control over the PSB companies. However, the expected representation of the audience is much less well positioned to bring its particular interests to bear in the decision-making process compared with the weight attributed to the members of the Presidency appointed by the Parliament. In order to get a clearer view of the actual degree of participation of the audiences, it is worth examining the balance between the Presidency (representing the state) and the civil members. This balance between the presence of the state and of the audience in the Boards of Trustees cannot be described merely by the number of members (eight to twelve members in the Presidency and twenty-one to twenty-three additional civil members in the Board). One should also note that the work of the Presidency is continual, but the Board has usually only up to four plenary sessions per year. Beyond this the one-year term of office proved insufficient in practice for civil members to become familiar with the operation of the PSBs. In addition, civil members working only on a part-time basis have significantly fewer chances of influencing the decisions of the Board than the professional members of the Presidency.

The funding of PSB in Hungary

The total income of PSBs in 2008 amounted to 56,895 billion HUF (approximately 190 million euros). This amount was provided via various sources. These sources and their shares in the financing are summarised in Figure 19.4.

The most definitive characteristic of the funding system is that it is not linked to the public service remit or public service tasks at all. Although in the case of ad hoc funding from the central state budget the Parliament defines the purposes of the subsidy (for example, costs of digital switchover or costs of reorganisation of MTV) the main element of the funding, the so-called 'licence fee', is not attached to any particular public service purpose. This makes the current system of financing PSBs questionable from the point of view of Community competition law.

The centre of the system of financing PSBs is the Broadcasting Fund (BF). The BF was established by the 1996 Broadcasting Act as a pool of resources serving the purposes of public service and the financial background of the administration of the media segment (Broadcasting Act §§77–8). It is managed by the ORTT, the media authority of Hungary. The intention behind establishing the BF was to create a system of media financing that would be completely independent from the central state budget. Thus, the purpose was to minimise external intervention in the functioning of the PSBs and the regulatory authority by financial means.

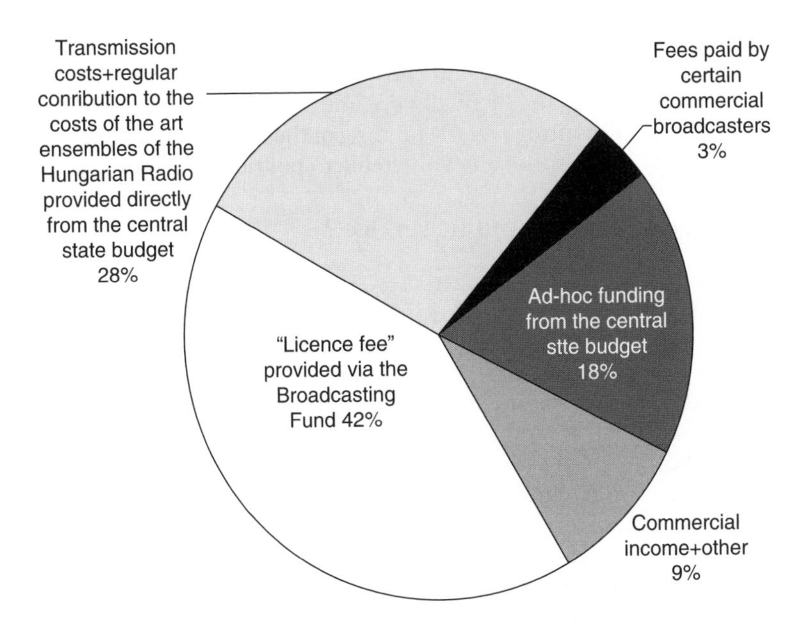

Figure 19.4 The sources of income for Hungarian PSBs (2008)
Source: Act CLVII of 2007, Act CLXIX of 2007.

However, the practice of financing PSBs in the past decade shows that this regulatory objective was not fully met. It can be seen from Figure 19.4 that a substantial proportion of funding was given to PSBs (more precisely to the MTV) on an ad hoc basis directly from the central state budget. This implies that it is at the discretion of the parliamentary majority to determine financial decisions concerning PSBs.

A further diminution of the role of the BF as an independent guarantor occurred in 2002 when the model of the licence fee was in fact abolished by the government. Until the summer of that year public funding constituted a significant contribution to the operational costs of the PSBs. The level of the licence fee, which was to be paid by all households equipped with one or more television sets, had been established by Parliament in the Acts on the annual state budgets. The money collected formed the income of the BF, but in the summer of 2002 the government decided to abolish the licence fee and introduced a funding scheme under which PSBs' income would be derived from the central state budget 'instead of the citizens'. The reasons behind the decision were not connected to media policy considerations. Instead, the aim of the then newly elected government was to fulfil its corresponding electoral promise made during the campaign and to boost its popularity with regard to the forthcoming elections of local governments.

While it is true that the introduction of this new system of financing did not lead to practical shortcomings and the financing of PSBs did not decrease, nevertheless the role of the BF finally became formal and the parliamentary majority gained full control over the funding of the PSBs. In effect, the system of funding became exceptionally vulnerable to political influence.

'State broadcasting' or 'public service'?

The question whether the Hungarian system, as described above, can be regarded as a modern framework for PSM can ultimately be answered by examining the relationship of the PSBs and the public itself, for if this relationship is strong and healthy we can assume that the broadcasters are of a genuine public service nature. If, however, this relationship is formal, weak or non-existent then the difference between the state-owned broadcasting entities and commercial broadcasters is merely a matter of ownership. Bearing this in mind we can find the following links between the Hungarian PSBs and their audiences:

- *At the level of the remit – none.* It should be clear by now that no public consultation rounds or empirical surveys have been used for the purposes of clarifying the public service remit.
- *At the organisational level – the involvement of the representatives of the civil society in the governing bodies.* However, it can also be inferred from the analysis of the organisational structure that the participation of civil society in the Board of Trustees of the public foundations owning the PSBs is problematic. Furthermore, it is obvious that civil members of the Boards of Trustees are in a much weaker position from both an operational and a practical point of view than delegates of the state in the same supervising bodies. Therefore it is not an exaggeration to conclude that the presence of some civil organisations in the governing bodies of the PSBs does not establish a real and living relationship between these broadcasters and Hungarian society.
- *At the level of financing – the role of the licence fee.* The licence fee model was viewed by Hungarian society and the PSBs in a considerably different way than, for example, in the case of British society and the BBC, for the obligation to pay the licence fee was generally unwelcomed by the Hungarian households. The proportion of the non-paying population was exceptionally high and by 2002 it reached approximately 60 per cent of all households (Prime Minister's Office 2007, vol. II: 89). Generally, the establishment of this way of funding failed to provide Hungarian citizens with a real sense of ownership of the PSBs.
- *At the level of financing – ratings.* High reach of audiences is defined as a *sine qua non* for public service (Nissen, 2006b: 24.). It is important to have mechanisms providing feedback to PSBs on their impact on society

on the basis of data available on the actual use of their services by citizens. However, no such mechanism was introduced by the Hungarian regulator. Ratings have substantial impact only concerning commercial activities (advertising and sponsorship) of the PSBs, but this domain is clearly outside their public service role.

To sum up, average Hungarian citizens probably do not feel that PSBs serve them better than commercial rivals. This is a serious strategic shortcoming of the system. The real independence of PSBs cannot be secured exclusively via institutional and financial guarantees. The ultimate guarantee of independence from politics and market powers lies in the recognition of the citizenry. If the society recognises the value of public service it will be ready to protect PSBs from external pressures. If, on the other hand, no particular value is attributed to public service programmes by the citizenry then PSBs will be under threat. In this sense the Hungarian PSBs are still 'state' enterprises, rather than real public service broadcasters. Establishing a regulatory framework that defines these companies as media service providers that form integral parts of the society and enjoy genuine independence from the state still remains a task for the regulator.

Recent developments and future prospects

Having examined the regulatory system applied to Hungarian PSB we can identify three major sources of shortcomings:

- Insufficient definition of the public service remit.
- Poor PSB connection with the society.
- Questionable solutions of funding from the point of view of the EC state aid regulation.

These problematic issues have been known to decision-makers for quite some time. In the past few years there have been several attempts at the structural reform of the regulatory framework governing the system of Hungarian PSM. The most recent examples were the elaboration of a draft National Audiovisual Media Strategy and the accompanying publication of an attached Regulatory Concept Paper and a draft Bill on Media Services. Both documents were prepared under the aegis of the Prime Minister's Office which is the ministry responsible for matters of audio-visual regulation. The draft Bill also reflects the work of representatives of the political parties represented in the Parliament and serves as a basis for their discussions. It is important to note that because of the political landscape characterised by the extremely hostile attitudes of the political parties towards each other and the high level of consensus needed by the constitution to pass legislation related to PSBs it seems uncertain whether these documents will actually lead to new regulation. However, they reflect the current regulatory way of thinking.

Both the draft Bill and the Regulatory Concept Paper foresee fundamental changes in regulating PSM. In an attempt to give a short account of these changes we first notice that both documents aim at creating a system of funding compatible with the corresponding EC competition rules. Both of them envisage a funding model based on the contractual relationship between the PSBs and the funding body. As a consequence the documents require the detailed definition of the public service remit. Moreover, in the views expressed in the Regulatory Concept Paper and in the Bill, this remit shall be extended to new media (Bill on Media Services, 2009, §54; Commissioner for Regulation of the Media, 2007: 33–6). In general, on questions of remit and funding the regulator seems to be conscious of the negative consequences of not addressing issues of PSM at all in the rapidly developing media ecology of the past decade. So, if legislation comes to the fore PSBs might expect an increasing level of legal certainty and an extended and better defined room for manoeuvre in formulating their policies.

In terms of public media's forging real contacts with society the prospects of success are slim. At the level of the definition of the public service remit the draft Bill and the Regulatory Concept Paper unquestionably promise progress. As a new element, public consultation rounds would be introduced at the phases of the preparation of the public service financing contracts and of preparing the strategies for PSM by the governing and supervising body (Bill on Media Services, 2009, §68; Commissioner for Regulation of the Media, 2007: 36). However, at the institutional level we cannot predict similar developments. Although the Regulatory Concept Paper clearly stated the need for the real representation of the civil society in the institutions governing and supervising the system of PSB (Commissioner for Regulation of the Media, 2007: 39) the first version of the draft Bill left no room for such involvement and proposed a governing and supervising body composed exclusively of members delegated by the Parliament. As a reaction to criticism though, the second version of the Bill might introduce a consumer panel with consultative functions (Bill on Media Services, 2009, §52). It still seems that the regulator is reluctant to make real attempts to cease the supremacy of the state at the institutional level of PSB and to seek real and effective means of involving the society itself.

References

Act XX of 1949 on the Constitution of the Hungarian Republic.

Act CLVII of 2007 on the budget of the ORTT for 2008.

Act CLXIX of 2007 on the budget of the Republic of Hungary for 2008.

AGB Nielsen, Hungary (2009) *Development of Television Viewing Time (ATV) and Audience Shares (SHR%) on the Hungarian Market 1997–2008, Total Population (4+)*, http://cs.agbnmr.com/Uploads/Hungary/stat_shr_negyedeves.pdf, accessed 3 April 2009.

Bill on Media Services (2009) http://www.parlament.hu/aktual/szakmaitervezet.pdf, accessed 27 February 2009.

Broadcasting Act (2006) I of 1996 on Radio and Television Broadcasting.

Commissioner for Regulation of the Media (2007) 'National Audiovisual Strategy – Concept Paper for Legislation', http://www.meh.hu/szolgaltatasok/dtv/nams20070426.html, accessed 27 February 2009.

Council of Europe (2007) 'Recommendation CM/Rec(2007)3 of the Committee of Ministers to Member States on the Remit of Public Service Media in the Information Society' (Strasbourg: Council of Europe).

Decision of the Constitutional Court 37/1992 (VI. 10) AB.

Decision of the ORTT 242/2008 (I. 30).

Nissen, C. (2006a) 'No Public Service without Public and Service: Content Provision between the Scylla of Populism and the Charybdis of Elitism', in *Making a Difference: Public Service Broadcasting in the European Media Landscape* (Eastleigh: John Libbey Publishing).

Nissen, C. (2006b) *Public Service Media in the Information Society*, Media Division Director General of Human Rights (Strasbourg: Council of Europe), February.

Prime Minister's Office (2007) 'National Audiovisual Strategy: Paper for Professional Debate', http://www.meh.hu/szolgaltatasok/dtv/nams20070426.html, accessed 27 February 2009.

Schulz, W. (2005) 'Public Service Broadcasters' Strategies in the Information Society' (Hearing Council of Europe), presentation, http://www.coe.int/t/dghl/standardsetting/media/restricted/MC-S-PSM/MC-S-PSB(2005)Schulz_en.pdf, accessed 23 March 2009.

20
Future Directions for US Public Service Media

Walter S. Baer

Introduction: the structure and financing of US public television and radio

Public Service Broadcasters (PSBs) in the United States, like their European counterparts, face formidable challenges resulting from rapid technological change, audience fragmentation and declining TV viewership. Old media models everywhere are breaking down in the new environment characterised by user-generated content, collaborative production and editing, and multiple distribution alternatives.

The US system of public service broadcasting differs markedly from those in Europe and most other countries. Unlike continental Europe, where broadcasting began in the 1920s as state owned and operated enterprises, the US chose to grant local broadcast licences to non-government entities – primarily to commercial firms to provide advertising-supported programming, but also to colleges, universities and churches. In contrast, the UK created a hybrid model in which the BBC, a non-government organisation supported by public funds, became the broadcasting monopoly. Only in the late 1930s did the US set aside a portion of the then-experimental FM spectrum for 'educational radio' followed by an allocation of the UHF spectrum for educational television after World War Two.

A major expansion of US non-commercial broadcasting came with Congressional passage of the Public Broadcasting Act of 1967, spurred by a blue-ribbon Carnegie Commission report that proposed 'a new institution for public television' with a broad civic and cultural scope (Carnegie Commission, 1967). Congress set out broad goals on the new legislation 'to encourage the growth and development of public radio and television broadcasting ... for instructional, educational and cultural purposes ... [and] programming that involves creative risks and that addresses the needs of unserved and underserved audiences, particularly children and minorities' (PBA, 1967). The 1967 Act established a non-profit, non-government Corporation for Public Broadcasting (CPB) to provide financial support to local public television and radio stations, but without the sustained funding or authority to build the strong national network that the Carnegie Commission had envisioned. Instead:

> Driven by fears of a politically liberal broadcasting service . . . and the concerns of commercial rivals, Congress deliberately created a decentralized national service that was anything but a 'system'. Congress provided only a small minority of what public broadcasters would need through federal funds, and that through a regular appropriations process. The choice of appropriations, rather than an endowment, guaranteed that public broadcasting's content would be perpetually under political scrutiny. (Aufderheide and Clark, 2008: 3)

Today, public broadcasting comprises more than 1000 independently run, non-commercial radio and television stations throughout the United States that receive funds from the CPB. They are operated by a diverse mix of non-profit community organisations, colleges and universities, and state and local municipal authorities. The CPB funds, on average, only 14 per cent of their total revenue. Audience subscriptions and business contributions (primarily for underwriting programmes for which they receive on-air credit) provide greater amounts, and the rest comes from state and local governments, colleges and universities, foundation grants and a variety of other sources (Figure 20.1). More than 350 US public television stations own and operate the Public Broadcasting Service (PBS), a private non-profit corporation founded in 1969 that provides programme distribution and other services to its members. Similarly, National Public Radio (NPR), founded in 1970, is a private non-profit corporation owned by its nearly 700 member radio stations. The entire US public broadcasting enterprise had annual revenue of $2.9 billion in 2007, less than half that of the BBC.

Both PBS and NPR enjoy high levels of public confidence (Ford Foundation, 2007), but public television and radio have different structures and different circumstances to contend with in the changing media environment. Public television has seen its audience share decline steadily over the past twenty-five years, due in good part to increasing competition from cable, satellite and internet TV to which more than 85 per cent of American households now subscribe. These multi-channel systems offer an abundance of content, from children's shows to science and nature series to sophisticated drama, which was once available only on PBS. As a consequence, public television's average Nielsen rating in evening prime time has fallen to below 1.3 per cent in 2008 (CPB, 2009b), and even some of the highest rated shows are having difficulty attracting the corporate underwriting support they require. Moreover, PBS produces no programming itself; instead, nine major producing stations compete for resources and coordinate with PBS to sell their programmes to other member stations. US public television

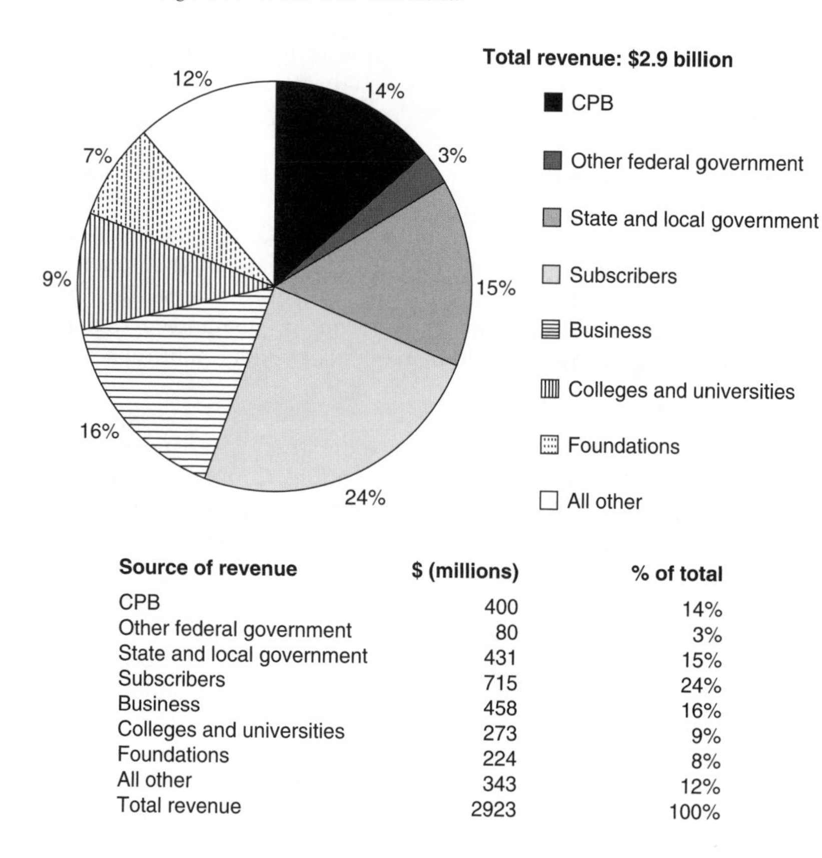

Source of revenue	$ (millions)	% of total
CPB	400	14%
Other federal government	80	3%
State and local government	431	15%
Subscribers	715	24%
Business	458	16%
Colleges and universities	273	9%
Foundations	224	8%
All other	343	12%
Total revenue	2923	100%

Figure 20.1 US public broadcasting revenue by source (fiscal year 2007)
Source: CPB (2009a).

in fact has a quite Byzantine structure for financing programme production, which is chronically underfunded.

Public radio, while not without it own problems, is in better shape. Its national audience has increased ten-fold since 1980 and, with commercial radio listeners declining, its audience share reached an all-time high of 5.2 per cent in 2008 (Figure 20.2). Unlike PBS, NPR has a strong programming group with production centres in Washington, DC and Los Angeles. Its morning and afternoon news programmes attract large audiences during rush hour drive times (radio's prime time) that serve to anchor stations' daily programme schedules. Production costs are of course much lower for radio than for television, and NPR programming benefits from a 2003 bequest of $230 million from an individual estate. American Public Media, owner-operator of forty-two public radio stations and the second

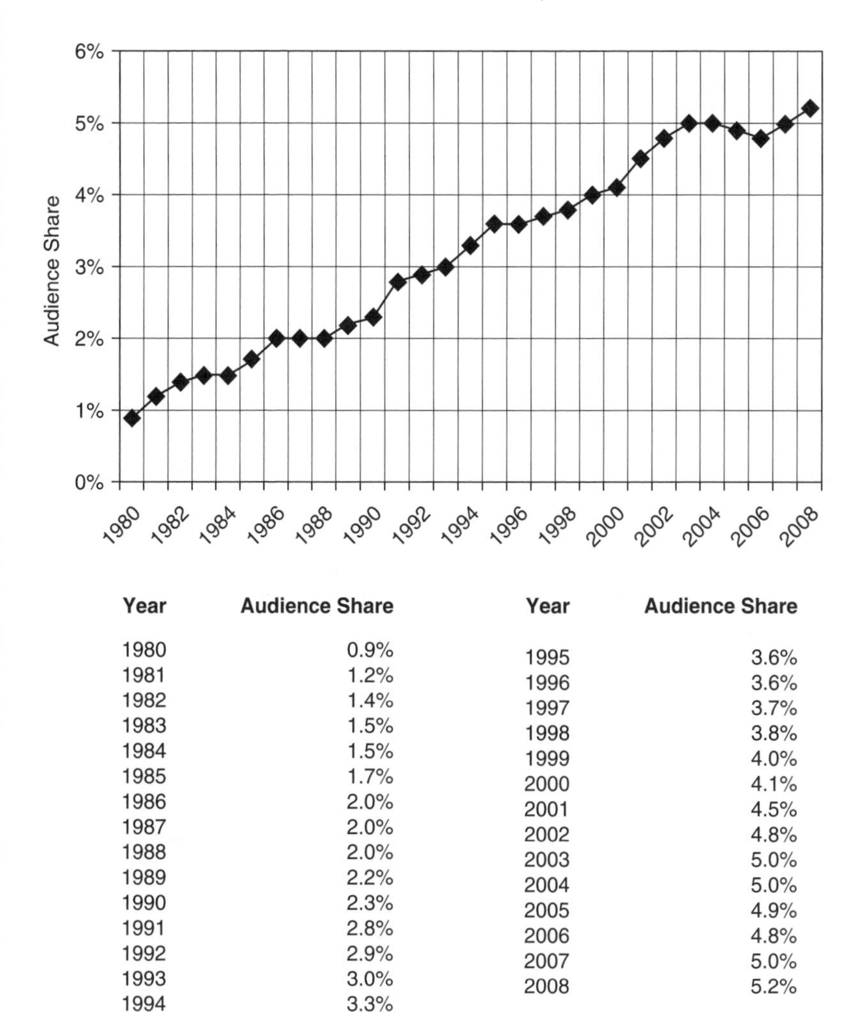

Year	Audience Share
1980	0.9%
1981	1.2%
1982	1.4%
1983	1.5%
1984	1.5%
1985	1.7%
1986	2.0%
1987	2.0%
1988	2.0%
1989	2.2%
1990	2.3%
1991	2.8%
1992	2.9%
1993	3.0%
1994	3.3%
1995	3.6%
1996	3.6%
1997	3.7%
1998	3.8%
1999	4.0%
2000	4.1%
2001	4.5%
2002	4.8%
2003	5.0%
2004	5.0%
2005	4.9%
2006	4.8%
2007	5.0%
2008	5.2%

Figure 20.2 US public radio's nationwide audience share (1980–2008)
Source: Arbitron data compiled by Radio Research Consortium (2008).

largest producer/distributor of public radio programming, also has a sizeable endowment unmatched in the public television community.

Public television and radio face serious financial problems, both short-term and long-term. The near-term problem is how to cope with the economic downturn that affects their two largest sources of income shown in Figure 20.1, audience subscriptions and corporate underwriting.

A CPB financial risk assessment in January 2009 forecast substantial revenue declines for US PSBs. It also stated that 21 per cent of the public television stations with annual revenues below $10 million were 'fragile', and some could go off the air (CPB, 2009b). Already in 2008, many stations had laid off staff and cut programming expenses (Janssen, 2008b). But beyond the current economic crisis, US public broadcasting must deal with the tectonic shifts in the media landscape brought about by the internet and related digital technologies.

Bringing public broadcasting online

In 2009, all US television stations were expected to make the long-awaited technical transition from analogue to digital broadcasting, enabling them to broadcast both high-definition (HD) TV and multiple channels of non-HD video and data services. Multi-casting opens up opportunities for PSBs to advance their public service missions by offering more educational, civic, children's and other local content to their communities, as well as programmes in languages other than English. Filling these additional channels will not be inexpensive, however. While public stations have raised the funds needed for the technical conversion ($1.5 billion), developing and paying for HD and multi-cast programme content will likely burden station resources for some time to come.

Moreover, PSBs need to move beyond their ageing core audiences and connect to a new generation that has grown up with computers and the internet, mobile phones and social networks. Younger people live in a different media world from that of traditional broadcasting, one that is centred on participatory and social media platforms such as Flickr, MySpace, YouTube, Facebook and Twitter. They want access to information, entertainment and peer networking when and where they choose, and they want to be media participants rather than just passive viewers and listeners.

US PSBs fully recognise that to engage 'the people formerly known as the audience' (Rosen, 2006), they must provide content and services that are available online and on mobile devices, available on-demand, and interactive. But charting a path for public service broadcasting to embrace the new digital and social media has not been easy. During 2005, a high-level citizen panel worked to develop a national agenda for 'Public Service Media in the Digital Age' similar to what the Carnegie Commission had done four decades before (Digital Future Initiative, 2005). Their report, however, resulted in little subsequent action. Other efforts by groups within public broadcasting tried to build system-wide consensus for digital content distribution (DDC Working Group, 2007; PSP Initiative, 2005), but did not resolve the underlying tensions and rivalries among the stakeholders and thus did not achieve the critical mass needed to move forward.

PBS and NPR online initiatives

Nonetheless, PBS and NPR have successfully developed and implemented a number of popular online, on-demand and interactive projects over the past several years:

- PBSKids.org, is a favourite US website for young children and their parents. It features free interactive games, songs, colouring pages and other activities using characters from PBSKids programming.
- *Frontline*, a highly regarded PBS documentary series for more than twenty-five years, reports that programmes streamed from its website now represent a substantial fraction of total viewing. *Frontline* also creates background Web pages for each programme, including a full transcript and timeline, extended interviews, readings and links, and viewer comments.
- NPR's website provides hundreds of free podcasts on a wide range of topics produced by NPR and more than fifty of its member stations. It also offers a series of more eclectic 'alt.NPR' podcasts tailored for young adult audiences.

Other online initiatives are at more developmental stages:

- More than 1200 television shows previously shown on PBS are available for on-demand viewing from Hulu.com, a for-profit online video service owned by NBC Universal and News Corp. Watching shows streamed on Hulu is free to anyone with a computer and broadband internet connection, but each PBS video is preceded by a thirty-second 'public service announcement' and PBS promotion.
- PBS has partnered with BitTorrent to make some recent shows from Nova, History Detectives, Scientific American and other series available for downloading and purchase on BitTorrent's peer-to-peer content distribution platform.
- NPR, in collaboration with member stations, offers hourly national and local news headlines to iPhone, Blackberry and other smartphone users over their mobile phone networks. The CPB-funded Independent Television Service has also commissioned a series of short videos by independent producers for mobile devices.

KCRW online

Reflecting the station-focused structure of US PSB, moreover, much innovation is occurring at the individual station level. The online experience of KCRW public radio, the NPR affiliate in Santa Monica, California, provides a good illustration. KCRW broadcasts an eclectic mix of music, news, arts and cultural programming that reaches 350,000 listeners per week in Southern California. The station was early in embracing the internet – its website went live in 1995, it began streaming programming online in 1999 and podcasting in 2003 – and it currently has one of the largest online operations in US public radio. As of February 2009, KCRW online includes (Dewan, 2009):

- Managing a dynamic website and deep, 25,000-page Web archive that attracts 625,000 visits per month.
- Streaming three channels online 24/7: Live (simulcast), Music and News, with average monthly listenership of 1.6 million hours.
- Making the three streams and on-demand programming available to iPhone and iPod Touch users.
- Podcasting twenty-seven programmes with 1.3 million downloads per month.
- Managing KCRW social networks on Facebook, MySpace and Twitter.

Ongoing efforts are under way to further integrate the live streams and Web archives with additional original content created by KCRW DJs, hosts, producers and staff. KCRW also seeks to extend its online presence by engaging its community/social networks through blogs, discussion forums and user-generated content.

KCRW's online expansion seems to be proceeding well. The online audience has grown steadily and is now about one-quarter the size of the radio audience. Listener interest in the content KCRW produces extends far beyond its broadcast service area; more than half the online audience is outside Southern California and 16 per cent are from outside the United States. But online operations require constant tending and investment as digital programming, tools and platforms evolve; and so far, the fundraising methods that work on-air have been less effective online. Over time, KCRW must find ways to foster communities of loyalty and support on the internet that match those it has built with traditional radio.

From interactive towards participatory public media

The projects outlined above are all 'interactive' in the sense of giving users more control over when, where and how they access PSB content. Many enable users to interact with extensive Web-based archives, enter contests or play games online, explore links to related websites, send in comments and join moderated discussion forums. But these examples are less 'participatory' than the immensely popular Web platforms like YouTube, Facebook and Wikipedia, whose users create most of the content, collaborate in tagging and editing content, and communicate freely among themselves without moderators or other filtering.

Web-based, participatory media sites change the traditional create–filter–publish model to one of create–publish–filter. Users who register and agree to a set of 'Community Standards' (that typically forbid pornography, hate speech, copyright infringement and spam) are able to post content to a website where it can be seen or heard by other users without pre-editing. Users who find something that has been posted offensive can flag it for review by the site's staff, who have the authority to remove the item if it violates the Community Standards or the site's Terms of Use. Such editing after-the-fact generally works well in removing the most abusive postings but of course does not guarantee the quality or appropriateness of what remains publicly available. Content ranking, recommending and social bookmarking tools on such websites as Del.icio.us, Digg and Reddit also provide a form of collaborative filtering that bypasses conventional editing by media organisations.

Steeped in broadcasting traditions of professional reporting and editing, PSBs are not very comfortable with create–publish–filter models for user-generated content. US PSBs in particular worry that their government funding could be at risk if offensive language or content finds its way onto their websites that are open to the public.

However, some media ventures formed in the past several years strive to involve their audiences in more participatory ways while still maintaining most aspects of editorial control. They do this by inviting users to submit content to a website in a relatively unstructured manner, and then using a variety of filtering mechanisms to select what will be presented to larger public audiences. The four projects described below – two from public radio and two commercial – offer participatory models that appear relevant to PSB evolution in the United States.

Public Radio Exchange (PRX)

Launched in 2002, Public Radio Exchange is 'an online marketplace for distribution, review, and licensing of public radio programming' as well as a 'growing social network and community of listeners, producers, and stations collaborating to reshape public radio' (PRX, 2009). Its mission 'is to create more opportunities for diverse programming of exceptional quality, interest, and importance to reach more listeners'.

Any adult can register as a PRX producer and upload audio 'pieces' to the www.prx.org website that are intended for use by non-commercial radio and to which they own the intellectual property rights. Producers choose the licence terms and prices for use of their pieces. The pieces are catalogued and made available for listening and peer review by PRX staff, as well as by other registered producers, radio station staff and listeners.

Member station staff can browse by length, format, topic or review status; and listen to the full length of any piece. PRX also regularly emails newsletters highlighting relevant topics and pieces that have received highly favourable reviews. For a piece that seems of interest, a station programmer can click on a 'Licence/Download' link to see its licence terms and price. If these are acceptable, the programmer simply clicks on the 'Licence this piece' button to download a broadcast-quality (MP2) copy. PRX then automatically debits the station's account and credits the producer's account.

PRX is an independent non-profit corporation funded by the Corporation for Public Broadcasting, other government agencies and private foundations. As of April 2008, producers had posted more than 20,000 audio pieces, and more than 1000 producers had licensed pieces to stations, earning more than $650,000 in royalties (Janssen, 2008a). The PRX model could in principle work for independent television as well as radio producers, although the copyright clearance issues are far more costly and complex for television than for radio.

Current Media

Current Media is a hybrid model of participatory media and traditional advertising-supported commercial television, primarily targeted to 18–34-year-olds. It operates a for-profit cable and satellite television channel, Current TV, and an internet site 'produced and programmed in collaboration with its audience. Current connects young adults with what is going on in their world, from their perspective, in their own voices' (Current Media, 2009).

Any registered user can produce and upload short video programme segments ('pods') to the Current.com website, where they are rated by other users and Current staff. Pods are typically one to eight minutes in length and can be on any topic so long as they meet Current's Community Standards. Current offers user training and tutorials, and for experienced user-producers, access to an assignment desk for new work. Viewer-created news segments are tagged as 'collective journalism' and considered by a designated editorial team.

A user whose pod is selected by the Current editorial staff to be broadcast on the Current TV cable/satellite channel receives a one-time payment in the $500–$1000 range. Such viewer-created content constitutes about one-third of the channel's programming; the rest comes primarily from in-house staff producers.

Users also produce some of the advertising carried on Current TV. Viewer-created ad messages (VCAMs) were not part of Current's original business plan, but the concept was picked up from 'fake ads' circulating on the internet, and the first VCAM was broadcast in May 2006. VCAM assignments are now listed on Current.com; submissions are rated by other users, and sponsors make the final selections. Users receive $2500 for each VCAM shown on Current TV and between $5000 and $20,000 more for placements elsewhere. Current's head of advertising sales said in a 2007 interview that Current TV viewers prefer VCAM ads to those created by ad agencies by a nine-to-one margin: 'It's in their voice, it's more authentic, they relate to it more and it's more believable to them' (Lafayette, 2007).

Current Media is a for-profit company initially formed in 2002 with Al Gore as chairman. Current TV was launched in August 2005 and, as of January 2009, reaches more than 40 million households in the US, and more than 15 million households in the UK, Ireland and Italy via cable and satellite. The Current.com website was launched in October 2007. The company's revenues come from distribution fees for each subscriber household that receives Current TV and from advertising, but it is not yet profitable.

In 2007, Current TV won an Emmy for 'Outstanding Creative Achievement in Interactive Television Service' and a Webby Award for 'best television network website'. In November 2008, Current TV and CBC, Canada's national public broadcaster, formed a partnership to launch Current Canada, 'a cross-platform media company uniquely focused on engaging young adult audiences through participatory and interactive initiatives on TV and on the Web' (Current Media, 2008). No similar collaborations with US pubcasters have been announced.

iReport

While user-generated content is central to Current TV's business model, other US commercial broadcasters use it in more constrained ways. An important example is iReport, created by the 24-hour cable news network CNN as a focus for news stories contributed by users. It also 'attracts significant traffic in its own right while ... serving to promote the overall CNN brand' (Hampel, 2008).

iReport began in August 2006 as a section of the CNN.com website where users could upload text, audio, photos and videos for consideration by CNN reporters and editors. It moved eighteen months later to a separate website, iReport.com, so that it could be clearly distinguished from CNN.com. CNN emphasises that 'iReport.com is a user-generated site. That means the stories submitted by users are not edited, fact-checked or screened before they post.'

Although iReport welcomes any user submissions that do not violate the posted Community Guidelines, an Assignment Desk lists topics of particular interest to CNN television producers and writers. Users must register before they can post stories, or rate or comment on other users' stories. Uploaded stories appear on iReport.com almost immediately and are categorised by time of posting, contributor, user-supplied tags and 'newsiest' – a proprietary formula that combines freshness, contributor activity and the story's popularity and ratings on the website.

A CNN writer or producer who finds a story of potential interest on iReport will arrange for a staff member to contact the contributor and go through the same vetting and fact checking process that CNN uses for other source material. User-generated stories that are vetted and used in CNN television news coverage are then tagged with a label 'On CNN' and can be seen on CNN.com as well as on iReport.com. Unlike Current TV, however, iReport contributors are not paid for stories that are selected for television broadcast.

iReport appears successful in attracting user-generated content and building an online community, primarily of young adults who might otherwise not be drawn to CNN. More than 200,000 people have registered on iReport since February 2008, and nearly 240,000 stories have been submitted (iReport, 2009). CNN routinely incorporates vetted iReport stories in its television news (more than 1300 stories in January 2009) but diligently separates the 'unedited, unfiltered' submissions on iReport.com from what appears under the CNN brand. At this stage, iReport appears to have had little impact on mainstream CNN editorial practices; but over time it could well do so.

:Vocalo

:Vocalo.org is a new, participatory community website and radio station serving a diverse multicultural population in north-west Indiana, near Chicago. It reverses the traditional PSB structure by making the internet its primary platform and broadcasting mostly local, user-generated content.

:Vocalo was created by Chicago Public Radio (CPR) with the goal of fostering community engagement among people who were not public radio listeners. CPR's research showed that its audience was overwhelmingly white (91 per cent), while the communities it served were racially and ethnically much more diverse. African-Americans and other minorities were highly critical of both NPR and CPR programming, saying 'Chicago Public Radio is not about Chicago' (Malatia, 2007). Determined to develop a local service that reflected community interests, CPR dedicated its Indiana FM transmitter to :Vocalo and hired local staff and on-air hosts with strong community ties but generally little radio experience. Operations began in April 2007.

:Vocalo encourages local listeners and others to register as users, create their own original audio content (subject to Community Guidelines), and upload it onto the website. Online tutorials and free training workshops help users improve their audio techniques. Users can choose whether to share their content only with :Vocalo staff, share it with other registered users who can rate and comment on it, or make it publicly available. Staff choose which material to broadcast. :Vocalo's on-air programming does not follow a conventional schedule but consists of short audio pieces, the majority of which are user-generated and cover a wide range of storytelling, interviews, music performances, personal diaries, commentaries and rants. Staff-produced programmes and commercial music make up the remainder. :Vocalo does not carry any CPR programming, not even NPR newscasts.

While :Vocalo's editorial model is clear and rather revolutionary for US PSBs, its financial model is not. Its expenditures in its first two years have been supported largely by CPR supplemented by grants from foundations, the Corporation for Public Broadcasting and a few Chicago-based corporations. CPR, like nearly every other US public broadcasting entity, has seen its listener subscriptions decline in 2008 and has laid off staff, including two :Vocalo on-air hosts. CPR's CEO is highly committed to the project, but the extent to which :Vocalo can continue to attract grant monies or develop other sustainable sources of funds remains unclear.

US public service media in the new digital media ecology

Public service broadcasters in the United States are gradually reconstituting themselves as digital public service media, joining what Yochai Benkler calls the 'networked public sphere' in which 'individuals, acting alone or with others ... [are] active participants in the public sphere as opposed to its passive readers, listeners, or viewers' (Benkler, 2006: 212). As in past communications and media transitions, digital and social networks will not supplant broadcasting (or print), but online platforms over time will become central. Distinctions between radio and television will likely fade as digital media integrate text, pictures, sound and video across multiple distribution platforms. Many observers also predict that mobile devices will provide the primary connections to the internet for most people by 2020 (Anderson and Rainie, 2008).

Managing the technology side of this transition is the easiest part; changing well-established practices and organisational structures to align with the new media ecology is a good deal more difficult. For example, US PSBs rely on on-air fundraising to get financial support from listeners and viewers, but such direct online appeals have brought little response so far. Online users have been notoriously unwilling to pay for online content, commercial or non-commercial, whether it is breaking news or analysis, magazine articles, or radio and television programmes. And recent proposals for developing financial support online for journalists through individual contributions (Kershaw, 2008) or micropayments (Isaacson, 2009) seem at least premature. Still, there are some notable successes in having users pay directly for online content, such as music downloads from iTunes, electronic books from Amazon.com, financial news from the *Wall Street Journal* website, and Radiohead's 2007 'pay what you want' arrangement for downloading a new music album. As PSBs work to build participatory communities online, they will need to try a wide range of approaches to generate financial support, to assess what works and what does not, and to share the results so that a set of best practices for online fundraising can emerge.

As much as they need new business models, public service media need to develop new models for adopting digital and social networks to foster civic engagement and revitalise democratic processes (Wilson, 2008). In the past several years we have seen an explosion in the use of online social media for citizen reporting, freewheeling discussions and debates, advocacy positions, and political organising at all levels from neighbourhood to community to national to global. Although the views expressed are often raucous and opinionated, there is already a great deal of constructive energy in the US networked public sphere.

What US public service media can add is to serve as trusted online conveners of community voices on public issues, as well as reliable content producers. Today, Minnesota Public Radio regularly emails thousands of listeners who have signed up to become part of its Public Insight Network, asking for their comments on current issues and inputs for news stories (MPR, 2009). Looking forward, public media must take further steps to engage their publics by employing the participatory tools and platforms that digital natives use routinely. They, like all media, will become network-centric, and they must be willing to give up some measure of traditional control over content production and editing in order to embrace more collaborative practices. This will demand substantial structural and cultural changes within public service media organisations, but such changes will be necessary if US public service media are to stay relevant and thrive in the media ecology of the twenty-first century.

References

Anderson, J. and L. Rainie (2008) 'The Future of the Internet III', Pew Internet & American Life Project, 14 December, http://www.pewinternet.org/pdfs/PIP_FutureInternet3.pdf, accessed 27 February 2009.

Aufderheide, P. and J. Clark (2008) 'Public Broadcasting and Public Affairs', Berkman Center for Internet & Society at Harvard University, http://cyber.law.harvard.edu/pubrelease/mediarepublic/downloads.html, accessed 27 February 2009.

Benkler, Y. (2006) *The Wealth of Networks* (New Haven and London: Yale University Press).

Carnegie Commission (1967) *Public Television, a Program For Action: the Report and Recommendations of the Carnegie Commission on Educational Television* (New York: Harper & Row).

CPB (2009a) 'Public Broadcasting Revenue by Source, FY 2007', Corporation for Public Broadcasting, 5 January, http://www.cpb.org/stations/reports/revenue/2007PublicBroadcastingRevenue.pdf, accessed 27 February 2009.

CPB (2009b) 'Economic Outlook: Impact on Public Broadcasting', Corporation for Public Broadcasting, Board Presentation, 26 January, http://current.org/pbpb/surveys/CPB-EconomicOutlook-Jan26-2009.ppt, accessed 27 February 2009.

Current Media (2008) 'CBC and CURRENT TV announce partnership to launch new participatory media company', media release, 10 November, http://i.current.com/pdf/CURRENT_AND_CBC.pdf, accessed 27 February 2009.

Current Media (2009) 'Frequently Asked Questions', http://current.com/s/faq.htm, accessed 27 February 2009.

DDC Working Group (2007) 'The Digital Distribution Consortium (DDC) Overview', February, http://www.integratedmedia.org/files/Media/022007_872_0078508.pdf, accessed 27 February 2009.

Dewan, A. (2009) (Director of New Media, KCRW), personal communication, 19 February.

Digital Future Initiative (2005) 'Final Report of the Digital Future Initiative: Challenges and Opportunities for Public Service Media in the Digital Age', New America Foundation, http://www.newamerica.net/files/archive/Doc_File_2766_1.pdf, accessed 27 February 2009.

Ford Foundation (2007) 'All Things Reconsidered', Ford Reports, 37(2), http://www.fordfound.org/impact/fordreports/publicsquare/all_things_reconsidered, accessed 27 February 2009.

Hampel, M. (2008) 'iREPORT: Participatory Media Joins a Global News Brand', Berkman Center for Internet & Society at Harvard University, http://cyber.law.harvard.edu/pubrelease/mediarepublic/downloads.html, accessed 27 February 2009.

iReport (2009) http://www.ireport.com/index.jspa, accessed 27 February 2009.

Isaacson, W. (2009) 'How to Save Your Newspaper', *Time*, 5 February, http://www.time.com/time/business/article/0,8599,1877191,00.html, accessed 27 February 2009.

Janssen, M. (2008a) 'Prize for PRX, a middleman appreciated by both sides', *Current*, 21 April, http://current.org/indies/indies0807prx.shtml, accessed 27 February 2009.

Janssen, M. (2008b) 'Even APM, WGBH trim their payrolls', *Current*, 22 December, http://www.current.org/economy/econ0823otherlayoffs.shtml, accessed 27 February 2009.

Kershaw, S. (2008) 'Crowdfunding: a Different Way to Pay for the News You Want', *The New York Times*, 24 August, http://www.nytimes.com/2008/08/24/weekinreview/24kershaw.html?_r=1& oref=slogin&pagewanted=print, accessed 27 February 2009.

Lafayette, J. (2007) 'Ads That Sell Themselves', *TV Week*, 27 June, http://www.tvweek.com/news/2007/06/ads_that_sell_themselves.php, accessed 27 February 2009.

Malatia, T. (2007) 'It's public radio but with nearly everything different, including the name', *Current*, 14 May, http://www.current.org/radio/radio0708malatia.shtml, accessed 27 February 2009.

MPR (2009) Public Insight Network, Minnesota Public Radio, http://minnesota.publicradio.org/publicinsightjournalism, accessed 27 February 2009.

PBA (1967) Public Broadcasting Act of 1967, Sec. 396. [47 U.S.C. 396], http://www.cpb.org/aboutpb/act/text.html, accessed 27 February 2009.

PRX (2009) The Public Radio Exchange, http://www.prx.org, accessed 27 February 2009.

PSP Initiative (2005) 'Executive Summary: the Public Service Publisher Initiative', 25 January, http://technology360.typepad.com/PSP_Executive_Summary.pdf, accessed 27 February 2009.

Radio Research Consortium (2008) 'Public Radio Nationwide Trend 2008', http://www.rrconline.org/reports/pdfCPBHistoryMSu6-12m2008.pdf, accessed 27 February 2009.

Rosen, J. (2006) 'The people formerly known as the audience', *Pressthink*, 27 June, http://journalism.nyu.edu/pubzone/weblogs/pressthink/2006/06/27/ppl_frmr.html, accessed 27 February 2009.
Wilson, E. J. (2008) 'The Emerging Media Eco-System and the Meanings of Democracy', presentation to Media Republic Forum, University of Southern California, 28 March, http://cyber.law.harvard.edu/interactive/2008/08/annenberg1, accessed 27 February 2009.

21
Identity Housekeeping in Canadian Public Service Media

Philip Savage

> 'The only thing that really matters in broadcasting is program content; all the rest is housekeeping'
>
> – Robert Fowler (Canada, 1965: 3)

Introduction

Canadian public broadcasting is closely linked with a larger project familiar to Europeans: telling a story about a collection of peoples spread over a northern continent. For half of Canada's 140-year history the public radio service of the Canadian Broadcasting Corporation (CBC), joined by TV and digital media, has provided the bulk of programming in the mixed private–public broadcasting system. The CBC was created to help form and then protect Canadian identity by broadcasting 'CanCon' (Canadian Content). The digital revolution only intensified the cultural project in the context of one of the world's highest levels of integration with a foreign – read American – market.

What is it about the CBC's programming that addresses the changing needs, desires and identities experienced by Canadians? In particular, how is CBC reflecting Canadian societal transformation while changing institutionally even as the technological, demographic, political and economic structures – the nation's prosaic housekeeping – change and Public Service Broadcasting (PSB) becomes Public Service Media (PSM)? Paradoxically, the CBC's oldest service, CBC Radio, leads at what PSM does best – developing new content to address dynamic notions of identity and audience reflection. As this spreads from regional radio operations throughout CBC-TV and CBC.ca, it provides new models for Canadian broadcasters and, possibly, for other national PSBs struggling with dynamic and varied national identities.

The political economy of Canadian broadcasting

PSB is seen as a central mechanism for the information and cultural exchange key to a democracy, which, according to Jürgen Habermas, relies on a vibrant public sphere for discussion among the broader civil society (Habermas, 1962). For about a century now mass media and especially electronic distribution have become the major sites for discussion, with the role of PSB an alternative to market-driven and increasingly consolidated commercial media which commodify the audience role rather than directly enhance citizen democratic potential (Mosco, 1996: 167–8).

In Canada, as in much of Europe, there are two main policy camps evaluating PSB's role – the more conservative 'economist' view and the social democratic 'culturalist' view. The economist view favours market forces as the key determinant of efficient allocation of any goods, including information and culture, and argues that only in extreme cases of 'market failure' should governments deploy regulation or public ownership of media. In the Canadian setting, economists have been particularly critical of governments who use market failure to justify nationalist goals. PSB is supported, they argue, as long as it follows narrow political goals of nation-building, including the maintenance of an 'orthodox' national identity (cf. Globerman, 1983). Culturalists point out that this happens less than argued, but agree that linking the economic goals to political orthodoxy is potentially anti-democratic and can limit the range of cultural viewpoints – especially alternative, 'disruptive' regional identities.

The culturalists argue that market failures do occur in countries like Canada where there is a high level of economic dependency. In fact the struggle for a level of indigenous control over cultural production, distribution and consumption, while often tagged with a 'cultural nationalist stereotype', at best should concern itself with enquiry into all forms of democratic expression under a broad goal of 'public communication' (Hogarth, 2000). This can include key aspects of a dynamic debate about identity (or rather, identities) but also allows for an ongoing discussion of 'culture failures' as well, namely: risk of cultural amnesia; lack of shared knowledge; and limitations on local artistic and intellectual development (Audley, 1994). But as both Audley (1994) and Hogarth (2000) have pointed out, whether supported for economic or cultural reasons, the resources going to Canadian PSB are in sharp decline, with a real impact on content throughout the Canadian media system.

Parliament provides CBC funding of 1 billion Canadian dollars (600 million euros) through an annual grant (politicians are free to vary the amount year by year).[1] With this, CBC provides an array of over twenty English-language and French-language (plus six aboriginal languages) conventional, specialty and satellite Radio/TV services as well as digital platforms.[2] Comparatively, CBC funding is low among OECD countries – 0.07 per cent of GDP in Canada versus Finland's 0.28 per cent and 0.23 per cent each for Denmark, Norway and the UK (Canada, 2003). Yet the overall broadcasting spend by Canadians is high. The federal broadcasting regulator, the Canadian Radio-television and Telecommunications Commission (CRTC) puts total revenues for television at 7.5 billion euros, while government funding for culture via federal, provincial and municipal governments is 5 billion euros. CBC thus represents 8 per cent of the total private–public broadcasting revenues and 12 per cent of public cultural expenditures.

Another trend is outsourcing, with public investment bypassing CBC and allocated to commercial broadcasters to achieve policy goals. In recent years, private TV broadcasters have achieved access indirectly to public funds of 600 million euros through:

- The Canadian Television Fund (CTF): 160 million euros.
- The federal tax credit initiative: 115 million euros.
- Provincial production subsidies: 135 million euros.
- CRTC simultaneous substitution rules: 190 million euros (or more) (Nordicity, 2008).

Service obligations remain loose for the private TV channels and policy achievements minimal. CRTC data show that only 8 per cent of viewing of entertainment genre programming on Canada's conventional private television services is to CanCon, even though most of the above funding is specifically designed to support domestic English-language TV 'drama and comedy' production, distribution and consumption (CRTC, 2008). The CBC proportion of viewing in this category is five times greater at 39 per cent – although only a fraction of its budget goes to English-TV drama and entertainment production (CRTC, 2008).

Canadian broadcasting legislation

The national Broadcasting Act (Canada, 1991) governs both public and private broadcasters in Canada (provinces constitutionally have limited broadcasting jurisdiction). The Act recognises a 'mixed' public–private broadcasting system in which policy and regulation of PSB and commercial players are developed to serve the public interest, with priority on paper given to the public broadcaster. As with budgets, reality differs and the implementation of policy and regulation has shifted to safeguarding large media conglomerates' commercial interests while most public service obligations fall to the CBC.[3] According to Sections 3 (l) and (m) of the Act CBC is required to provide programming on radio and television (no mention of new media) that should be 'predominantly and distinctively' Canadian, reflecting all its regions, in English and French, be made available universally throughout the country, reflect both the 'multicultural and multiracial' aspects of Canada, and not just 'contribute' to cultural expression, but also to a 'shared national consciousness and identity' (Canada, 1991).

These same programming goals apply to private broadcasters. On paper the only significant difference is that the CBC is expected not only to be *predominantly* Canadian, but also to provide *distinctively* Canadian programming. In reality, private broadcasters air Canadian-produced programmes with few Canadian references on screen which might diminish foreign sales, whereas CBC's *distinctively* Canadian programming tends to mean an open reference to Canadian characters, places, themes and stories in programming. Some describe it as *industrial* versus *cultural* production goals. The PSB with the cultural goals must ensure the 'system as a whole' accomplishes its broader programming versus 'sales' goals.

A 2003 parliamentary investigation reported that the current Act in itself continues to provide an adequate framework for an evolving broadcasting system, but that uneven policy implementation had exacerbated three key ongoing crises:

- Continued low production/consumption of English-language Canadian TV programming.
- Reductions in community/regional programming.
- Chronic under-resourcing of the CBC (Canada, 2003).

As in previous broadcasting reports, parliamentarians saw these as a historical theme – an ongoing political risk to cultural sovereignty. Indeed the report, entitled *Our Cultural Sovereignty*, began by requoting a 1932 Royal Commission on the threat that ubiquitous American radio content posed to Canada's very existence:

> the majority of the programs heard are from sources outside of Canada. It has been emphasized to us that the continued reception of these has a tendency to mould the minds of young people in the home to ideals and opinions that are not Canadian. (Canada, 2003: 23)

Culture, identity and broadcasting in Canada

British broadcasting scholar Richard Collins spent time in Canada and attempted to steer a course between economist and culturalist viewpoints. In part he developed a 'post-nationalist' argument in his 1990 book, *Culture, Communication and National Identity*. Collins traced the historical obsession of 'patriating' Canadian minds through broadcasting with significant relevance to identity struggles emerging in Europe. Like the economists, he criticised overly simple and deterministic notions of political sovereignty linked to broadcasting (Collins, 1990). As he saw it, the simple mantra was: Political Sovereignty requires Cultural Sovereignty, which requires Cultural Identity, which requires Indigenous Content Production, which requires Canadian Content Production.

For culturalists (and in particular cultural nationalists), Collins' proposed 'uncoupling' was an attack on the CBC, and not welcome (or worse, politically naïve) in an era of neo-conservatism and deep PSB cutbacks coming out of the 1980s. The cultural sovereignty model had at least helped legitimise public investment in culture and PSBs as a good investment in national cultural defence. However, other culturalists had already identified the model as problematic.

Also in 1990, University of Montreal Professor Marc Raboy wrote *Missed Opportunities*, tracing the history of broadcasting policy as a political instrument for reinforcing a narrow and centrally constructed notion of Canadian identity against regional challenges (most notably from Quebec as well as those seeking a looser political federation). The Act itself imposed a national unity 'patriotism' burden on CBC, even while CBC journalists walked a journalistic tight rope, covering provincial and federal elections in which regional political parties like the Parti québécois (PQ) or Bloc québécois (BQ) ran campaigns specifically against the orthodox notion of national unity.

Concurrently, argued Raboy, Canadian policy allowed for a type of private broadcaster protectionism which legitimised increased public monies for a small number of wealthy media conglomerates with guaranteed high profit margins (often 30 per cent plus annually) and unrestricted consolidation, based on a business model of imported American radio and TV programming (to garner high advertising revenues) with only minimum industrial commitments to CanCon. This national identity/commercial protectionism model shortchanged taxpayers, audiences and the development of democratic media.

Marc Raboy eventually helped draft *Our Cultural Sovereignty*, and move the identity discussion away from preserving 'the' national identity to developing clearer public service goals around culture and dynamic identities:

> for a broadcasting system that reflects what is distinctive about Canada, its racial and cultural diversity, its multitude of expressions and values. It will seek to speak on behalf of the public, indeed the multiple and diverse publics who cohabit this vast territory (Canada, 2003: 4)

Collins had argued similarly that it was time to rethink the Canadian broadcasting system, from 'an imperfect version of a nation-state of the good old-fashioned kind' and think of it as a 'pre-echo of a post-national condition'. Public broadcasting could be freed from narrow constraints of 'correct' and standardised interpretations of the Canadian experience. If it failed to do so, it would stultify and even hurt its intended goals such 'cultural pluralism, and political equality' (Collins, 1990: xii).

CBC Television: a popular CanCon entertainment strategy

CBC Television continues to try to develop popular national entertainment that engages audiences. However, the measures of success CBC-TV

Table 21.1 Canadian content consumption by media

Media	*% Canadian*
Newspapers	95
Magazines	41
Television (overall)	33
Books	30
Radio	30
Recorded music	16
Television (English-language drama)	5
Film	3

Source: CRTC (2008); Magazines Canada (2005).

uses encompass a narrow view of public involvement relative to that of CBC-Radio. TV largely uses commercial, ratings-based language of average minute audiences (AMAs) and Prime Time Share. This is not surprising given the structural connection to a commercial model; almost half of CBC-TV's 400 million euros programming budget comes from advertising. These days CBC-TV is particularly vulnerable.[4]

The current head of CBC's English services, Richard Stursberg, has pinned his and the CBC's overall success on making CBC-TV programming popular with large audiences through a 'Canadian entertainment strategy':

> English Canadians read English Canadian newspapers, they go to Canadian rock and hip hop concerts, they read Canadian novels, they buy the records of Canadian artists and in theatre, dance and classical performance, we have our global stars and we rave about them. But when it comes to the most popular forms of narrative – television and feature films – Canadians overwhelmingly prefer the stories of another country. (Stursberg, 2006: 2)

The cross-media comparisons bear out the truth of Stursberg's analysis. While Canadians overwhelmingly read Canadian-produced newspapers (95 per cent), and spend significant portions of their magazine (41 per cent), book (30 per cent) and radio time (30 per cent) with Canadian content, viewing of English-language Canadian TV entertainment is very low (5 per cent) (see Table 21.1). It is true that the large shortfall in Canadian TV drama viewing is partially made up for by CanCon viewing in news (80 per cent plus) and sports (50 per cent plus), and to a lesser extent in genres like children's programming. In this context PSB already plays a vital role. The English and French services of the CBC/Radio-Canada attract the largest single amount of viewing to Canadian television – both overall (80–85 per cent) and in

Table 21.2 Canadian content viewing by television channel

Canadian television channel sources	*% Viewing to Canadian*
Television overall	33
CBC/Radio-Canada	
CBC-TV – overall	80
CBC-TV – comedy/drama	39
Radio-Canada TV – overall	85
Radio-Canada TV – comedy/drama	61
Conventional private TV	
English-language – overall	33
English-language – comedy/drama	8
French-language – overall	72
French-language – comedy/drama	25

Note: Private TV does not include US channels and Canadian pay/specialty.
Source: CRTC (2008).

Table 21.3 CBC-TV share of TV prime-time viewing (7–11 p.m.)

Pre-entertainment strategy			*Post-entertainment strategy*		
2003–4	*2004–5*	*2005–6*	*2006–7*	*2007–8*	*2008–9*
7.1%	6.9%	7.5%	7.4%	7.8%	9.2%

Note: 2008–9 figures only to mid-season.
Source: CBC (2008a); Wilson (2009).

terms of entertainment programming (39–61 per cent) (see Table 21.2). But on the English side CBC-TV still feels the pressure to 'repatriate' Canadian entertainment.

CBC-TV has based its popular entertainment approach on significant new hours of CanCon continuing drama series and factual entertainment. In the CBC-TV prime (7–11 p.m.) schedule the amount of hours for this increased to 275 hours in the current season from 150 hours pre-2006 (see Table 21.3). According to Christine Wilson, CBC-TV Deputy Director of Programming, success came in 2008–9 with a share increase to 9.2 per cent compared with an average of 7.5 per cent prior to the start of the entertainment strategy.

While too early to show clear success, CBC-TV is exhibiting a combination of a more ratings-driven strategy and development of a less traditional view of Canadian stories and identity. There may be some slight widening in the notion of a 'Canadian identity' in its programming. Says Wilson: 'it would be hubris to think about anyone being the keeper of the Canadian identity; our role is to be a place where Canadians can come and see themselves' (Wilson, 2009). Instead, CBC-TV hopes to provide a broader reflection of Canada's diversity on the small screen: 'We think about the Canadian experience that ranges across all age groups and that has a lot of different ethnic faces and socioeconomic faces. We try to reflect all of that in our programming. Not in a kind of self-conscious way but in a way that speaks to Canadians' (ibid.).

One of the complaints from more traditional CBC supporters is that the new focus comes at the cost of information and regional programming. The CBC does not share in detail how TV programming dollars were redirected, but managers like Wilson will say that increased entertainment spending comes from stopping a 'high-impact' programming strategy of imported and Canadian made-for-TV specials and movies. It remains to be seen whether in this current recession the strategy continues; as of February 2009 the CBC contends there is no reversal on a numbers-driven entertainment strategy (ibid.).

CBC Digital – Web 2.0 versus programme extension

Digital *extension* of existing broadcasting content via the internet (though not always new content on new media) is the current CBC mantra. In 2007–8, CBC declared modest, 'housekeeping'-oriented goals for its digital platforms, that is, to provide a more efficient dissemination of radio and TV content. Indeed few at CBC.ca are willing to make the leap – at this point – to new media potential for cultural democracy strategies often associated with Web 2.0.[5]

This is apparent in the CBC's submission to a CRTC hearing on regulating/subsidising CanCon broadcasting material distributed on the internet.[6] The CBC argued that established broadcasters – especially PSBs – should have prime access to any public funds or industry incentives directed to Canadian new media content production, shoring up work done in this area by existing broadcasters, especially the CBC. The submission underlined the argument with specific reference to the Act (but with limited reference to Web 2.0 potential):

> We believe our efforts in new media contribute to the further development of a shared national consciousness and identity for Canadians and that our use of this platform also contributes to the preservation of our democratic values. (CBC, 2008b: para 13)

Democratisation and regional reflection aspects, however, are downplayed in planning and resources. In the financial and performance indicators sections of its annual report, the CBC digital strategy is described as a success in terms of programme extension and commercial potential (see Table 21.4).

According to Dan Hill, CBC Senior Director of Digital Programming, the goals are necessarily modest in light of political and economic realities since,

Table 21.4 CBC.ca performance indicators

INCREASE TRAFFIC	CBC.ca continued as the number one news/media website with 4.1 million average monthly unique visitors
IMPLEMENT WEB 2.0	CBC.ca drew over 27,000 registered members and over 10,000 comments on CBC.ca content with its user-generated content tools
	Since 1 June 2007, CBC has added over 5700 on-demand videos and served 6.2 million streams
INCREASE REVENUES	CBC.ca ad banner revenue rose 18 per cent during the fiscal year for a 2007–8 total of $3,119,000[7]

Source: CBC (2008b).

first, CBC is not mandated by the Act to provide new media services and, second, CBC has no additional public funding in this area. As a result, digital activities at CBC are commercially revenue-dependent or support existing broadcast programming (budgeted as promotion and distribution). This effectively limits experimentation in advanced interactive digital content: 'It's a luxury we can't afford to be in [especially during recession]. It was vanity showcase things that a sponsor might be interested in (and funded) to experiment and test in digital space. But that's gone away. From a market perspective advertisers no longer have budgets for things like that' (Hill, 2009).

CBC Radio: old model, new identities

CBC Radio, most particularly in its regional operations, which make up 60 per cent of its operations, is reinventing public broadcasting to deepen engagement, specifically around changing notions of reflection and identity. Indeed it puts a dynamic notion of identity at the centre of its programming strategy, especially in its largest urban centres and most notably in the example of CBC Radio One English programming in Toronto. The Greater Toronto Area, with a population of almost 5.5 million, is Canada's largest and most multicultural city – the most ethnically diverse in the world according to some local sources (Toronto, 2009). At CBC Toronto, thirty-five radio staff produce two weekday daily morning and afternoon three-hour shows, and two weekend morning shows, as well as a mix of daily newscasts, interviews, columns, sports and cultural coverage of the city (Paul, 2007). On weekends staff produce two three-hour morning programmes (with coverage to the entire province) and a one-hour arts programme on Saturday afternoon. The total local content represents fifty hours of programming each week, with the remaining 118 hours provided by the national CBC Radio

One network. However, while local programming represents only 30 per cent of weekly programming, due to its schedule placement in radio listening periods, it generates approximately two-thirds of total station listening.

The change to a new type of programming came in 2002 after intense reflection by station manager Susan Marjetti and her staff, in collaboration with community leaders:

> We started the redevelopment process with one simple question: when you turn on the radio to 99.1 FM Toronto do we actually sound like this city looks? And it was absolutely unanimous at the time that no, we do not. So is that acceptable for the public broadcaster who takes public monies from all Canadian taxpayers? And that launched a major philosophical discussion about the role of the pubic broadcaster. (Marjetti, 2009)

What emerged was deliberative thinking about programming that engages audiences through: a diverse programme content strategy; diversity staffing; and audience feedback, as measured on systematic quantitative and qualitative bases.

Programming

The *programme content strategy* led to three types of storytelling through radio that can be understood, according to the CBC radio programmers as: (1) coverage of a 'community about a community' that often leads to a shared understanding among the entire audience, for example, a story about the well-known Toronto Caribana Festival – specific to the Caribbean-Canadian community; (2) coverage 'transcending the specific community', for example, the story of a 12-year-old Chinese girl kidnapped from her bedroom in Toronto – a story of a child missing in the Chinese community but also any parent's nightmare that speaks beyond that community; and (3) coverage of a 'universal value or appeal', for example, tax increases for homeowners – it does not matter if you are Chinese-Canadian, Indo-Canadian or African-Canadian (Marjetti, 2009).

The notion of a highly 'professionalised' journalism (common to PSB) was also deliberately deconstructed. The CBC staff felt they could not reflect a culturally diverse and dynamic city only in traditional journalistic ways (Paul, 2007). Staff began consciously to use music, art and cultural genres to tell new types of stories on radio in traditional news and current affairs time slots. For instance, freelance columnists from diverse backgrounds were incorporated into the 'prime real estate' of the schedule, providing a 'new lens' that is not limited only by 'newsworthy' political and economic developments: 'Toronto is a masala of spices and you often get a better sense of the feel of Toronto through art than through journalism' (Marjetti, 2009).

Staffing

The *change in staffing* among the thirty-five local programmers was driven by the new content. CBC reports to Parliament indicate that 7 per cent of employees nationally were members of a visible minority group, but at CBC Radio in Toronto it was almost four times greater.[8] Marjetti says 'to change the outside you change the inside'. Diversity was emphasised particularly in on-air roles. The bulk of on-air voices – especially in key hosting roles – are from non-mainstream backgrounds (Paul, 2007). Marjetti (2009) also speaks of key producers like the Caribbean-Canadian senior producer for the morning programme who 'brings his perspective and his sensibilities to that story meeting table. He has built into his objectives – as do all the shows and all the programme leaders – goals of diversity and reflection, both on the hiring front and also on the on-air programming front.'

Audiences

There are about fifty stations available over the air to Toronto listeners – many American from across Lake Ontario – making it one of the most crowded radio markets in North America. CBC has a 10 per cent share of total radio listening and a weekly reach of 1.1 million people for at least fifteen minutes (BBM, 2008). By conventional ratings, therefore, CBC Radio One has become the market leader – indeed these figures represent a doubling in ratings performance in recent years. Yet they constitute for the CBC Radio staff only part of the audience success; the senior producer said in response to BBM's success: 'We're CBC – we don't have to cater to advertisers. Our mandate is to serve the public' (Paul, 2007).

CBC's audience research group conducts ongoing quantitative analysis, its own surveys and a range of programme testing with diverse Torontonians both through focus groups and auditorium testing. But radio programmers have developed additional strategies to hear directly from the audience, with a particular emphasis on under-represented ethnic voices. They often combine this with live town hall programming and on-air forums where they not only elicit programme material but also new programme feedback and ideas.

Early on in the programme development process, CBC Radio senior programming staff met with people known to represent a diverse and engaged segment of the public:

> In all we had about 36 different people come in over the course of six months from various walks of life, various social economic backgrounds, various ethnic backgrounds. All were people who live in this city, and some of them felt quite passionate about the CBC, but they also had issues with the CBC, that is, they did not think that the CBC was speaking to them. (Marjetti, 2009)

A year later the reconstituted group gave feedback on changes and these ideas were further integrated on-air. One of the participants, a leader within the Canadian Sikh community, remarked he had 'never seen so complete a turn around in one year' (ibid.).

Conclusion

Canadian PSM – and specifically the bulk of the CBC's shrinking 'housekeeping' budget allocation – appears fixed primarily on the popular 'entertainment strategy' of CBC English TV. Arguably, while trying to move beyond a narrow notion of Canadian identity, there are commercial pressures that limit its potential. Similarly on the new media front, there is currently very low funding for digital and new media experimentation in more open and democratic formats (along the lines of Web 2.0 or newer incarnations). While this is suggestive of another 'missed opportunity' to explore new dimensions of the electronic public sphere, experiments in 'glocalised' (locally produced, globally oriented) content on CBC Radio, while small in budget terms, are really at the centre of the CBC's dynamic cultural identity reshaping in Canada.

Whereas the political and economic struggles continue around maintaining sufficient public infrastructure to support a transformed CBC, these new and evolving radio forms of content provide a model that transcends narrow cultural nationalist views while at the same time directly confronting overly economic approaches to local content production, distribution and audience engagement. Specifically an analysis of the 'audience as public' model which appears to be driving the successful radio experimentation by CBC Radio in Toronto, suggests a transformative approach to public service content production; one that confronts and incorporates the multiple and changing identities as the basis for new models of public engagement in broadcasting and new media.

Raymond Williams argued that sometimes the oldest communication forms have a lasting value for cultural expression, even when economic and technological structural change seems to pass them by (Williams, 1981). A historical analysis of the experimentation with dynamic identity formation in local CBC Radio programming suggests that the Habermasian ideal of PSB public space can coexist with new forms of content in old media, and potentially provide models for new media. A concentration on local content in a global environment within nationally supported systems is possible – and there is no reason to believe that the political will and the economic wherewithal is absent from the Canadian PSM setting, at least in the long term.

Notes

1. CBC is able to bring the total annual operating revenues up to 1.2 billion euros, through advertising revenues and subscription fees to some of its TV and new media services (CBC, 2008a).

2. For a full listing see cbc-radio-canada.ca/submissions/pdf/services.pdf, accessed 12 April 2009.
3. In 2007 three major media corporations remained after a decade of mergers. In English-Canada, CTV Globemedia had acquired CHUM broadcasting to attain revenues in the range of 1.2 billion euros per year and CanWest Global had acquired Alliance-Atlantis to bring their revenues to about 1.8 billion euros (at the time of writing CanWest is in bankruptcy proceedings). In the French-language sector, Quebecor Media had revenues of about 1.5 billion euros (Alexander, 2007: 17). Both CTV and Quebecor are part of larger conglomerates.
4. CBC faces a 120 million euro budget shortfall for the 2009–10 fiscal year, representing almost 10 per cent of the total budget – due largely to CBC-TV's advertising dependence and the recessionary pressures (CBC, 2009).
5. The strongest advocate for Web 2.0 was former CBC Radio producer and Head of CBC.ca, Sue Gardner, but her ideas were not endorsed by CBC VP Richard Stursberg.
6. Broadcasting Notice of Public Hearing CRTC 2008-11, http://www.crtc.gc.ca/eng/archive/2008/n2008-11.htm, accessed 12 April 2009.
7. The CBC's internet advertising represents less than 3 per cent of the total estimated Canadian internet advertising in 2007 (CBC, 2008b).
8. By federal legislation, statistics are generated on the basis of 'self-declared' visible minorities in internal surveys conducted by government agencies (CBC, 2007).

References

Alexander, I. (2007) 'Canadian Public Television at a Crossroads', presentation to Communication Policy Class at McMaster University by Ian Alexander, Director of Strategic Projects, CBC, 8 March 2007 (Toronto: CBC).

Audley, P. (1994) 'Cultural Industries Policy: Objectives, Formulation and Evaluation', *Canadian Journal of Communication*, 19(3): 17–52, http://www.cjc-online.ca/viewarticle.php?id=246, accessed 3 June 2009.

BBM (2008) 'Radio Top-Line Reports, Toronto CMA, Fall 2008' (Toronto: BBM Canada), http://www.bbm.ca/en/radio_top_line.html, accessed 3 June 2009.

Canada (1965) *Report of the Advisory Committee on Broadcasting* (Ottawa: The Queen's Printer).

Canada (1991) *Broadcasting Act* (Ottawa: Royal Statutes of Canada), http://laws.justice.gc.ca/en/B-9.01/, accessed 10 April 2009.

Canada (2003) *Our Cultural Sovereignty: the Second Century of Canadian Broadcasting*, Report of the House of Commons Standing Committee on Canadian Heritage (Ottawa: The Queen's Printer), http://www2.parl.gc.ca/HousePublications/Publication.aspx?DocId=1032284&Language=E&Mode=1&Parl=37&Ses=2, accessed 1 April 2009.

CBC (2007) 'Federal Institution Submission (FY 2005-2006) on the Operation of Section 3(2) of the Canadian Multiculturalism Act', http://www.cbc.radio-canada.ca/docs/equity/annualreports.shtml, accessed 21 February 2009.

CBC (2008a) Annual Report 2007-08, Volumes I and II (Ottawa: CBC), http://www.cbc.radio-canada.ca/annualreports/2007-2008/index.shtml, accessed 18 February 2009.

CBC (2008b) 'Call for Comments on Canadian Broadcasting in New Media, Broadcasting Notice of Public Hearing CRTC 2008–11 – Comments of CBC/Radio-Canada, 5 December 2008' (Ottawa: CBC), http://www.cbc.radio-canada.ca/submissions/2008.shtml, accessed 18 February 2009.

CBC (2009) 'CBC Approves Budget: Cuts Expected', http://www.cbc.ca/arts/media/story/2009/03/17/cbc-budget-moore.html, accessed 3 June 2009.

Collins, R. (1990) *Culture, Communication and National Identity* (Toronto: University of Toronto Press).

CRTC (2008) *Broadcast Policy Monitoring Report 2007, Radio, Television, Broadcasting Distribution, Social Issues, Internet* (Ottawa: Canadian Radio-television and Telecommunications Commission), www.crtc.gc.ca/eng/publications/reports/PolicyMonitoring/2007/bp.m.r2007.htm, accessed 18 February 2009.

Globerman, S. (1983) *Cultural Regulation in Canada* (Montreal: The Institute for Research on Public Policy).

Habermas, J. (1962 [English translation 1989]) *The Structural Transformation of the Public Sphere: an Inquiry into a Category of Bourgeois Society* (Cambridge, MA: MIT Press).

Hill, D. (2009) (Senior Director, Digital Programming, CBC, Toronto), personal interview, 20 February.

Hogarth, D. (2000) 'Communication Policy in a Global Age: Regulation, Public Communication and the Post-nationalist Project', in M. Burke, C. Mooers and J. Shields (eds), *Restructuring and Resistance: Canadian Public Policy in an Age of Global Capitalism* (Halifax, Canada: Fernwood), pp. 212–25.

Magazines Canada (2005) 'Brief to the Standing Committee on Finance, 2005', http://www.cmpa.ca/, accessed 28 December 2005.

Marjetti, S. (2009) (Regional Director of Radio, CBC, Toronto), personal interview, 20 February.

Mosco, V. (1996) *The Political Economy of Communication* (London, Sage).

Nordicity (2008) *Canadian Television: Why the Subsidy?* (Ottawa: Nordicity Group Ltd.), http://www.nordicity.com/reports.html, accessed 10 April 2009.

Paul, L. (2007) 'Morning Glory', *Ryerson Review of Journalism*, Spring: 69–71.

Raboy, M. (1990) *Missed Opportunities: the Story of Canada's Broadcasting Policy* (Montreal: McGill-Queen's University Press).

Statistics Canada (2008) 'Government Expenditures on Culture: Data Tables, 2005/2006', Catalogue no. 87F0001X (Ottawa: Statistics Canada), http://www.statcan.gc.ca/pub/87f0001x/87f0001x2008001-eng.htm, accessed 1 February 2009.

Stursberg, R. (2006) Speech to the Economic Club of Toronto, 7 November (Toronto, CBC), http://www.cbc.radio-canada.ca/speeches/20061107.shtml, accessed 15 February 2009.

Toronto (2009) 'Toronto's Racial Diversity', http://www.toronto.ca/toronto_facts/diversity.htm, accessed 21 February 2009.

Williams, R. (1981) *Culture* (Glasgow: Collins/Fontana).

Wilson, C. (2009) (Deputy Programme Director, CBC-TV, Toronto), personal interview, 20 February.

22
Public Service Media in Australia: Governing Diversity

Gay Hawkins

Introduction

In October 2008 the Australian government's Department of Broadband Communications and the Digital Economy (DBCDE) launched an inquiry into the future of Public Service Media (PSM). The discussion paper informing this inquiry was titled 'ABC and SBS: Towards a Digital Future'. It invited submissions on the complex and wide-ranging issues facing public broadcasting in the twenty-first century, with the aim of enabling policy to be developed that would allow 'national broadcasting to thrive in a digital, online, global media environment' (DBCDE, 2008: 1). In seeking to review the functions and future of the Australian Broadcasting Corporation (ABC) and the Special Broadcasting Service (SBS), in a fully digitised environment, this inquiry identified several key issues. These were: how to harness new technologies to enhance charter objectives; how to expand the amount and diversity of Australian content; how to extend the impact of news and current affairs to better inform all Australians; and how to develop the capacity of these media in enhancing social inclusion and the governance of cultural diversity.

In this chapter I focus on the last of these issues. While there is no doubt that they are all central to the evolution of PSM, my claim is that issues of cultural diversity and social inclusion go to the heart of contemporary public media practices in Australia. Like most nations facing the demands of contemporary democracy, Australia is a highly pluralised and increasingly globalised society. As a consequence, the social fabric of Australia's cultural diversity has become, not only more complex, but also more pervasive. In seeking to govern this political and social reality all media play a critical role. Because they are central to private and public life, they are crucial in providing audiences with the resources and frameworks to recognise, engage with and negotiate various dimensions of difference. The specific issues for PSM are: What is their unique role in negotiating difference? How can they function as sites where diverse constituencies and perspectives can be recognised and represented? How can PSM pluralise homogeneous notions of national culture and identity? And how can they enhance an inclusive democracy via innovative media practices that extend possibilities for participation, public connection and debate?

Australian PSM offer interesting and unique answers to these questions. Due to their contrasting histories, organisation and charters, the ABC and SBS approach them in distinct ways. Using a comparative analysis it is possible to see how the challenge of cultural diversity is apprehended by a traditional PSB such as the ABC, with a standard charter to 'inform, educate and entertain all Australians', and by the much newer SBS, which was set up in the mid-1970s with an explicitly multicultural remit. Its pluralist inflection of the generic PSB remit is often dismissed as a narrow-cast diversion from the main broadcasting game at the ABC. The standard analysis is that SBS stands as the niche, poor relation in the contemporary media landscape in Australia; a service largely for 'ethnics' and other minorities, while the ABC is genuinely comprehensive, mainstream and national.

In terms of the future development of PSM in Australia this analysis is limited because it ignores the impact of massive changes in technology, audience practices and social composition. These changes offer both organisations a range of possibilities for innovative responses to cultural diversity. For example, thanks to the rise of digital media, the ABC is now able to resist some of the homogenising tendencies of traditional PSB and develop niche services and audiences via new media platforms. Whether this amounts to a significant pluralisation of the 'mainstream', and the national public sphere, is still to be decided but it is an important shift. In contrast, SBS is seeking to shake off its niche or narrow-cast status and turn to mainstream pluralism: to make multiculturalism everybody's business with the introduction of more populist content. Some argue that a desperate search for ratings is behind this 'mainstreaming' strategy. However, this analysis denies the wider political significance of SBS' new approaches to the changing dynamics of cultural diversity, and their implications for a radical renewal of public service communication. In the rest of this chapter I critically evaluate the implications of these changes within the ABC and SBS, paying close attention to the differences between pluralising the mainstream and mainstreaming pluralism.

PSM in Australia

Before analysing these shifts in detail it is necessary to briefly explain the organisation and functions of the ABC and SBS. This background makes it possible to contextualise the particular challenges they each face in governing cultural diversity. The ABC has to respond to these challenges from the position of being Australia's original and largest PSB. Established in 1932 it drew on models from the colonial homeland, Britain. Like the BBC it was seen as a citizen-forming institution, central to making Australians visible to themselves, for fostering national culture and interests, and for producing informed democratic subjects.

By the twenty-first century this paternalistic model of citizenship has been transformed into a neoliberal regime of choice, participation and decentred democracy. By this I mean that as the ABC has shifted from a PSB to a PSM it has become less monolithic in terms of representing a unitary national public or public good. Its national agenda is now heterogeneous and dispersed across a series of media institutions, working across multiple platforms, territories and constituencies. Today it runs a national analogue television network (ABC 1) that reaches 98 per cent of the population and two digital channels (ABC 1 and ABC 2) that reach 97 per cent of the population; these channels have a combined weekly audience of more than 12 million people. It also runs an international television service (Australia Network) reaching 21 million homes across the Asia-Pacific region, a range of national, state and regional radio services, three digital radio services transmitted on its digital television platform as well as an international radio network (Radio Australia). In terms of its online presence, ABC Online has been at the forefront of internet platform development in Australia and gets around 2 million unique users per month. In June 2007 ABC audiences downloaded more than 2 million ABC Radio programmes and 1.8 million TV and news videos. In July 2008 the ABC launched the full screen internet TV service iView (DBCDE, 2008: 51).

In contrast to the ABC, SBS was established to cater to the communication needs of ethnic and Aboriginal minorities ignored by mainstream media, and to promote awareness and understanding of diversity amongst all Australians. Set up as a multi-lingual radio service in 1975, it began television broadcasting in 1980. Today it runs two TV (analogue and digital) channels that cover more than 95 per cent of the Australian population. In the main capital cities these channels get an average weekly audience of around 5 million people. SBS Radio broadcasts in sixty-eight languages to all capital cities and some regional centres. It attracts around 800,000 listeners per week. SBS Online averages around 469,000 unique users per month (DBCDE, 2008: 52). This distinctive media organisation remains one of the most significant achievements of successive Australian governments' multicultural policies. According to Ang et al. (2008: 4), SBS has not simply responded to social change; it has also generated it by showing the key role that media can play in nurturing social engagement with the age of diversity.

In terms of funding and regulation both the ABC and SBS are statutory authorities, functioning as independent corporations with their own legislation and charters. They get the bulk of their funding from government although they are also able to earn additional revenue through business activities including merchandising and, in the case of SBS, advertising. As the much bigger and more complex institution ABC gets approximately A$850 million per year, as opposed to SBS which gets approximately A$191 million. Currently, both PSM are extending their digital TV services to provide the same coverage as their analogue services, in preparation for the switch-off of analogue signals in 2013. They are also lobbying the federal government hard for extra funds necessary to extend their digital and online services and to become central players in the digital cultural economy. The ABC is way ahead in this game. It has set the benchmark for innovative multi-platform media in Australia, leaving both the commercial channels and SBS way behind.

As these PSM head into a changing digital environment, with the expansion of a high-speed broadband network in Australia, the challenges of responding to cultural diversity are set to intensify. Not only will audiences have access to a far greater range of digital TV channels, the proliferation of internet-based platforms means that their traditional authority and influence will be contested. As mentioned, how these media should function in a multi-channel, multi-platform environment is currently under review. The key question here is how will these developments impact on their capacity to govern difference both now and in the future? Three issues are crucial to this process: strategies for recognition and social inclusion, extending participation, and cosmopolitan versus national public culture. I will consider each in turn.

Recognition and social inclusion

Recognition and social inclusion are regarded as key strategies in the enactment of contemporary multicultural policy. They focus on the need to pluralise national culture by acknowledging other identities and perspectives and fostering diverse public spheres. A commitment to pluralism and inclusion does not imply assimilation. Rather, it involves the active promotion of various forms of diversity in ways that acknowledge both the right to be different *and* included in national public culture.

Even though SBS has a statutory obligation to realise multicultural policy via distinct media services, the ABC is also expected to provide programmes that 'reflect the cultural diversity of the Australian community' (ABC Act, 1983). This means that both PSM have had to develop techniques for recognition and social inclusion. The key issue is what their different approaches to diversity reveal about the limits and possibilities of PSM. Equally important is how each organisation sees its obligations to recognition and social inclusion *in relation to the other*. Did the creation of SBS let the ABC off the hook in terms of transforming its representations of Australia as multicultural, and imagining its audience as linguistically and culturally diverse?

In many senses the answer to this question is yes. The creation of SBS was partly a response to the ABC's persistent monoculturalism and the difficulty this organisation has had in adapting to massive changes in the composition of the Australian population. Several government reports into the state of the ABC in the 1970s noted its reluctance to acknowledge diversity; hence the decision to establish SBS, to set up a separate public broadcaster with a specific multicultural brief. Even though the ABC was keen to run this new service as another channel, on the basis of its track record this option was considered out of the question.

This was a reasonable assessment. Historically, the ABC has generally managed difference using two primary strategies: ignore it or render it a problem. While it has always acknowledged special populations in the interests of responding to the market failures of commercial media, the special categories have traditionally been 'women', 'rural' and 'children'. These populations have been well served by the ABC. However, ethnically diverse audiences have not been considered special or in need of distinct programming; for a long time they were largely ignored and thereby rendered invisible by the national broadcaster. This is not to deny the fact that the audience for the ABC was inevitably culturally diverse, it is just that these populations simply did not see or hear themselves on ABC TV or radio. This strategy of social exclusion gradually came unstuck during the 1980s when the ABC could no longer ignore the rise of multiculturalism as the new national imaginary. In this period it began programming an innovative range of local productions focusing on the migrant experience, indigenous culture, and other diversities from queerness to disability. It also established Aboriginal and multicultural units to develop relevant local content. Much of this programming could not be described as typical 'multicultural programming' in the sense that it sought to serve the special needs of culturally diverse audiences. Rather, it was aimed at wider audiences and sought to make different cultures and experiences visible within mainstream culture.

However, by the end of the 1990s these initiatives had largely collapsed. The election of a conservative government saw significant budget cuts and a retreat from programming innovations. There was also a concomitant increase in imported content largely from the BBC. Representing the rich cultural diversities within Australia was pushed aside in favour of relentless Englishness (Hawkins, 1997: 14).

There is no doubt that in the 1980s there was a shift towards a more pluralised understanding of Australia and Australians on the ABC, and a real attempt to recognise populations that had been largely invisible with the creation of new content. These strategies were important to the wider construction of a multicultural national identity in Australia, helping audiences engage with difference. They also disrupted narrow and complacent senses of 'common culture', but they had their limits. Two problems emerged from this focus on recognition and social inclusion via programming. First, it ran the risk of making a fetish of difference, of representing diverse populations as exotic or worthy. This reduces multiculturalism to an empty celebration of difference and can deny the civic contracts and relations between diverse publics. Second, while diversity may have filtered into the margins of the television schedule it did not really have any impact on the overall look and feel of the ABC. Thirty years after this heady period of discovering diversity the organisation still has no on-air presenters with accents and few from non-Anglo-Celtic backgrounds. Subtitles are still a rarity and little local drama content focuses on the politics of diversity as a central theme. The wider representational politics evident on the ABC retain a persistent whiteness.

This assessment may sound mean-spirited and politically correct. That is not the intention. Rather, it is to show how difficult it is for the ABC to radically shift its normalising assumptions about majority culture. Due to reduced or static funding under the conservative Howard Liberal government (1996–2007), and the concomitant pressure to attract bigger audiences in prime time, exploring the dynamics of diversity was considered risky and irrelevant. This shows how seeking to appeal to an imagined mainstream is inherently homogenising. On ABC television populist logics rule and it is a populism that has an increasingly stingy and limited commitment to pluralism.

Related to this is the effect of ten years of conservative government and its serious retreat from official promotion of multiculturalism. In this political context the ABC has not felt obliged to maintain, let alone further develop, its earlier commitment to recognition and social inclusion. These commitments have been partially displaced to new media platforms and ABC 2. The key point is that in its dominant mass media platforms, the ABC's obligation to the promotion of diversity remains underdeveloped. Yet, throughout this period, the cultural make-up of Australia has become more multi-layered and complex. Australia's largest cities are now hyper-diverse and it is impossible to consider the culture as white, Anglo-Celtic or mono-cultural (Ang et al., 2008: 49). This reality presents serious challenges for the ABC. There is a widespread sense that its current television programming, particularly in relation to locally made and imported drama, seems either parochial or English, and irrelevant to large sections of the population.

In stark contrast, SBS' strategies for recognition and social inclusion have been innovative and extensive. The standard analysis is that SBS is simply doing its job, and is able to embrace such strategies because, unlike the ABC, it is not burdened with obligations to an imagined mainstream or majority culture. This assumes that SBS is an ethnic broadcaster that serves mainly ethnic audiences and that its remit is exclusively narrow and based on special media provision for minorities. On the contrary, like the ABC, SBS is a *public* broadcaster whose key role is to broadcast cultural diversity for the benefit of the nation as a whole. As Nigel Milan, a former managing director of SBS declared 'SBS's ultimate responsibility is to nation building, to showing multicultural Australia to itself; to tell the stories of Australia in the languages of Australia and to unite the nation through understanding and acceptance of cultural diversity' (Ang et al., 2008: 2). What is so striking about this formulation is the way in which this multicultural national imaginary implicitly challenges monocultural notions of the mainstream or 'the public' at play in the ABC. Ingrained in the ethos of SBS is the fact that the public is characterised by plurality not unity and that it consists of multiple histories and perspectives relatively unfamiliar to one another. SBS assumes that what Australians have in common *is* diversity and the role of PSM is to create spaces where the connections and differences between particular groups and perspectives can be explored and negotiated.

In this context SBS' recognition strategies have been innovative and increasingly sophisticated. Central to these has been a throughgoing commitment to multi-lingualism and an acknowledgement of Australia's linguistic diversity. This has taken different forms according to the media platform. In radio, where SBS began, the initial approach was to broadcast programmes in key languages, presented and produced by ethnic communities. This represented a form of 'ethno-multiculturalism' based on catering to the special needs and interests of migrants and ethnic communities (Ang et al., 2008: 19). The primary objective was cultural maintenance. By helping ethnic groups maintain their language, homeland identifications and traditions, they were able to sustain their different cultural identities. Gradually, this approach shifted from servicing ethno-specific communities to a more professionalised and *multi*cultural ethos. Radio is still broadcast in sixty-eight languages but the emphasis now is on content that helps listeners engage with contemporary issues in Australia. Instead of a focus on homeland issues or community politics there is more content about current affairs. Rather than nurture diasporic yearning and ethnocentrism, the objective is to help audiences to be able to participate in public culture in their own language. This has the effect of both creating diverse public spheres and encouraging migrants to become more confident as citizens in a multicultural Australia.

Broadcasting in diverse languages is an exercise in public validation and legitimation that involves complex registers of recognition. Not only does it contribute to migrant audiences' sense of well-being and involvement in Australian culture but it also symbolically acknowledges many languages beyond English as part of the national linguistic culture. It remains one of SBS' most powerful gestures of social inclusion.

Dealing with linguistic diversity on TV has involved very different strategies. Because TV is considered a comprehensive medium the focus has been on making imported programming (roughly 90 per cent of SBS' content), from an incredibly wide range of countries, accessible to all audiences. This rich array of foreign-language content was not seen as a medium for ethnic insularity or cultural maintenance. Rather, it was programming designed to broaden the cultural awareness of *all* Australians beyond their ethnocentric comfort zones, hence the crucial role of subtitling in enabling cross-cultural communication.

While subtitling was common international practice in Europe, when SBS TV began it was relatively rare in Australia. Despite leading this important social change the prime-time schedule on SBS TV now offers little subtitled material. This shift began in the mid-to-late 1990s when SBS began to seek out larger audiences in the interests of making cultural diversity more accessible. Subtitled content was seen as a barrier to mass appeal and the development of 'popular multiculturalism' (Ang et al., 2008: 19). English is now the dominant language on screen in the evenings. There has been some protest about this shift with critics arguing that it signals the mainstreaming of SBS and an abandonment of its social remit and charter objectives.

This accusation has some weight; however, it does ignore the fact that making multiculturalism popular, or mainstreaming pluralism, can be an equally powerful strategy for social recognition and inclusion. In the last few years SBS has embarked on an ambitious initiative to screen more locally made content in the interests of making multiculturalism relevant to 'all Australians'. This has seen the screening of an impressive array of content that has explored the dynamics of everyday diversity. Rather than framing difference as a special issue or worthy, SBS takes it as the norm and this has enabled it to reflect a much more contemporary Australia and to explore the deeper and more complex issues surrounding hyper-diversity. While this content is predominantly in English, the dominant language, it often captures the reality of linguistic hybridity by showing characters moving into other languages or speaking in heavily accented English.

A commitment to multi-lingualism and linguistic pluralism is only one of SBS' many strategies for social inclusion but it is perhaps the most significant. It also highlights the differences between the ABC and SBS. In the ABC there have been sporadic but somewhat weak attempts to include different cultures and experiences in the prime-time schedule. But this has mainly occurred in drama programming and has never impacted on the general cultural ambience of the broadcaster where subtitles are extremely rare and 'proper' English dominates. On SBS both local and imported drama reflects extraordinary linguistic and other diversities. So too does the on-air culture where other languages, accents and identities are everywhere present. On SBS the recognition of diversity is normal and this makes its prime-time content far more innovative than the ABC's. Because SBS brings together different viewpoints and experiences into a common public sphere it resists homogenising the mainstream and constructs it, instead, as a dynamic space for engaging with difference.

Extending participation

User-led participation and innovation are regarded as key social and cultural values of PSM in the multi-platform era. The concept of a networked public sphere is central to the development of more engaged citizens who are able to participate in and initiate public debate. In Australia, the ABC is seen as the leading innovator in extending citizen participation. Not only was it the first media organisation to engage in serious research, development and experimentation with interactive online services, it now offers a range of multi-platform media that situate PSM at the heart of an emergent digital commons. As Flew et al. (2008: 13) argue, now is the crucial time for government to support the ABC and SBS to develop 'online architectures of participation' that will enable the 'creative output, information, opinions and stories of the Australian public to be distributed and discussed'.

How these architectures of participation will be used to develop innovative responses to cultural diversity is yet to be seen. They present enormous potential for moving beyond recognition and the representation of diverse viewpoints to user-generated content that lets people tell their own stories in their own terms. There is no question that these developments have allowed the ABC to diversify its online content and allow multiple voices to be heard in ways that television has been largely unable to do. This reflects the capacity of Web 2.0 and other digital media to foster more engaged and active forms of citizenship. However, the tendency within the ABC is still to see this content as add-ons, as peripheral to the central and authoritative role of expert media practitioners, whether they are journalists or producers, who implicitly remain as gatekeepers and curators of this content. There are also few explicit strategies to seek out culturally diverse participation and user-generated content.

As already outlined, SBS has a strong tradition of participation in relation to its radio service, which has always used extensive community-generated content. In relation to multi-platform media, SBS' services are small and underdeveloped especially in comparison to the ABC's. They are also often information or read only rather than read/write. One area of enormous potential for SBS is the development of more diverse sources of news and information. Since its inception SBS TV has pioneered internationally oriented news bulletins based on the extensive exploitation of various satellite feeds coming into Australia. This has led to the development of one of the most cosmopolitan and highly respected news services in Australia (Ang et al., 2008: 176). However, access to these diverse news sources is now readily available on subscription and internet news services and this poses a serious challenge to SBS' unique free-to-air bulletin. Flew et al. (2008: 23) argue that 'in the context of global concentration of news and information sources bloggers are increasingly seen as valuable contributors to news reporting ... SBS could develop – particularly in the online space – a potential "meta-news-aggregator"'. This would capitalise on SBS audiences' extensive information networks both within ethnic communities, and beyond to homelands and every conceivable international issue. SBS has a long record of productive diversity; it would continue this by using audiences' community and international links to add depth to the news service and enhance its democratic reach and potential. This form of multi-platform participation could also feed directly into SBS' evening news bulletin, one of its most highly rated programmes, further enhancing SBS' commitment to mainstreaming pluralism.

Conclusion: cosmopolitan versus national public cultures

The differences between the ABC and SBS in responding to the complex realities of multiculturalism are stark. While many media analysts in Australia still cling to the fantasy that this task is primarily the role of SBS, this conveniently lets the ABC off the hook. It also further entrenches the assumption that the primary function of PSM is to foster mainstream culture in the national interest. As I have shown, invocations of the mainstream are inherently homogenising and normalising and this inevitably pushes diversity to the margins. Australia's cultural diversity is pervasive; there are multiple publics and interests, and national life, more than ever, involves a constant series of interactions between diverse constituencies. PSM have a key role to play in this process of social inclusion and engagement across difference. The ABC's various strategies aimed at diversifying the mainstream, limited as they are, have played an important role in this process. However, there is no doubt that SBS' achievements are far more significant. Not only do they show how PSM can be radically reimagined as multicultural, they also signal how a cosmopolitan approach to media provision can extend the reach and impact of a national institution and be a source of wider media and social change.

SBS' commitment to linguistic diversity, its extensive use of international content from all around the world and its cosmopolitan news service, are a few of many examples one can cite to show how it has dramatically extended the national ethos of PSM, and challenged other media in Australia to become less parochial. These examples show how this PSM has not only made available programming never before seen on Australian television, from non-English language movies to experimental documentaries, but has also mediated this content with a very distinct set of orientations to 'the foreign'. Within the general rubric of multiculturalism, a whole set of new television, radio and now multi-platform interfaces have been created that allow SBS audiences and users to have an international perspective on the world. These interfaces also allow them to feel implicated in multiple public spheres and diverse communities both within and beyond the nation. Audiences are also able to participate in transnational identifications and see the connections and interrelationships between their attachments to 'home' and those of others.

These examples show how it is often very difficult to sustain any kind of fixed or essentialised distinction between the national and the foreign. From its inception SBS has not necessarily opposed the foreign to the national, the foreign exists alongside or *mixed up with* the other work SBS does in addressing multicultural Australia. What SBS' approaches to cultural diversity often facilitate is a series of *eccentric* connections with the world that disrupt any kind of hierarchy from the local to the national and then the international. Often, the presence of these cosmopolitan orientations and identifications displace the centrality of the nation as the primary mode of identification in relation to the world or to the political events of other nations. Or these expansive perspectives reveal nationalism's limits, the ways in which it can blind audiences to the contingency of their identity.

Finally, SBS continues to shame other Australian media, the ABC included, about their extraordinary parochialism. This is more than just a question of pronunciation of foreign names, or a fear of subtitling anything, or the use of diverse international content; it is about the limits and possibilities of PSM. What SBS shows is that there are diverse publics and uses of media organised around plural geographies and identifications. Understanding Australia is increasingly problematic without a recognition of how it is thoroughly interconnected with other public spheres, political processes and communities. And, in recognising this, SBS' exploitation of the foreign and its commitment to mainstreaming pluralism reveals the cosmopolitan possibilities of PSM.

References

Ang, I., G. Hawkins and L. Dabboussy (2008) *The SBS Story: the Challenge of Cultural Diversity* (Sydney: UNSW Press).

Australian Broadcasting Corporation Act (1983) (Canberra: Australian Government).

Department of Broadband Communications and the Digital Economy (DBCDE) (2008) *ABC and SBS: Towards a Digital Future* (Canberra: Australian Government).

Flew, T., S. Cunningham, A. Bruns and J. Wilson (2008) *Social Innovation, User-Created Content and the Future of the ABC and SBS as Public Service Media*, QUT Digital Repository at http://eprints.qut.edu.au/.

Hawkins, G. (1997) 'The ABC and the Mystic Writing Pad', *Media International Australia*, 83: 11–17.

Special Broadcasting Service Act (1991) (Canberra: Australian Government).

23
New Zealand on Air, Public Service Television and TV Drama

Trisha Dunleavy

Introduction

This chapter focuses on television, exploring the 'public service' contribution of New Zealand on Air (NZoA), this country's public broadcasting agency. Originally called the New Zealand Broadcasting Commission, NZoA was created as part of the neoliberal-styled restructuring and deregulation of New Zealand broadcasting in 1988–9. Its statutory mission was to disburse public funding to broadcasting projects deemed able to 'reflect and develop New Zealand identity and culture' (New Zealand Broadcasting Act, 1989: 17). Designed to operate in a deregulated broadcasting system in which the public network had been stripped of its PSB remit and related funding, NZoA represented a new approach to PSB, a radical alternative to the tradition of a single PSB network.

Since 1989, NZoA has been the primary institutional facilitator of 'public service' TV programming (PSTV) in New Zealand. Already unusual among international PSB providers because it is an agency rather than a broadcaster, three other elements of NZoA's operation have combined to make it a very effective component of a PSTV environment that is challenged as much by limited population size as by insufficient public investment. First, NZoA funding is allocated on a contestable basis in a highly transparent process that is designed to match public funding with strong ideas. Second, this money is disbursed to producers rather than networks so that it buys programmes rather than services. Third, it is used to support productions destined for private as well as public networks, although the necessary prerequisite for both is that the completed programmes will air on free-to-air (FTA) broadcast channels.

While the above elements of NZoA's approach have been important to its perceived success in the twenty years it has operated, its potential influence in PSB terms has always been undermined by a fourth characteristic: that NZoA's level of annual funding has never been sufficient to pay for the range of activities in which it is involved, this obliging it to *support* rather than to fully fund productions in partnership with other investors. Television's importance to NZoA's overall PSB contribution is underlined by its receipt of two-thirds of NZoA's annual government grant, the latter being NZ$116.1 million (51 million euros).[1] Although this level of funding places real limits on the volume and range of PSTV programmes that it can facilitate, public and political responses to NZoA suggest that it is considered to have 'punched well above its weight' in terms of the 'public-value-per-public-dollar' invested.

Consistent with its mission to facilitate programming that can 'reflect and develop New Zealand identity and culture', NZoA has allocated funding to maintain TV production in the categories of drama and comedy, documentary and information, children and young people, and special interest programming, all of which have remained important to PSTV objectives in New Zealand and for which NZoA has been responsible in the last twenty years.[2] NZoA has evidently succeeded in this mission, with the resulting programmes testifying to its facilitation of increased cultural diversity and creative innovation.[3] Although any one category could be examined to elucidate how NZoA functions to facilitate PSTV in New Zealand's under-resourced PSB landscape, this chapter foregrounds drama, as one that has been unusually challenging because of its high costs and risks.

New Zealand's television landscape and PSTV programming

New Zealand television's greatest challenge has been reconciling the limitations of a small national audience of just 4.2 million people with the voracious appetites of a relatively expensive medium. With New Zealand-made programmes often lacking commercial viability and imported TV shows remaining comparatively inexpensive and abundant, television output has consistently been dominated by American, British and Australian programmes, leaving domestic productions comprising only 30–35 per cent of the total.

A small population, coupled with limited political enthusiasm for and public investment in non-commercial broadcasting, has influenced New Zealand's PSTV potentials in two other ways. One is that New Zealand's public television network, TVNZ (as with its ancestor, the NZBC) has always operated on a 'mixed' economic footing rather than as a non-commercial entity. This has meant that TVNZ, still the leading network with 51 per cent audience share (AGB Nielsen Media Research, 2009), has needed to reconcile PSB objectives with the pursuit of advertiser-funded commercialism, with the latter consistently overriding the former. The second is that, since its 1988–9 transformation, TVNZ has been more than 90 per cent reliant on advertising revenue, thus ensuring that commercialism drives its decisions and reducing the distinctions between so-called 'public' and 'private' networks that might otherwise apply.

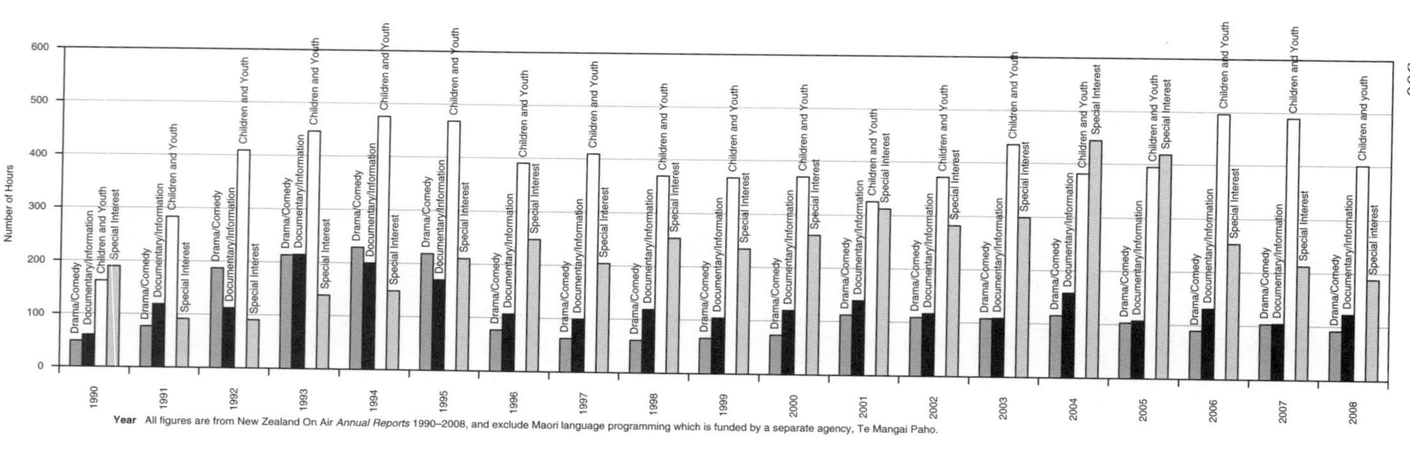

All figures are from New Zealand On Air *Annual Reports* 1990–2008, and exclude Maori language programming which is funded by a separate agency, Te Mangai Paho.

YEAR	Drama / Comedy	Documentary / Information	Children and Young People	Special Interest
1990	49	60	162	189
1991	77	119	283	91
1992	187	112	410	90
1993	213	214	447	138
1994	229	200	476	148
1995	218	169	469	210
1996	77	107	391	247
1997	62	99	410	204
1998	60	119	369	253
1999	66	104	367	234
2000	73	120	370	261
2001	112	139	325	312
2002	109	116	372	282
2003	108	109	436	298
2004	115	158	382	445
2005	103	107	395	419
2006	89	131	497	253
2007	103	104	490	212
2008	92	123	401	188

0	E
39.49	D
49.49	C
54.49	C+
59.49	B-
64.49	B
69.49	B+
74.49	A-
79.49	A
84.49	A+

Figure 23.1 Hours of NZ on-air funded TV programming in four generic categories (1990–2008)

In this context, PSTV objectives in New Zealand have been closely aligned with efforts to maintain a desirable range of New Zealand-made programmes in FTA schedules. However, there is an important distinction to be made between New Zealand's PSTV categories and the interrelated, though broader area of New Zealand-produced TV programming (or 'local content'). New Zealand's PSTV categories comprise drama, comedy, documentary and information, children and youth, Maori language, arts, music, and ethnic minorities, all of which involve programmes that combine a social or cultural significance with a degree of commercial fragility. Yet this same combination of features is not true of all New Zealand-produced TV programming. Mainstream programmes in the local content categories of news, sports, light entertainment and popular factual, for example, have been less subject to direct competition from imports and also tend to garner the country's largest TV audiences. As such, they enjoy greater commercial viability and a stronger, more consistent schedule presence than is possible for the majority of New Zealand's PSTV programmes.

As the above range of PSTV categories implies, New Zealand television's 'public service' objectives in programming have centred upon two main aims: (a) the maintenance of a basic supply of programmes for audience groupings that television struggles to cater for, and (b) the attempt to counter 'market failure' in at-risk, yet culturally beneficial TV genres. While New Zealand's PSTV categories have changed very little since television was first introduced, they have been subject to increasing commercial pressure on FTA schedules in the last fifteen years due to a proliferation of channels, more aggressive inter-network competition (particularly between leading rivals TVNZ, TV3 and Sky) and the consequent fragmentation of an already small national audience.

The restructuring of New Zealand television: the context for NZoA's creation

The impulse that made the creation of a stand-alone PSB agency seem logical and necessary was the 1988 decision to remove 'public service' objectives and public funding from TVNZ. Television's 1988–9 restructuring was tasked with providing the institutional and legislative framework for the medium's transition from an era of state monopoly (1960–89) to one of direct competition between TVNZ and incoming private networks. However, entailing a definitive rejection of the regulatory rigour of the past, this restructuring was philosophically consistent with the neoliberal transformation of the New Zealand public sector occurring more broadly. From 1984 to 1999, New Zealand governments prioritised a 'rolling back' of the New Zealand state (Kelsey, 1998), an outcome that public sector restructuring achieved in two ways. Pursued in tandem, one was the reformation of public companies as commercially focused 'state-owned enterprises' (SOEs), and the other, the removal of their 'public' functions and related funding. Completed by late 1989, the transformation of TVNZ underlined that within this programme of state sector 'reform', broadcasting was treated no differently from other areas of public activity rather than being considered a special case.

One aim of TVNZ's SOE transformation was to place it in a position of competitive neutrality with its incoming private network rivals, the first being TV3 (1989–) and the second, the subscription-based Sky (1990–). Helping to justify the government's conception of TVNZ as a network that should operate commercially was that it was already reliant on advertising revenue. Whilst TVNZ and its public network predecessors had always combined advertising revenue with public funding, TVNZ's dependence on the former increased rapidly after 1980 as the latter declined. With TVNZ around 80 per cent reliant on advertising revenue by 1988 the distance between its 'mixed' funding tradition and its SOE reorientation as a 'strictly commercial' company was smaller than it might otherwise have been. By mid-1989 TVNZ had been stripped of its earlier PSTV obligations, NZoA had begun operating, and the public income that TVNZ used to receive now belonged to NZoA. Underpinning this new structure was the requirement for increased transparency. Whilst neoliberal ideology accepted the validity of public funding for PSTV outcomes, it demanded that these be more clearly identifiable than had been possible whilst such outcomes had been the responsibility of a 'mixed' model TVNZ.

Hence, NZoA was established to disburse public funding for remaining PSB activities, some of which were permanently lost following TVNZ's SOE transformation. The government's intentions for television became clear after it rejected the 1986 Chapman Report on Broadcasting which had advocated a strengthening of TVNZ's public role. Although New Zealand broadcasting was ripe for restructuring in the late 1980s, the far-reaching changes it sustained were informed not by the PSB orientation of the Chapman Report but by the 'free market' thrust of the Rennie and Stevenson Reports which followed it in 1988. Whilst the Rennie Report provided the blueprint for an SOE structure and 'strictly commercial' focus for TVNZ, the role of the Broadcasting Commission (which was renamed NZoA in 1989) was outlined in the Stevenson Report. Anticipated by details in the 1988 Stevenson Report, NZoA's statutory functions were outlined in the 1989 New Zealand Broadcasting Act, the most significant piece of legislation to follow television's restructuring and deregulation.

Making a little go a long way: NZoA's funding allocation strategies

Given the broad nature of its remit, the size of NZoA's annual funding purse was geared to be among the most important arbiters of its effectiveness in 'public service' terms. In the financial year 2007–8, as mentioned, this was NZ$116.1 million (51 million euros), of which some NZ$74.3 million (32.6 million euros) went to television.[4] While the very creation of NZoA underscored a neoliberal vision for broadcasting that would require the remaining PSB activities to justify every dollar of public funding invested in them, another explanation for such a limited funding supply to NZoA was the 1989 Broadcasting Act's conception of it as offering financial support to, as opposed to 100 per cent financing for, the PSTV programmes for which it would be responsible. Notwithstanding the limits on its income, that NZoA is perceived to have performed in PSTV terms, owes much to the inventiveness of the strategies and policies for funding allocation that it developed around the priorities and directions outlined in its founding legislation.

The legislative clause stipulating that NZoA should 'make funding available on such terms and conditions' as it 'thinks fit' (New Zealand Broadcasting Act, 1989: 18) underlines that it retained the ability to devise its own funding criteria. The discussions below focus on three 'rules' which, originating as a brief clause in the 1989 Act, have since governed NZoA's funding processes. Together, these initiated a significant departure from earlier ways of thinking about PSTV programming, including who would have the right to screen it and what objectives it would aim to serve. As a founding member of NZoA's Board recalled, 'it consciously developed a vision' for a contemporary and forward-looking New Zealand PSB that valued accessibility (important to which would be TV programmes with broad appeal) and catered to the full range of age groups (Horrocks, 2009).

Considerations of potential audience size

Propelling NZoA in the above directions was firstly the requirement that it consider 'the potential size of the audience to benefit' (New Zealand Broadcasting Act, 1989: 18), which brought two important repercussions. First, NZoA has been obliged to support TV programmes for mainstream audiences rather than concentrate exclusively on minorities. Second, this rule has encouraged NZoA to reconcile the level of funding invested in a given TV project with the size of the audience anticipated for it as a completed programme. Although this audience size rule held the potential to marginalise minority programmes as beneficiaries of NZoA funding, this was countered by two other provisions in the Act, these requiring NZoA to provide programmes for 'minorities in the community including ethnic minorities' (ibid.: 18) and to ensure the 'availability of a balanced range of programmes' (ibid.: 19). Accordingly, while more of NZoA's annual funding goes to general audience programming, a proportion is deployed to provide programming for minority audiences and special interest groups.

The requirement for a broadcaster commitment

A second rule was that NZoA's allocations should prioritise programmes that 'if produced, would be broadcast' (ibid.: 19). While the Act did not prevent NZoA from granting funding without this guarantee, in the context of limited funding with which to experiment, NZoA interpreted it so as to ensure that its TV funding would not be allocated to productions that networks were unwilling to screen. Although this rule has always threatened to reduce risk-taking and innovation by NZoA-supported programmes to only that which the TV networks that air it are willing to tolerate, the programmes themselves suggest that, as far as is possible for advertiser-funded networks, this threat has been minimised. Important to this has been NZoA's significant presence in the funding mix and the contributions it is therefore able to make to negotiations about new productions which, without NZoA's involvement, would be confined to the host network as 'buyer' and production company as 'supplier'.

An important example of NZoA's intervention in network commissioning was its role in the establishment of TVNZ's soap opera, *Shortland Street.* Launched in 1992 and continuing, this programme's unrivalled longevity, five-nights-a-week format, and 7 p.m. timeslot has given it easily the highest, most sustained audience profile of any TV programme that NZoA has supported. NZoA had offered a $3 million (1.3 million euros) tender, then one-third of the total cost for a year of episodes, to entice either TVNZ or TV3 to respond with a soap proposal.[5] NZoA chose TVNZ's *Shortland Street* over TV3's *Homeward Bound,* with the new soap commissioned for TV2, the 'younger' half of TVNZ's then two-channel pair. Although TVNZ/TV3 competition for the project implied their mutual eagerness to broadcast an NZoA-supported soap, network unease about the sizeable risk that this entailed was evident in the resistance of senior TVNZ executives who needed convincing that a local prime-time soap was even viable. It 'was an enormous battle', according to Ruth Harley, then Chief Executive of NZoA. 'People didn't want to make it. There was huge scepticism – scepticism being a kind word for it – about whether it could be done' (McDonald, 1995). The problem was not TVNZ's own investment in the production but the 'opportunity cost' the network incurred by 'stripping' an untested local soap in a lucrative 7 p.m. slot. That NZoA persisted and TVNZ eventually relented, demonstrates that, even though a broadcaster retains the right to refuse a PSTV proposal, NZoA's negotiating power can be significant and even more so when its financial contribution is greater than the one-third share it offered in this case.

Exemplifying the impacts at network level has been the consistently 'edgy' drama and comedy programming commissioned by TV3, all of which was facilitated by a high proportion of NZoA funding. As successful instances of creative innovation within their own sub-genres, leading examples have included: *Mataku* (2002–4), a *Twilight Zone*-styled anthology drama series of

Maori 'ghost stories'; *Outrageous Fortune* (2005–), a renewable drama series whose 'West Auckland crime family' concept took this important sub-genre in a new direction; and *bro' Town* (2004–9), an animated sitcom that probes class and racial tensions from the perspective of four schoolboy characters, three of whom are Samoan.

Contestable funding

The third rule that has been important to NZoA's funding of PSTV production and to its perceived success in this, originated in the Act as an instruction that it 'invite competitive proposals' (New Zealand Broadcasting Act, 1989: 20); this subsequently became known as 'contestable funding'. Even more unusual in PSTV terms was that NZoA's contestable funding would be available to programmes destined for private as well as public networks. The principles of competitive neutrality, competitive tendering and transparency had informed the procedures of a majority of the public corporations and agencies that were created as a result of the neoliberal transformation of New Zealand's public sector. When applied to broadcasting, these same principles dictated that PSTV programmes should no longer be the exclusive domain of public networks: if TVNZ were going to operate in a position of competitive neutrality with its broadcast rival TV3, then both should be able to host NZoA-supported programmes. The 1989 Act had avoided directing NZoA as to how competitive tendering would be achieved in television, leaving the agency to devise its own eligibility criteria and policies. With the repercussions of this third rule being revealed by NZoA's first funding decisions in 1989, contestable funding would thenceforth allude to two policies for its allocations to TV production, both yielding positive outcomes in PSTV terms.

First was that the programmes that NZoA supported could air on TVNZ, TV3 or another broadcast network. Although, as the 1989 Act was being drafted, only one private network (TV3) yet existed, the conceptual openness of this third rule allowed it to remain relevant as the broadcast television sector expanded. Two later additions to the broadcast TV system were TV3's second channel C4 (1996–) and new private network, Prime (1998–). Importantly, these additions were not accompanied by increases to NZoA's annual income. NZoA's anticipation of an increased volume of calls on its funding as television's broadcast sector grew, seemed to inform the policies that it formed around 'contestability', which came to include additional criteria about which networks would qualify to screen NZoA-supported programmes. NZoA's legislative confinement to 'broadcast' programmes (ibid.: 18) combined with its obligation to consider audience size to shape the criteria it has used since the 1990s. NZoA funding has supported programmes destined for FTA networks with national reach and higher levels of audience share (NZoA *Statement of Intent 2008–11*: 3). Accordingly, the priority hosts for NZoA-funded programmes (and although Prime and Maori TV networks also qualify) have been TVNZ and TV3, whose three leading channels hold a combined peak-hour share of 69.5 per cent (AGB Nielsen Media Research, 2009).

The second policy, a NZoA initiative rather than a formal directive, was that NZoA funding would be disbursed directly to production companies or producers rather than to the network broadcaster of the finished programme. Although this policy has not prevented NZoA funding being allocated to producers working on an 'in-house' basis at TVNZ or TV3, an increasing majority of NZoA's television funding has been allocated to independent producers. Whereas by 1998 some 75 per cent of this funding went to the independent production sector, by 2008 the proportion was an overwhelming 89 per cent.[6] Echoing the industrial impact of the 'PSB publisher' remit given to Britain's Channel 4 (Harvey, 2000), the emphasis on outsourcing as opposed to the in-house production of PSTV programmes since the creation of NZoA, has stimulated the expansion of New Zealand's 'indie' sector and greatly enhanced the range of perspectives and approaches available to the resulting programmes.

NZoA and the case of television drama

Even before the creation of NZoA, New Zealand-produced TV drama had qualified as an 'endangered species'. However, this position for drama has been more overt during New Zealand television's multi-network era than it was through its decades of public monopoly. In the context of the 'strictly commercial' objectives imposed on TVNZ from 1989 and the continuing reluctance of both TVNZ and TV3 to meet drama's high costs by themselves, PSTV justifications for drama have been this genre's lifeline. Although hours-on-screen totals can obscure the trends of the last twenty years, the increased number of national broadcast networks operating (these now including TVNZ, TV3, Prime and Maori TV) has not been matched by any similar increase in the number of TV dramas produced each year. Moreover, annual output in the 'high-end' area of drama series, serials and one-offs has gradually reduced since 1995. Although TVNZ and TV3 together hold 72.9 per cent of the national audience (AGB Nielsen Media Research, 2009) neither has been willing to commission new drama programmes without the 'risk capital' provided by public funding. Accordingly, TV drama production has survived in New Zealand only by virtue of NZoA's ability to remain its primary investor.

This situation is borne out by TV drama's greater reliance on NZoA funding in 2009 than in 1992 when *Shortland Street* was established with only a one-third contribution from NZoA. The Stevenson Report (1988) had envisaged that NZoA would provide 'establishing grants' for TV production, this assuming not only the participation of other investors in the funding mix but also a complete withdrawal of NZoA support once a programme had demonstrated its success. However, as reduced private investment in TV production collided with increasing economic pressure on New Zealand networks, the 'establishing grant' approach needed to be replaced by a more flexible 'subsidy' model, this involving an increase to NZoA's share of the total investment in PSTV programmes. By 2009, NZoA's funding contribution to a given TV production averaged 64 per cent of the total cost and could extend as high as 92 per cent for programmes aimed at specialist or minority audiences (NZoA *Annual Report* for the year ending 30 June 2008: 45–7).

Given NZoA's remit to 'reflect and develop New Zealand identity and culture', drama has been a key performance indicator for it, due to the capacity this genre has shown to foster cultural identity whilst simultaneously helping to defeat 'cultural cringe'. Together with earlier assertions about drama's higher cost than other TV production forms, the above considerations explain why drama has consumed the largest proportion of NZoA's annual funding allocations to television. Of the total of NZ$74.3 million (32.8 million euros) that NZoA spent on television in the 2007–8 financial year, some NZ$23.8 million (10.5 million euros) was allocated to drama programmes, with the related but separate budget categories of 'comedy' and 'children's drama' receiving an additional NZ$6.4 million and NZ$4.6 million, respectively (ibid.: 45). Although NZoA has no editorial control over and nor can it determine the ultimate timeslot for any TV programme that it funds, its position as the majority investor in TV drama has given it sufficient weight to counterbalance the commercial priorities of the networks with its own PSTV imperatives. When asked what qualities NZoA seeks when assessing drama proposals, Jo Tyndall (then Chief Executive) highlighted the following five characteristics:

- 'Strong expressions of New Zealand culture' and 'stories that say something about us as a nation'.
- 'An accurate reflection of the changing society', incorporating cultural and ethnic diversity.
- Enough experience in the creative team to deliver a good programme.
- Conceptual innovation.
- 'Strong broadcaster support' and a timeslot that is appropriate to the investment cost (Tyndall, 2003).

Drawn from the 'high-end' area of TV drama, two recent examples can illustrate how the above criteria converge upon the individual programme. One is *Outrageous Fortune* (TV3, 2005–), a drama series now in its sixth season and the other, *Piece of My Heart* (TV One, 2009), a feature-length one-off. Infused with elements of comedy and melodrama, *Outrageous Fortune* follows an extended family that is held together by its plain-speaking matriarch, Cheryl West. Set in the Bogan heart of a sprawling urban and largely working-class West

Auckland,[7] the conflict that has fuelled *Outrageous Fortune*'s fertile central narrative is that despite this family's notoriety in the sub-criminal fringe in which it resides, Cheryl has determined that it will clean up and 'go straight'. Having amassed over eighty-nine episodes, *Outrageous Fortune* has become New Zealand's most enduring, successful hour-long drama to date, conferring enviable brand benefits on TV3 in the process. The second example is *Piece of My Heart*, a telefeature that was produced for TV One's longstanding *Sunday Theatre* slot. With its historical (circa 1960) story told in flashbacks by its main character, Flora, *Piece of My Heart* explores the survival struggle of two pregnant teenagers who, sent to a Magdalene-styled nursing home after 'shaming' their families, are then tricked into surrendering their babies for adoption. A contrast with *Outrageous Fortune* in form as well as in tone, *Piece of My Heart* exemplifies what John Caughie (2000) termed 'serious drama', the indicative characteristics being its seriousness of purpose, conceptual originality and the uncompromising quality of its artistry and performances.

Conclusion

This chapter's investigation of NZoA – the leading facilitator of New Zealand's PSTV programming in the last twenty years – may be relevant to other countries in which such programming is threatened by continuing channel proliferation, market fragmentation and related challenges to the dominance of PSB funding by public networks. Even though its BBC network remains the world's unrivalled exemplar of PSTV objectives, provisions and outcomes, one such country is the UK, for recent debates about the future of British PSTV have lingered upon two problems in which ITV, Channel 4 and Five networks are all implicated. One is whether public funding for TV production should be made available for programming destined for these non-BBC networks, all of which have a PSB role. The other is how the concerns expressed by these networks about their ability to continue to offer the range of programme forms specified by their PSB status and licences can be addressed.

NZoA's inventive system and perceived success offers important food for thought given the kinds of problems that national TV systems are facing in an era of unprecedented channel proliferation. The PSB ideal is that noncommercial public networks remain the centrepiece of national TV systems. However, not all countries can afford them and it is increasingly difficult for a single network to meet the diverse PSTV needs of a contemporary public. Conspiring with longer-standing problems of delivering PSTV in a small national market, the 1988–9 deregulation of New Zealand television forced its policy-makers to confront these challenges early. Faced with the likely annihilation of remaining PSTV programming if it could not succeed, NZoA was strongly motivated to develop new approaches to the allocation of its public funding. This chapter has examined three of the strategies which have distinguished NZoA's efforts to make a limited funding supply go as far as possible to maximise the PSTV outcomes. What the NZoA experience can offer to countries grappling with similar problems to those described here is not only an additional way to allocate public funding for PSTV outcomes in an environment of increasing channel capacity and audience fragmentation, but also a 'tested and proven' model through which public funding can be allocated to PSTV programmes produced for commercial networks.

Notes

1. In the period 1989–99 NZoA was funded by a universal public broadcasting fee (PBF) paid by citizens. Since the 1999 abolition of this PBF system, NZoA has been funded from a direct annual government grant.
2. Until 1993, Maori programming was the fifth PSTV category for which NZoA was responsible. From that year onward, responsibility for this category passed to the newly created Maori broadcasting agency, Te Mangai Paho.
3. See Figure 23.1 for figures illustrating NZoA's facilitation of PSTV programmes by genre and annual hours produced from 1990–2008.
4. NZ On Air, *Annual Report* for the year ending 30 June 2008: 4, 17.
5. This was an early example of the 'competitive tender' strategy that NZoA went on to use more broadly, aiming to encourage network commissioning of programmes in PSTV genres not yet 'represented' by a New Zealand-produced example.
6. NZ On Air, *Annual Report* 2002–3: 49 and NZ On Air, *Annual Report* for the year ending 30 June 2008: 6.
7. 'Bogan' is a pejorative colloquialism in Australia and New Zealand. Roughly equivalent in meaning to the American 'white trash' and British 'chav', it recognises a demonstrable dearth of 'cultural capital'.

References

AGB Nielsen Media Research (2009) 'TV Channel Share Data for the First Quarter'.

Caughie, J. (2000) *Television Drama: Realism, Modernism, and British Culture* (Oxford: Oxford University Press).

Chapman Report (1986) 'Broadcasting and Related Telecommunications in New Zealand', Report of the Royal Commission of Inquiry.

Dunleavy, T. (2005) *Ourselves in Primetime: a History of New Zealand Television Drama* (Auckland: Auckland University Press).

Dunleavy, T. (2008) 'New Zealand TV and the Struggle for "Public Service"', *Media, Culture and Society*, 30(6): 795–811.

Harvey, S. (2000) 'Channel Four Television: From Annan to Grade', in E. Buscombe (ed.), *British Television: a Reader* (Oxford: Oxford University Press), pp. 92–117.

Horrocks, R. (2004) 'Turbulent Television: the New Zealand Experiment', *Television and New Media*, 5(1): 55–68.

Horrocks, R. (2009) (Member of NZoA Board 1989–2000), personal communication, 16 April.

Kelsey, J. (1998) *The New Zealand Experiment: a World Model for Structural Readjustment?* (Auckland: Auckland University Press).

McDonald, F. (1995) 'Awesome', *New Zealand Listener*, 6 May: 32–5.

McMahon, K. (2008) 'Grade Demands End to ITV Indie Quota', *Broadcast*, 10 October: 1.

Murdock, G. (1997) 'Public Broadcasting in Privatized Times: Rethinking the New Zealand Experiment', in P. Norris and J. Farnsworth (eds), *Keeping it Ours: Issues in Television Broadcasting in New Zealand* (Christchurch: New Zealand Broadcasting School), pp. 9–33.

New Zealand Broadcasting Act (1989).

New Zealand on Air (NZoA) *Statement of Intent 2008–11.*

New Zealand on Air (NZoA) (1989–2008) *Annual Reports.*

Parker, R. and K. McMahon (2008) 'ITV Aims to Tear Up Terms of Trade', *Broadcast,* 8 August: 1.

Rennie Report (1988) 'The Restructuring of the Broadcasting Corporation of New Zealand on State-Owned Enterprise Principles', Report of the Steering Committee, July.

Stevenson Report (1988) 'Report on Implementation of Broadcasting Policy Reform', Official Committee on Broadcasting, August.

Tracey, M. (1998) *The Decline and Fall of Public Service Broadcasting* (Oxford: Oxford University Press).

Tyndall, J. (2003) (Chief Executive of NZoA), personal interview, 11 November.

Index

NB: Page numbers in **bold** refer to figures and tables